SPECIATION OF FISSION AND ACTIVATION PRODUCTS IN THE ENVIRONMENT

Proceedings of the Speciation-85 Seminar organised by the Commission of the European Communities in collaboration with the National Radiological Protection Board, Chilton, UK, held in Christ Church, Oxford, UK, 16–19 April 1985

Programme Committee
R. A. BULMAN, *NRPB, Chilton, UK*
J. R. COOPER, *NRPB, Chilton, UK*
M. J. FRISSEL, *RIVM, Bilthoven, The Netherlands*
C. MYTTENAERE, *Directorate-General for Science, Research and Development, CEC, Brussels, Belgium*
H. SMITH, *NRPB, Chilton, UK*

Local Organising Committee
R. A. BULMAN, *NRPB, Chilton*
J. R. COOPER, *NRPB, Chilton*

Scientific Secretariat
C. MYTTENAERE, *CEC, Brussels*
H. SMITH, *NRPB, Chilton*

SPECIATION OF FISSION AND ACTIVATION PRODUCTS IN THE ENVIRONMENT

Edited by

ROBERT A. BULMAN

and

JOHN R. COOPER

National Radiological Protection Board,
Chilton, Oxfordshire, UK

for

Directorate-General for Science, Research and Development,
Commission of the European Communities

ELSEVIER APPLIED SCIENCE PUBLISHERS
LONDON and NEW YORK

ELSEVIER APPLIED SCIENCE PUBLISHERS LTD
Crown House, Linton Road, Barking, Essex IG11 8JU, England

Sole Distributor in the USA and Canada
ELSEVIER SCIENCE PUBLISHING CO., INC.
52 Vanderbilt Avenue, New York, NY 10017, USA

WITH 92 TABLES AND 123 ILLUSTRATIONS

British Library Cataloguing in Publication Data

Speciation of fission and activation products
in the environment.
1. Radioisotopes—Environmental aspects
2. Chemistry, Analytic
I. Bulman, Robert A. II. Cooper, John R.
III. Commission of the European Communities.
Directorate-General for Science, Research and
Development
363.7′384 TD196.R3

Library of Congress Cataloging in Publication Data

Speciation-85 Seminar (1985: Christ Church, Oxford)
Speciation of fission and activation products in the
environment.

'Proceedings of the Speciation-85 Seminar, organised by the Commission of the European Communities in collaboration with the National Radiological Protection Board, Chilton, UK, held in Christ Church, Oxford, UK, 16–19 April 1985'—P. ii.
Includes bibliographies and index.
1. Transuranium elements—Analysis—Congresses.
I. Bulman, Robert A. II. Cooper, John R. III. Commission of the European Communities. Directorate-General for Science, Research, and Development. IV. Great Britain. National Radiological Protection Board.
V. Title.
QD172.T7S64 1985 546′.44 85-27520

ISBN 0-85334-422-1

Publication arrangements by Commission of the European Communities, Directorate-General Information Market and Innovation, Luxembourg

EUR 10059

Printed in Great Britain by Galliard (Printers) Ltd, Great Yarmouth

INTRODUCTION

For several years now there has been the recognition that the environmental impact of toxic elements is determined not by their gross concentration but by their chemical form. By identification of their speciation it becomes possible to make predictions about the movement of toxic elements through the biosphere.

In the last few years numerous reviews, reports and a few books have been published on speciation studies. In only a few of these publications has the speciation of the transuranic actinides been discussed. In general, the nature of the speciation of stable elements in the environment can be used to predict the speciation of radionuclides such as ^{54}Mn, ^{60}Co, ^{65}Zn, ^{90}Sr, ^{106}Ru, ^{125}Sb, ^{129}I and many others.

In this seminar the programme committee set out to draw together scientists who were working on the various parts of the environmental pathway of fission and activation products. The success of these aims should be judged by the diversity of the papers presented.

It is our opinion that these Proceedings should be viewed as an interim report of the speciation of fission and activation products in the environment. Our knowledge of the speciation of these radionuclides in various parts of the pathway back to man is still poor. New techniques in analytical chemistry will improve our knowledge of the speciation of the elements. With improvements in the understanding of the speciation of radionuclides in the environment, the task of establishing the acceptability of nuclear power will be made easier.

As editors of these Proceedings we have aimed for a uniformity in the presentation of the papers. As the Proceedings are produced from camera-ready copy the onus for presentation of error-free scripts has lain with the authors. In some cases we have returned manuscripts for alterations and reduction in length where this has been too excessive but in other cases where there are variations in style, for instance in the style of references cited, we have chosen to stick to the target date for publication rather than face additional delays which would result from returning manuscripts.

We wish to thank all the participants for attending the seminar and helping to make it a friendly meeting. We also thank our colleagues on the Organising Committee for their support and the Commission of the European Communities for generous financial support.

Robert A. Bulman
John R. Cooper

CONTENTS

Session 3: Gastrointestinal Uptake
(*Chairman:* HYLTON SMITH)

Session 4: Speciation of Iodine
(*Chairman:* JOHN R. COOPER)

Session 5: Freshwater Environment
(*Chairman:* ADRIEN CREMERS)

Session 6: Marine Environment
(*Chairman:* THOMAS SIBLEY)

Session 7: Speciation in Plants and Micro-organisms
(*Chairman:* MARYKA BHATTACHARYYA)

ABBREVIATIONS

The gastrointestinal uptake factor for some radionuclides has been identified on some occasions in these Proceedings as f_1.

K_D (or Kd) has been used to express equilibrium concentration of sorbed radionuclide divided by equilibrium concentration of radionuclide in solution.

ANALYTICAL TECHNIQUES FOR IDENTIFICATION OF CHEMICAL SPECIES

R M BROWN, J S HISLOP AND C J PICKFORD

Chemical Analysis Group, Environmental & Medical Sciences Division
AERE Harwell, Oxfordshire, OX11 ORA, United Kingdom

ABSTRACT

The speciation of radionuclides in the environment is still at a very early stage of development and many of the existing methods are semi empirical. The situation is somewhat improved for inactive species and for species at higher concentrations. This paper reviews certain of the techniques applicable to speciation and considers these under the headings of direct and indirect methods. Examples of the application of speciation techniques mainly in the non-nuclear area are given along with an assessment of their applicability to radionuclides in the environment.

1. INTRODUCTION

The range of topics covered by the papers to be presented to this conference highlights the diversity of interpretation placed by different authors on the term "speciation". This varies from determination of isotopic composition, through whether an element or nuclide is present as an adsorbed species or is in true solution, to the full identification of the molecular compounds present. This range, therefore, covers physical, physico chemical and chemical form. The purpose of this paper is to consider those techniques which provide information on chemical form (including valence state).

An assessment of speciation is important because the chemical toxicity of an element or nuclide and its fate in the environment is governed principly by its chemical form and not by its atomic identity. (This is unlike its radiological toxicity which is a characteristic only of the isotopic identity). Who would have predicted for example that the dominant chemical form of arsenic in fish would be the non-toxic organic compound, arsenobetaine, or that nature would detoxify inorganic mercury present in fresh water systems by methylation. Equally, that the resulting methyl mercury would concentrate in fish and pose a greater hazard to man than the original inorganic form.

As far as the chemical form of fission and activation products in the environment is concerned, very little information is available and what there is has rarely been substantiated to confirm its accuracy. A number of factors have contributed to this, not least being the low concentrations of nuclides present in many systems. In addition the difficulties in maintaining the integrity of the chemical form during the collection of a representative sample and during the analytical process are often overlooked.

As the study of speciation of radionuclides is at such an early stage of development, it was considered appropriate by the organisers of this conference for workers in the field to be aware of the range of potential techniques available for chemical speciation and to illustrate how certain of these techniques have been applied to speciation of non active elements in environmental samples. Some of these stable elements may be used as models for fission and activation products. Not all of the techniques discussed here may be applicable to fission and activation products, particularly if present at low concentrations. In many cases, however, e.g. at point of release of liquid or gaseous effluent, knowledge of chemical form at higher concentrations may be relevant in predicting the subsequent fate in the environment. Several excellent reviews exist[1,2] including a very recent one relating to nuclides in ground waters[3].

The range of matrices of interest is also likely to be extensive. Water is of considerable concern, particularly sea and surface water, although there is increasingly interest in speciation of nuclides in sub-surface ground waters which have been in contact with nuclear waste repositories. A related matrix is soil pore water since this may affect the uptake of activity into plants. Similarly the chemical form in plants, animals and fish will affect the uptake in man himself. At all stages of this biological chain the chemical form of the element or nuclide may change, sometimes in an unpredictable manner, and it is for this reason that a range of practical chemical speciation techniques is required. While computational methods, which will be described later at this conference, can provide a valuable insight into relative stabilities and probabilities of existence of particular chemical species under stated conditions some advance knowledge of these must initially be assumed. It is also in general not possible to include biological processes in these computer models; these are often the most important processes occuring in the environment.

TABLE I Possible chemical forms of metals in natural waters(1)

Chemical form	Possible examples	Approximate diameter (nm)
particulate	solids	>450
simple hydrated metal ion	$Zn(H_2O)_6^{2+}$	0.8
simple inorganic complexes	$Zn(H_2O)_5Cl^+$	1
simple organic complexes	Cu-glycinate	1-2
stable inorganic complexes	PbS, ZnCo3	2-4
adsorbed on inorganic colloids	Cu^{2+}-Fe_2O_3, Cd^{2+}-MnO_2	10-500
adsorbed on organic colloids	Pb^{2+}-humic acid, Zn^{2+}-organic detritus	10-500

The range of possible chemical forms of nuclides in the environment is extremely large and the distinction between chemical and physico chemical may not always be clear cut. Table I shows possible chemical forms of metals in natural waters and highlights the nature of the problem[1]. Similar schemes have been postulated for radionuclides. Considering only organic complexes in waters, the data given in Table II shows some of the species which may be present in aqueous wastes from reprocessing of nuclear materials[3].

TABLE II Potential sources of complexing agents from aqueous radioactive process wastes[(3)]

Ligands used directly in processes:

TBP	HDEHP	Tartaric acid
DBBP	NTA	Hydroxyacetic acid
EDTA	DTPA	Sugar
EDTA	Citric acid	

+ Solvents used in processes
+ Impurities in and degradation products of ligands and solvents

Many of the techniques used for speciation studies, some of which will be detailed in later papers, are what could be described as generic speciation procedures. These involve the partitioning of a nuclide into operationally defined fractions. Unfortunately in many cases the operations may not be defined sufficiently and re-establishment of conditions may not be achievable by workers in different laboratories or by the same workers on different types of samples. In the case of soils this may be due e.g. to differences in buffering capacity or ion exchange properties.

In the present paper, for convenience, we have divided chemical speciation techniques into two groups:-

(i) those techniques that give data on speciation directly, ie those that give a response specific to a molecule or oxidation state, and

(ii) those that give data indirectly, ie those that rely on separation followed by element or nuclide specific detection.

2. DIRECT METHODS OF SPECIATION

Direct methods can be divided into three distinct groups, chemical techniques, molecular absorption and emission techniques and instrumental techniques.

2.1 Chemical Methods

Chemical methods of speciation rely upon fundamental differences in chemical behaviour to measure or separate different forms of an element. Most published procedures involve several different steps or techniques to achieve speciation, and the distinction between chemical methods and instrumental or combined procedures is not clear cut. For the sake of convenience, chemical methods have been divided into the following sub groups:-

2.1.1 Methods Based on Size

Techniques that can be included within this category include simple filtration, to separate dissolved from particulate species, including colloids. This group of procedures is widely used in the radionuclide area, because of the tendency for trace levels of these elements to occur as fine particulates, or adsorbed onto sediments[4]. Filtration techniques, involving use of pore sizes down to 0.22 μm do not nominally change the speciation of elements. Separations of this kind have been primarily used in the heavy metal area to separate hydrated ions from colloids[5,6].

Other size-based techniques include dialysis[7] and ultrafiltration[8]. Gel permeation chromatography has been widely used[9,10] to separate macro-molecules according to size or permeability, although it is usually combined with instrumental procedures for measurement of the separated species. Published applications of this technique have been mainly confined so far to the non-radionuclide field.

2.1.2 Methods Based on Solubility

Although differential solubility of trace elements or radionuclides when extracting with solvents of differing polarity is a standard procedure[11] it can hardly be called speciation, but at best "fractionation". It is open to the criticism that results obtained using this technique are purely empirical, and that considerable changes may have taken place in the speciation of the sample during extraction.

Liquid-liquid extraction is a well established procedure for most elements however[12], and is used to concentrate or separate trace elements from complex matrices. Because of the necessity to extract, or form specific complexes before extraction, it is inherently speciation dependent, and the literature contains many methods for the separation of e.g. Fe^{3+} from Fe^{2+}, Cr^{3+} from CrO_4^{2-}, MeHg from Hg^{++} etc. Due allowance must be made for the kinetics and equilibrium constants of the reactions involved, however, to avoid changes in speciation during extraction. Theoretical modelling may play an important role in this context. The stabilities of genuine "organometallic" compounds are much greater than those of simple inorganic or organic complexes, and these may, therefore, be separated or speciated with much more confidence that no change has taken place.

2.1.3 Methods Based on Differing Reactivity

Many chemical reactions are species dependent. The classical reactions of Fe^{2+} with oxidising agents are typical of this type of procedure. Such reactions are also possible with transuranic elements, but this is only usually carried out as a means of assay at high concentrations. Co-precipitation with fluorides is also valence state dependent, and this reaction forms the basis of a speciation scheme for Pu III, IV and V[13]. Various forms of Sb, Se and As have been speciated in ground waters by their valence state dependent reactions with sodium borohydride[14], liberating gaseous hydrides which are usually, but not exclusively measured instrumentally. Adjustment of pH produces changes in speciation of some elements which may be used to confirm valence state. Pu is a good example of this latter type of behaviour: in the stable nuclide area, metals such as Fe etc form a variety of halogen complexes dependent on valence state, as the pH and halogen concentrations are raised. Once again, such reactions have been characterised at high concentrations, and the actual behaviour of elements at environmental concentrations may be completely different.

2.1.4 Ion Exchange Methods

Ion exchange theoretically allows anionic and cationic species to be separated in waters. In practice problems often arise and it is widely reported that the sum of the cationic and anionic species for a particular element may often exceed the total elemental concentration[15]. This is a consequence of the ready interchangeability of species when in contact with resins. However, practical schemes have been proposed[16] to separate metals into anionic, cationic and neutral (unabsorbed by ion exchangers) species using this kind of procedure. Marchand[17] published a particularly interesting paper in which he separated species in seawaters using a cation exchange scheme: he found that tracers added to the seawater for elements such as Fe and Cr exhibited different speciation to the natural elements present. Clearly this kind of situation is potentially important in nuclear environmental studies if assessment of the uptake of radionuclides by living organisms is being made.

2.1.5 Chelating Resins

Chelating resins have many of the advantages and disadvantages of liquid-liquid extraction, although they may often be more convenient to use, particularly when incorporated into filter papers. As with cationic or anionic resins, there is now evidence that spikes or tracers added to seawater may not be retained in the same way on chelating resins such as Chelex 100 as naturally present trace elements[18]. This may be a consequence of either the valence state, or the presence of colloids or organic complexes. Florence and Batley[19] suggest that metal species are often present as species adsorbed onto colloids or very small particles. This suggestion has also been proposed by other authors, and clearly demonstrates the complexity of speciation measurements or calculations that would be required in order to predict accurately the behaviour of radionuclides or stable elements at very low concentrations in environmental matrices.

2.2 Molecular Absorption and Emission

2.2.1 UV/Visible Spectrometry

Stable molecular species and radicals often absorb UV, visible or IR radiation giving rise to characteristic absorption spectra. In some circumstances, re-emission of UV or visible radiation may also occur, giving rise to fluorescence or to phosphorescence. Molecular absorption occurring in the UV or visible region is inherently species dependent. Obvious examples of this use of speciation from basic chemistry include:- Cr^{III} and Cr^{VI}, Fe^{II} and Fe^{III} and Mn^{II} and Mn^{IV}. However, the sensitivity of measurement of such a system is not sufficient to allow direct measurement of environmental concentrations of even common elements, let alone radionuclides present at very low levels. Addition of a ligand to form a species dependent coloured complex may considerably increase the sensitivity of detection of many ions. Routine methods exist for quantitative determination of many elements, including radionuclides, at relatively high concentrations, such as might exist in direct leachates of stored waste material.

Recently, new techniques have become available which improve this situation somewhat. Preconcentration of metallic ions such as uranium onto an ion chromatographic[20] or an HPLC column, followed by post column derivatisation provides for very sensitive determination of certain ions. Concentrations down to sub-ppb may be reached by this route.

2.2.2 Infrared Spectrometry

Infrared spectrometry is widely used in organic chemistry to identify molecular species. It is not often used to identify inorganic molecules, except at higher concentration, but there is no reason why the technique would not be applicable to compounds separated by chromatography. Fourier transform IR is more advantageous in this respect because of the short scanning times required, allowing for real time measurements and the ability to carry out repeated scans to improve signal-to-noise ratios and hence sensitivity. IR techniques are not usually applicable to aqueous media however, so that prior extraction and pre-concentration may be required.

2.2.3 Fluorescence

Molecular fluorescence is a technique that is much more sensitive than absorption since it relies on the measurement of emission signals above a low background, rather than the difference between an incident and transmitted light beam which is limited in sensitivity by the noise level present on the source. Uranium is often routinely determined by fluorimetry[21,22] giving limiting sensitivities down to about 10^{-10} g levels. A fluoride matrix is generally used for measurement, which imposes constraints upon the technique as far as speciation studies are involved. The use of laser sources gives higher absolute sensitivities for uranium[22] but again requires considerable modification of the sample matrix, and thus probably alters the element speciation.

Although plutonium is also potentially determinable by fluorimetry at environmental levels, the other transuranic elements are not. However some other elements which are of interest in environmental radionuclide studies, such as the lanthanides, fluoresce readily in certain valence states and thus potentially may be determined or speciated by this route. Lanthanide ions also phosphoresce, so that methods of determination based on this property might well be applicable.

2.2.4 Laser Spectrometry

Laser based techniques such as laser induced photoacoustic spectroscopy[23] and laser induced thermal lensing spectroscopy[23] have been proposed as techniques suitable for trace actinide speciation studies. Unfortunately, the former technique is currently limited in sensitivity to 10^{-6} to 10^{-8} M for uranium (i.e. 240 to 2.4 ppb) which may not be sufficiently sensitive for trace speciation studies and the latter technique appears neither sensitive nor specific enough at the present time.

The ready availability of lasers has also revitalised the technique of Raman spectroscopy[(56)] which complements IR spectroscopy and may be used to identify the chemical form of species in solution, including aqueous solutions, down to concentrations of 10^{-4} M. Of particular note is the development of laser raman microprobes which can be used to identify the chemical form of species present as small particulates. Purcell and Etz[(57)] have used the technique for identification of uranium oxide and nitrate in particulates with dimensions of 5-15 μm.

2.3 Other Instrumental Methods

2.3.1 X-ray Diffraction

X-ray diffraction (XRD) has been used by Biggins and Harrison[25] for the identification of lead species in street dust, after magnetic and density separation. They concluded that the most frequently observed compound was $PbSO_4$, but equally that only a minor proportion of the lead present in a street dust exists in a crystalline form suitable for XRD analysis. XRD is only capable of determining the chemical form of major constituents of samples but significant data can be obtained from sub-mg weights of sample.

2.3.2 Mössbauer Spectroscopy

Mösssbauer spectroscopy can be used for identification of chemical form of nuclides which exhibit the Mössbauer effect by recoilless emission and absorption of γ radiation. The technique can only be applied to solid samples and to a limited number of nuclides. Several of these, e.g. ^{127}I, ^{129}I, ^{83}Kr, ^{124}Xe and ^{131}Xe are of interest in the nuclear industry. The technique is unlikely to be suitable for environmental materials unless significant biological preconcentration occurs. The technique has been mainly applied to determination of the chemical form of ^{57}Fe including that in soil[26] and in atmospheric particulates[27,28] and of ^{119}Sn

2.3.3 Electrochemical Methods

This approach has been widely applied to the measurement of chemical form of the elements and a review of the application of voltametric procedures in trace metal chemistry of natural waters and atmospheric precipitation has recently been published by Nürnberg[29]. While potentially offering the capability of identifying the form of individual chemical species by comparison with synthetic standards, electrochemical procedures have tended to be used to classify groups of compounds having similar electro-chemical behaviour. Thus, anodic stripping voltametry is used to distinguish between labile and non-labile metals[1], but this can be considered only as a generic speciation procedure as behaviour will be critically dependent on the electrolyte conditions defined. Using polarographic procedures the relative stabilities of various complexes can be measured[29]. Alternatively the complexing capacity of a natural water may be determined by titrating with a suitable heavy metal.

Electrochemical procedures do not readily fit into the somewhat simplified classification we have defined of direct methods and combined methods. The extremely high sensitivity of voltametric methods for determination of elemental concentrations in solution makes them useful for detecting nuclides following the separation of their species. In this regard the use of adsorption voltametry (or cathodic stripping voltametry) in which a metal in solution is measured with high sensitivity (10^{-10}M) in the presence of a suitable organic complexing ligand has particular potential. Using catechol the technique has been applied to the determination of uranium by van den Berg[30].

2.3.4 Mass Spectrometry

Mass spectrometry (MS) has many of the characteristics the analyst requires for obtaining data on chemical form. Organic mass spectrometers inherently produce molecular information and are very sensitive, being capable of measuring quantities of materials in the 10^{-9} g range. With the recent advent of the fast atom bombardment ion source[31], the analyst is no longer restricted to recording spectra of volatile materials. MS, however, records spectra of the major sample components, unless a preseparation process such as GC or HPLC has been used.

2.3.5 Nuclear Magnetic Resonance

Like Mössbauer spectroscopy NMR is an extremely powerful technique for identification of chemical form but is restricted to a limited number of nuclides and generally is limited in sensitivity. Nuclides such as ^{205}Tl have, however, been measured[32] in solution at concentrations down to 5 x 10^{-4}M. A number of nuclides of particular relevance in the nuclear area, are appropriate to study by NMR, including ^{3}H and ^{99}Tc.

3. INDIRECT METHODS OF SPECIATION

3.1 General Considerations

The second group of analytical techniques capable of producing data on speciation (the indirect methods) can be considered under the alternative heading 'combined techniques'. These use the combination of separation by chromatography or other means with element or molecule specific detection. The organic chemist has been using combined techniques for speciation for many years, with combinations such as gas chromatography-mass spectrometry (GC-MS) and more recently high pressure liquid chromatography-mass spectrometry (HPLC-MS) and gas chromatography-fourier transform infrared spectrometry (GC-FTIR). (The organic chemist would have little use for a total carbon concentration figure in a sample). This philosophy of combining separation and specific detection has recently been adopted by inorganic chemists, and is beginning to produce some very elegant speciation data. Several comprehensive reviews already exist[33-37].

3.2 Specific Combined Speciation Techniques

As in organic chemistry, two forms of instrumental chromatography have been interfaced on-line to specific detectors, namely GC and HPLC. GC separates molecular species according to differences in molecule polarity, or volatility. Samples are normally introduced as a solution in an organic solvent. All species to be detected have to be volatilised, and the detector has to be able to cope with a gaseous eluent. For the majority of environmental samples, this leads to two problems. Firstly most environmental samples have an aqueous component, leading to a requirement to extract a representative sample into a suitable organic solvent prior to GC separation, and secondly the majority of species likely to be of interest are involatile at typical GC temperatures.

HPLC separates according to a number of physical and chemical parameters, including partititon between two different solvents (the mobile and the stationary phases), and molecular size. In all forms of HPLC, the mobile phase plays an integral part in the separation; changing the solvent has considerable effect on the separation. (In GC the mobile phase (carrier gas) is merely a medium to promote physical movement; changing from one gas to another has little or no effect on the separation). In addition, HPLC can readily be used on aqueous samples with little or no sample pre-treatment, and species do not need to be volatilised. In practice, HPLC is often the preferred separation method for environmental speciation studies.

As indicated earlier, there are few techniques capable of producing direct molecular information on trace elements in environmental matrices. There are even fewer that are capable of being interfaced on-line to GC or

HPLC. Techniques such as Atomic Absorbtion Spectrometry (AAS), in both its flame and graphite furnace modes, and plasma-emission spectrometry, in its inductively coupled, direct current and microwave induced forms (ICP, DCP and MIP respectively) have been used as single or multi element specific detectors. These combinations produce "element specific" chromatograms. Essentially two separations are being achieved; separation of species of an element from each other, and separation of that trace element from the bulk of the matrix. Molecular identification is achieved in the same way as in conventional GC or HPLC - comparison of retention time with that of a standard. Table III summarises the chromatographic and detection possibilities that have been interfaced to each other.

TABLE III Analytical techniques that have been interfaced together

Chromatograph	Detector
GC	Molecular - MS
	IR/FTIR
HPLC - reverse phase	NMR/FTNMR
normal phase	
ion exchange	Elemental - AAS (Flame or Furnace)
gel permeation	Plasma (ICP, DCP, MIP)
	Electrochemistry
	Isotopic - ICPMS
	Radiometric methods

The molecular detectors listed are used extensively in organic chemistry, but have the disadvantage of not separating the element of interest from the bulk of the matrix.

3.3 Practical Considerations

Interfaces between chromatographs and detectors can be quite simple. For example, typical HPLC eluent flow rates of 1 to 2 ml min $^{-1}$ equate with solution uptake rates for flame AAS and ICP nebulisers, so that the combination of HPLC to flame AAS simply requires two pieces of tubing to be joined. Interfaces between GC's and specific detectors are more complex because of the need to keep them heated to avoid condensation of the sample vapour. Combined techniques such as GC-MS and HPLC-MS require even more complex interfaces because of the different operating pressures of the two halves of the system. In general the simpler the interface, and the smaller the dead-volumes it contains, the less likely one is to lose chromatographic resolution. This is a particular problem in HPLC-ICP. The nebuliser and spray chamber required to convert the liquid HPLC eluent into an aerosol cause a marked loss in resolution. On the other hand, GC's can be successfully interfaced to plasma emission spectrometers, where a nebuliser and spray chamber are not required.

The ideal chromatographic detector would be used on-line, producing a continuously varying signal in real time, i.e. sampling the eluent and measuring the species of interest at a rate of 1 Hz or greater. Of the possible detectors mentioned earlier (Table III) two are not capable of making measurements sufficiently rapidly. Graphite furnace AAS involves subjecting samples to a thermal cycle with a period of around 60 seconds. Techniques involving measurement of radioactivity generally require even longer, unless one sacrifices the specificity achieved by energy dispersion, and uses a simple ratemeter. In both these cases, it would be normal to collect discrete fractions, and analyse each in turn off-line, producing a time-integrated signal. Elegant solutions to the graphite furnace AAS problem have been reported[38], these only bring the fraction collection and analysis on line; the signal is still time-integrated. Collecting fractions does however have one distinct advantage over on-line detection. If the fraction size is large enough (>0.25 ml), it is possible to use several different analytical techniques on each fraction, to obtain more data.

3.4 Examples

It is not possible in a paper such as this to illustrate all of the potential combinations of chromatography and detectors listed earlier.

The following examples give an illustration of what has been achieved with combined techniques, both in terms of technique development and application to real environmental problems.

The first paper to illustrate the potential of combined techniques was published by Kolbe et al[39] in 1966. They used the GC-AAS coupling to demonstrate the separation of tetralkyllead compounds in petrol. Segar[40] interfaced a GC to a GFAAS, to take advantage of the improved detection limits achievable with the graphite furnace, and achieved limits of detection of 10^{-8} g for tetramethyllead. AAS was first reported as a metal-specific detector for ion exchange liquid chromatography in 1973 by Manahan and Jones[41] who separated chelating ligands as their copper complexes. Pankaw and Januner[42] separated and preconcentrated Cr^{III} and Cr^{VI} in natural waters by the same method. Jones and Manahan[43] coupled HPLC to AAS for the separation of organochromium compounds. As they pointed out in a later publication[44], combined HPLC-AAS leads to poorer limits of detection (relative to AAS itself) because of the dilution occuring in the HPLC. Brinckman et al[45] coupled HPLC and GFAAS and investigated the suitability of the technique for separating a number of organo-metallic compounds of arsenic, lead, mercury and tin. They also indicated the advantage of collecting discrete fractions from the HPLC, in that further analytical measurements can be made. Fraley et al[46] used an ICP as a single element detector for HPLC, separating EDTA and NTA complexes of Cu, Zn, Ca and Mg to demonstrate the potential. Irgolic et al[47] went one step further and used a simultaneous ICP as a multi-element specific detector. This group demonstrated the separation of a series of arsenic compounds from each other, and the separation of arsenate, selenite and phosphate anions which co-eluted. Morita and his co-workers[48] have also demonstrated the separation of arsenic compounds by HPLC-ICP, including separation of the two principle forms found in marine fish and crustaceans, arsenobetaine and arsenocholine. Browner et al[49], have considered the placement of the spray chamber, and effect of solvent flow rate, on a HPLC-ICP coupling.

While all of the examples quoted have demonstrated the potential of combined methods, they all concentrate on separating components in synthetic solutions, often at concentrations significantly higher than encountered in the environment. Real applications are now beginning to appear in the literature. Holak[50] has speciated mercury in fish by HPLC-

AAS, finding all the mercury to be present as methylmercury chloride. Chau et al[51] have determined tetraalkyllead compounds, and Pb^{2+}, in water from Lake Ontario. Gardiner et al[52] have used column gel permeation, coupled to a DCP, to speciate protein-bound Cu, Fe and Zn in human serum and intravenous infusion fluids. O'Neill et al[38] have applied HPLC-GFAAS to the determination of organocopper species in soil pore water, while the present authors have applied a similar technique to cadmium and manganese species in soil pore water, and to the speciation of arsenic in several marine organisms, radioactive inorganic arsenic was used to trace the course of the bioconversion, using both GFAAS to determine total arsenic and γ-spectroscopy to determine active arsenic[54]. Means et al[55] have used a similar combination of HPLC fractionation and GFAAS/γ activity detectors to speciate ^{60}Co and U in leachates from the Oak Ridge Laboratory. They showed that 90-95% of the ^{60}Co and 70% of U was associated with molecular weights around 200-300. The remainder of each was associated with higher molecular weights ($\geqslant$700). They did attempt to further identify the main species concluding that low molecular weight EDTA complexes were the probable species.

4. CONCLUSIONS

It has not been practical in this paper to review in any detail all methods applicable to speciation of stable or radioactive elements in the environment. The range of techniques discussed at this conference and the potential problems associated with their use highlights the relatively early stage of development of these procedures. In many instances developments have taken place by applying experience gained in one discipline to problems encountered in another. It is hoped that by indicating where advances are taking place in the speciation of stable nuclides, benefits may be derived in the radioactive area.

REFERENCES

1. T.M. Florence and G.E. Batley, Crit. Rev. in Anal. Chem., 9, p219 (1980).
2. T.M. Florence, Talanta, 29, p345 (1982).
3. P.M. Pollard, AERE-R 11496, HMSO London (1985).
4. M.B. Lovett and D.M. Nelson, Techniques for Identifying Transuranic

Speciation in Aquatic Environments, Proc. of meeting organised by IAEA and CEC, Ispra, p27 (1980).
5. J. Lecomte, P. Mericam and M. Astruc, Symposium of the Proc. of the Int. Conf. on Heavy Metals in the Environment, p678, Amsterdam (1981).
6. U. Förstner and W. Salomons, Env. Tech. Letters, 1, p494 (1980).
7. P. Benes and E. Steinnes, Water Research, 9, p741 (1975).
8. P. Benes, E.T. Gjessing and E. Steinnes, Water Research, 10, p711 (1976).
9. C. Steinberg, Water Research, 14, p1239 (1980).
10. S.F. Sugai and M.L. Healy, Mar. Chem., 6, p291 (1978).
11. R. Chester and S.R. Aston, Techniques for Identifying Transuranic Speciation in Aquatic Environments, Proc. of meeting organised by IAEA and CEC, Ispra, 4 p173 (1980).
12. A.K. De, S.M. Khopkar and R.A. Chalmers, Solvent Extraction of Metals, VNR Ltd, London (1970).
13. D.M. Nelson and M.B. Lovett, Nature, 276, p599 (1978).
14. M. Thompson, B. Pahlavanpour and L.T. Thorne, Water Research, 15, p407 (1981).
15. M. Astruc, J. Lecomte and P. Mericam, Env. Tech. Letters, 2, p1 (1981).
16. R.H. Filby, K.R. Shah and W.H. Funk, Proc. 2nd Int. Conf. Nucl. Methods in Environ. Res. NITS, Springfield, Va (1974)
17. M. Marchand, J. Cons. Lut. Explor. Mer., 35, p130 (1974).
18. J.P. Riley and D. Taylor, Anal. Chim. Acta., 40, p479 (1968).
19. T.M. Florence and G.E. Batley, Talanta, 23, p179 (1976).
20. Dionex Application Note, No. 48, Dionex Corporation, Sunnyvale, CA 94086 (1983).
21. G. Phillips and G. Milner, Chem. of Actinides, 10 Pergamon Press (1983).
22. S. Klainer, T. Hirschfeld, H. Bowman, F. Milanovitch, D. Perry and D. Johnson, LBL-11981 (1980).
23. W. Schrepp, R. Stumpe, J. Kim and H. Walther, Appl. Phys., B32(4), p207 (1983).
24. J. Beitz and J. Hessler, Nucl. Technol., 51, p169 (1980).
25. P.D.E. Biggins and R.M. Harrison, Env. Sci. Technol., 14, p336 (1980).
26. C.A.M. Ross and G. Longworth, Clays and Clay Minerals, 28, p43 (1980).
27. B. Dzienis and M. Kopcewicz, Tellus, 25, p213 (1973).
28. B. Kopcewicz and M. Kopcewicz, J. de Physique, 37, pC6-841 (1976).
29. H.W. Nürnberg, Anal. Chim. Acta., 164, p1 (1984).
30. C.M.G. van den Berg, Anal. Proc., 21, p359 (1984).
31. M. Barber et al, J. Chem. Soc. Chem. Comm., p325 (1981).
32. J.F. Hinton, G.L. Turner, G. Young and K.R. Metz, Pure & Appl. Chem., 54, p2359 (1982).
33. F.J. Fernandez, Atom. Abs. Newsletter, 16, p33 (1977).
34. F.J. Fernandez, Chromat. Newsletter, 5, p17 (1977).
35. J.C. VanLoon, Anal. Chem., 51, p1139A (1979).
36. I.S. Krull and S. Jordan, Int. Laboratory, p13, (Nov./Dec. 1980).
37. J.W. Carnahan, K.J. Mulligan and J.A. Caruso, Anal. Chim. Acta., 130, p227 (1981).
38. L. Brown, S.J. Haswell, M.M. Rhead, P. O'Neill and K.C.C. Bancroft, Analyst, 108, p1511 (1983).
39. B. Kolb, G. Kemmner, F.H. Schleser and E. Wiedeking, Z. Anal. Chem., 221, p166 (1966).
40. D.A. Segar, Anal. Lett., 7, p89 (1974).
41. S.E. Manahan and D.R. Jones, Anal. Lett., 6, p745 (1973).
42. J.F. Pankow and G.E. Janauer, Anal. Chim. Acta., 69, p97 (1974).
43. D.R. Jones and S.E. Manahan, Anal. Lett., 8, p569 (1975).
44. D.R. Jones and S.E. Manahan, Anal. Chem., 48, p1897 (1976).
45. F.E. Brinckman, W.R. Blair, K.L. Jewett and W.P. Werson, J. Chrom. Sci., 15, p493 (1977).

46. D.M. Fraley, D.A. Yates, S.E. Manahan, D. Stalling and J. Petty, Appl. Spectr., 35, p525 (1981).
47. K.J. Irgolic, R.A. Stockton and D Chakraborti, Spectrochim. Acta., 38B, p437 (1983).
48. M. Morita, T. Uchiro and K. Fuwa, Anal. Chem., 53, p1806 (1981).
49. B.S. Whaley, K.R. Snable and R.F. Browner, Anal. Chem., 54, p162 (1982).
50. W. Holak, Analyst, 107, p1457 (1982).
51. Y.K. Chau, P.T.S. Wong and O. Kramar, Anal. Chim. Acta., 146, p211 (1983).
52. P.E. Gardiner, P. Brätter, V.E. Negretti and G Schulze, Spectrochim. Acta., 38B, p427 (1983).
53. R.M. Brown, C.J. Pickford and W.L. Davison, Int. J. Environ. Anal. Chem., 18, p135 (1984).
54. R.M. Brown and N.J. Portsmouth, unpublished information.
55. J.L. Means, D.A. Crerar and J.O. Duguid, Science, 200, p1477 (1978).
56. D.L. Gerrard, Anal. Chem., 56, p219R (1984).
57. F.J. Purcell and E.S. Etz, Microbeam Analysis, p301 (1982).

A CRITICAL EVALUATION OF SEQUENTIAL EXTRACTION TECHNIQUES

P.NIREL, A.J.THOMAS and J.M.MARTIN

Laboratoire de Géologie, Ecole Normale Supérieure
46, rue d'Ulm - 75230 PARIS Cédex 05

ABSTRACT

The chemical speciation method of Tessier & al (1) has been applied to artificial substrates to test both its specificity and reproducibility for various trace elements, especially the stable isotopes from various fission and activation products found in the aquatic environment. Measured speciation is significantly different from expected results for most of the studied elements. Partial readsorption of extracted elements depends on the nature of the matrix analyzed and results in under-estimations in the first fractions.

INTRODUCTION

Sediments may represent an efficient sink for a wide variety of pollutants, but over large time scales they are not always permanently fixed and may be recycled through biogeochemical processes (2).
It is well known that the determination of total concentrations only, does not allow to assess the availability of pollutants to the environment or to living organisms (3).Since a few years sequential extraction procedures are used to estimate the type of associations between pollutants and particulate matters, because they permit the determination of various chemical fractions related to a gradient in association stability (4, 5).

Among the various methods which have been recently developed, Tessier's method (1) is of particular interest. It has been so far widely used so that numerous comparisons can be made with published results.Its multi-fraction approach allows applications to a variety of substrates, with a good reproducibility. A wide range of trace-elements and pollutants are found in the environment but the method has only been tested on a restricted number of elements, especially metals (Cd, Co, Cu, Ni, Pb, Zn, Fe and Mn)

The sequential procedure is briefly summarized below (F1 to F5):

(a) Fraction 1: extraction with 1N $MgCl_2$ at pH 7.0 ("exchangeable metals")
(b) Fraction 2: the residue from (a) is leached with 1M sodium acetate adjusted to pH 5.0 with acetic acid ("metals bound to carbonate")
(c) Fraction 3: the residue from (b) is extracted with 0.04M NH_2OH,HCl in 25% (v/v) acetic acid ("metals bound to Fe-Mn oxides")
(d) Fraction 4: the residue from (c) is extracted with 30% H_2O_2 ajusted to pH 2.0 with HNO_3 and then with 3.2M CH_3COONH_4 in 20% (v/v) HNO_3 ("metals bound to organic matter")
(e) Fraction 5: the residue from (d) is digested with a mixture of nitric, hydrofluoric and perchloric acids ("residual metals").
It is admitted , however, that the "specificity" of this method is practically defined by the operational protocol itself (1, 6).

The purpose of this study was: (i) to apply this method to a greater number of chemical elements, especially those which radioactive isotopes, originating from bomb testing and nuclear industry, are of environmental significance (Ce, Co, Cs, Sb, Zr), and (ii) to verify if the elements extracted by the destruction of a given phase remain, as expected, in a soluble form after extraction.

Physical phases according to Tessier's definition were thus simulated by preparing artificial substrates spiked with stable equivalents of these artificial radionuclides and also As. These substrates, as well as a mixture of them simulating a sediment matrix, were therefore analyzed before and after the selective extraction procedure, so that elemental budgets could be obtained.

ANALYTICAL TECHNIQUES

Elemental determinations were made by instrumental neutron activation analysis (INAA) using a simplified monostandard method (7). 200 to 1000 mg samples were irradiated in polyethylene vials during 90mn in a well thermalised neutron flux (1.2 E+13n/cm2.s). After a few days of decay they were counted with a high-purity germanium detector (32% relative efficiency and 1.8 KeV resolution) coupled to a 8K multichannel analyzer placed inside a 1 cubic meter low-level lead shield 15 cm thick.

TABLE I: Selective extractions performed on individual substrates and the mixture (+). Unlikely source of element in the corresponding fraction of the mixture (0).

	Substrate						Mixture
Reagents	Montm.	CaCO3	Fe ox	Mn ox	H.A	Quartz	
MgCl2	+					0	+F1
NaOAc	+	+				0	+F2
NH2OH,HCl	+	0	+	+		0	+F3
H2O2	+	0	0	0	+	0	+F4
Acids	+	0	+	0	+	+	+F5
% in mixt.	13.7	4.6	4.5	4.5	4.6	68.10	

A number of trace-elements naturally occurring in the irradiated substrates could also be determined. However, due to the time required to perform chemical extractions, several relatively short-lived activated isotopes were no more detectable with sufficient accuracy after several weeks and thus could not be measured in all samples. Reported errors are one standard deviation.

Artificial substrates coresponding to the five fractions were prepared according to Meguellati & al (8) and represented by montmorillonite (Montm.), calcium carbonate, iron and manganese oxides, humic acid :Fluka AG (H.A) and quartz respectively. All substrates, except quartz, were spiked with As2O5, Ce(NO3)3, Co(NO3)2, CsNO3, ZrCl4 (containing Hf) solutions, either by direct coprecipitation (CaCO3, oxides) or a 24h immersion (montmorillonite, humic acid). Solid phases were recovered by filtration and oven dried. An artificial matrix was prepared by mixing these substrates as indicated in table I. Homogeneization was ensured by grinding and shaking during 4 days. Chemical extractions were performed according to table I. The complete sequence was only applied to the montmorillonite and the mixture, whereas the residues recovered after partial attack of Fe-oxide and humic acid were completely mineralized.

TABLE II: Percentage of elements removed from the individual substrates by selective extractions and complementary attacks (nm = not measurable on the substrate before extraction; nd = not detected after extraction).

	attack	As	Ce	Co	Cs	Hf	Zr	Ca	Cr	Fe	Sb	Sc	Sm	Th	mean
				spiked											
Montm.	F1	3	58	45	2	< 1	< 1	25	12	50	nd	2	19	< 1	
	F2 to F5	nd	13	30	66	61	76	nd	nd	60	nd	98	nd	98	
	Total		71	75	68	62	77			110		100		99	83
CaCO3	F2	88	93	97	nm	84	98	96	nm	nm	69	nm	nm	nm	
	rinsing	nd	1	1		3	3	nd			nd				
	Total		94	98		87	101								95
Fe-ox	F3	33	101	100	nm	50	90	nm	91	44	36	nm	87	nm	
	rinsing	48	2	2		34	39		7	64	56		26		
	Total	81	103	102		84	129		98	108	92		113		101
Mn-ox	F3	53	100	101	88	70	89	nm	nm	nm	77	nm	nm	nm	
	rinsing	nd	1	1	1	16	17				19				
	Total		101	102	89	86	106				96				97
H.A.	F4	80	81	88	77	30	33	nm	67	70	nm	37	59	22	
	residue	nd	3	8	11	33	69		11	17		48	nd	74	
	Total		84	96	88	63	102		78	87		85		96	87

RESULTS AND DISCUSSION

Individual substrates

They have been submitted to the corresponding treatment of Tessier's method and sometimes to a complementary attack of the eventual residue (table II). The overall efficiency of the recovery for the various elements averaged 90%.

The CaCO3 substrate was completely destroyed and all measurable elements were solubilized and correctly recovered. Dissolution of the

Fe-oxide was incomplete. However a good recovery was observed for Ce, Co, Cr and Sm, whereas a significant percentage of the other elements remained associated with the residue. On the contrary, no solid residue could be observed after the attack of the Mn-oxide, but only a few elements were well recovered (Ce, Co, Cs). Subsequent washing of the vial with 12 N HCl was necessary to recover the missing 16-20% of the other elements. The humic acid substrate contained 10-15% solid impurities prior to the analysis. The chemical recovery was satisfying for As, Ce, Co but part of the other elements (especially Hf, Sc, Zr) remained associated with the residue. In the case of the montmorillonite, the recovery of spiked elements varied from <1% (Zr) to 58% (Ce). A similar range was observed for the other elements.

Obviously,the quality of the artificial substrates may be suspected and contribute to explain part of the observed anomalies. Drying at 110°C of the spiked Fe-oxide has possibly resulted in a partial cristallization of this substrate, although no convincing evidence of the occurrence of cristallized iron-bearing minerals could be obtained by conventional X-ray diffraction analysis. As, Hf, Sb and Zr seem to have been preferentially trapped by such a "resistant" iron phase. An alternative explanation is a relative unadequacy of the reagents used during the attack. The latter hypothesis is supported by the results obtained with the Mn-oxide showing that after an identical chemical extraction, a significant but lower percentage of the same elements was also lost with Tessier's procedure. With regards to the montmorillonite sample, pretreatment with diluted HNO_3 according to Meguellati & al (8) has probably damaged the crystal lattice to a certain extent, so that an unexpected high amount of iron was removed by the $MgCl_2$ attack. However this interpretation is not consistent with the very bad recovery of most of the spiked elements which were supposed to be located in exchangeable sites, and readsorption processes can be envisaged.

In conclusion, although several artifacts related to the preparation of the artificial substrates cannot be definetely ruled out, this selective extraction procedure is likely to result in an insufficient recovery of As, Sb, Hf and Zr in the oxides and possibly also in the montmorillonite and the humic acid fractions. In spite of this problem, a reasonable rate of recovery (>80%) was found in the oxides and the humic acid fractions for several elements such as metals (Cr, Co), an alkaline element (Cs) and a rare earth element (Ce), and in the carbonate fraction for all measurable elements.

Mixture of the substrates

The former data obtained by analyzing individual substrates should allow to predict the result of a complete extraction sequence performed directly on a mixture of these substrates. We have thus compared this expected speciation, computed on the basis of the experimental recoveries previously described weighted by the abundance of the substrates in the mixture, to the experimental measurements (table III).

The ratio between measured and calculated speciation is very variable, showing discrepancies of one order of magnitude in the first fractions, and only a factor of 2 in F4 and F5. A more practical examination of the results can be done by considering the difference between measured and calculated percentages which is plotted for each fraction in figure 1. In order to avoid a misleading representation of the data when the measured and calculated elemental budgets do not coincide accurately, the calculated data have been corrected for this difference (Cs, Zr, Hf and Cr).

This comparison however may be biaised to a certain extent by the methodology which has been employed. As shown in table I, the complete extraction sequence has not been performed on each of the individual components of the mixture, except montmorillonite. Thus the potential source of elements extracted from the mixture with each fraction cannot be always clearly identified. For instance, exchangeable elements in F1 may originate not only from the spiked montmorillonite but also from other substrates.

Nevertheless the insufficient extraction in F1 already observed with the montmorillonite is confirmed. The significant differences between measured and calculated speciation (table III) appear negligible on an absolute basis (figure 1) due to the small percentage of montmorillonite in the mixture.

TABLE III: Total concentrations in the mixture and percentage of elements extracted in the various fractions (calculated and measured).

		Ce	Co	Cs	Hf	Zr	Cr	Fe	Sc	Th
total	ppm	1212+-10	144+-1	89+-2	153+-5	4940+-200	10+-2	26940+-730	0.84+-0.03	5.3+-0.3
F1	calc	16.50+-0.50	3.12+-0.09	1.03+-0.04	0.14+-0.01	0.17+-0.03	1.3+-0.4	2.90+-0.12	0.5 +-0.1	0.26+-0.02
	meas	0.89+-0.02	0.70+-0.01	5.52+-0.18	0.94+-0.04	1.26+-0.15	7.7+-1.7	0.29+-0.02	0.8 +-0.1	0.75+-0.10
F2	calc	10.30+-0.10	9.91+-0.07	16.6 +-0.5	4.9 +-0.2	5.3 +-0.2	0.9+-0.3	0.04+-0.01	0.83+-0.07	0.26+-0.09
	meas	1.70+-0.05	1.20+-0.02	39.3 +-0.6	2.3 +-0.1	2.8 +-0.1	9.5+-2.3	1.00+-0.04	3.6 +-0.2	1.9 +-0.6
F3	calc	46.0 +-1.2	76 +-2.1	19.1 +-1.2	13.1 +-1.6	17.6 +-1.4	49 +-10	45 +-2	23 +-1	0.59+-0.02
	meas	58.3 +-1.8	84 +-2	16.3 +-0.6	5.6 +-0.3	3.8 +-0.5	73 +-25	72 +-3	47 +-2	<=7.6
F4	calc	12.0 +-0.3	4.41+-0.16	25.2 +-0.7	8.8 +-0.4	13.9 +-0.6	14.1+-2.9	0.77+-0.06	26 +-1	11.0 +-1.7
	meas	26.4 +-0.8	6.9 +-0.2	19.4 +-0.7	23.9 +-1.0	24.5 +-1.2	16.6+-7.1	14.7 +-0.6	21 +-1	8.5 +-1.8
F5	calc	1.61+-0.04	1.56+-0.03	9.02+-0.28	38 +-2	44 +-2	14 +-3	65 +-2	43 +-2	79 +-6
	meas	2.61+-0.08	1.53+-0.05	4.3 +-0.4	54 +-2.4	68 +-3	20 +-7	26 +-1	24 +-1	76 +-2
F1-5	calc	86+-1	95+-2	71+-1	65+-2	81+-3	79+-11	114+-3	93+-2	91+-6
	meas	90+-1	94+-2	85+-1	86+-3	100+-4	127+-27	114+-3	96+-2	87-95

Comparisons in F2 and F3 are discussed below for elements presenting a similar behaviour.

- Thorium and chromium : no significant matrix effect was detected and deviations from calculated speciation were smaller than 5%. This result is very uncertain for Cr due to large analytical errors.
- Iron and scandium : the sharp excess in F3, which is later compensated by a correlative deficiency in the next fractions, points to the occurrence of additional sources not taken into account in the calculations. The iron excess (7000 ppm) can only originate from the residue left after F3 attack on the iron oxide (16000 ppm), other sources being negligible. The iron content of the oxide alone and of the oxide in the mixture is similar, thus the homogeneity of this substrate cannot be suspected. Sc excess (0.2 ppm) cannot originate from the oxides, carbonate and quartz in which this element was not detected. Partial attack of montmorillonite and humic acid (in which total Sc content amounts to 0.7 ppm) may have supplied the excess Sc in the mixture, as well as the residue previously found. If this latter source is real, and in the case of iron, the anomaly found in F3 can be

ascribed to a matrix effect, the extraction efficiency in this fraction beeing dependant on the composition of the sample analyzed.

- Zirconium and hafnium : the deficiency in F3 confirms the conclusion derived from the attack on the cxides alone. The unadequacy of the reactives for these elements is poorly reproducible in different matrixes and results in the formation of an unsoluble coumpound which is later recovered by subsequent extractions.

- Cerium and cobalt : the common pattern is a systematic deficiency in the two first fractions. This result is not consistent with the single substrate experiment which has shown an excellent recovery in the carbonates. Here also a matrix effect must be envisaged. Furthermore the observed unadequacy of the reactives in F1 is confirmed, especially for Ce.

- Caesium : the origin of the excess in F1 (4 ppm) and F2 (20 ppm) cannot be easily identified. On the one hand the excess in F1 may result from a better recovery of spiked Cs in the montmorillonite (60 ppm), thus implying a some matrix effect; on the other hand partial extraction of Cs from the Mn oxide and the humic acid (total content: 24 ppm) may have also contributed to both excesses, and a methodological artifact cannot be discarded.

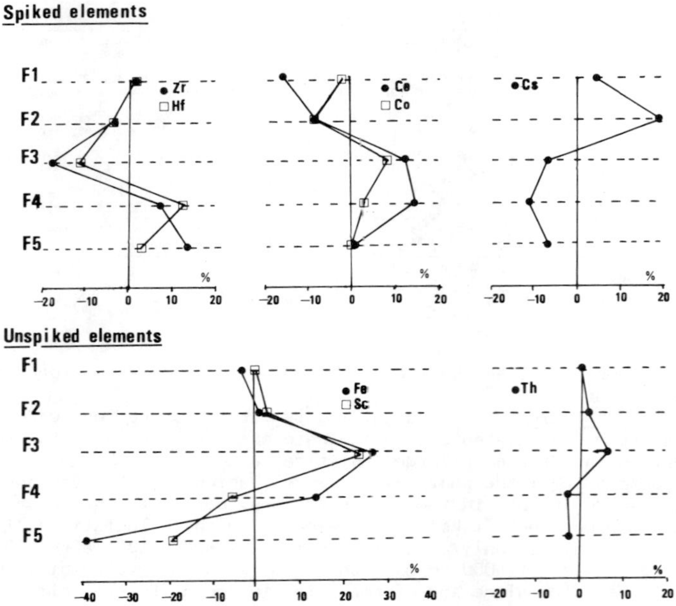

FIGURE 1: Differences (in %) between measured and calculated speciation in the mixture of artificial substrates. (Negative values correspond to underestimations).

In conclusion, results obtained with individual substrates were poorly reproduced in the mixing experiment, showing that the sequential extraction procedure is biaised by matrix effects. Similar inconvenients have been already reported by Robinson (9). Most of the over-estimations cannot be interpreted without ambiguity and methodological problems are probably implicated. Extraction deficiencies in the montmorillonite were replicated in the first fraction of the mixture (except Cs). Under-estimations of F2 imply a readsorption of the elements after destruction of the carbonate. This observation confirms the conclusions of Rendell & al (10). Insufficient extraction of Zr and Hf from the oxides was aggravated in F3 of the mixture. All these deficiencies are of course counter-balanced by excesses in the last fractions.

CONCLUSIONS

A number of inherent limitations must be kept in mind before using this method. Its application to elements not currently investigated may be hazardous if careful tests are not undertaken to check its efficiency. In particular under-estimations are to be expected in the first phases with Zr, Hf, Ce and Co. Environmental availability of artifical isotopes such as Zr-95, Ce-144 and Co-60 may therefore be poorly estimated.

Application to natural sediments will give reproducible results as far as the relative abundance of their organic and inorganic components does not show large varitions, but a reliable identification of a given fraction (even operationally defined) cannot be ensured in different matrixes. Comparisons of speciation between different sedimentary environments may be questionable.

Quantitative extrapolation of the results of the present study is difficult. The artificial substrates which were used are not fully representative of their natural equivalents, and the quantitative impact of transfers from a given fraction to the following ones is obviously dependant on the initial speciation of the sample analyzed.

ACKNOWLEDGMENTS : this study has been carried out under financial support of EDF (Electricité de France) contract N°E30 L 14/2 E 5931 and the CNRS :Land Sea Interaction Coordinated Group "I.C.O" and U.A 386.

REFERENCES

1. A. Tessier, P.G.C. Campbell, M. Bisson, Anal. Chem. , 51 , 7, p. 844-851 (1979)
2. E.A. Jenne, Symposium on Molybdenum in the environment , W. Chappel & K. Petersen Eds, (M. Dekker, Inc N.Y), 2 , p. 425-553 (1977)
3. J.M. Brannon, J.R. Rose, R.M. Engler, I. Smith, Chemistry of Marine Sediments , T.F. Yen Ed. (Ann Arbor Science), p. 125-149 (1977)
4. R. Chester, M.J. Hughes, Chem. Geol. , 2 , p. 249-262 (1967)
5. R. Van Valin, J.W. Morse, Mar. Chem. , 11 , p. 535-564 (1982)
6. U. Forstner, W. Salomons, NATO Workshop on Trace Elements Speciation in Superficial Waters and its Ecological Implications, Nervi (Italy) (1981)
7. G. Delcroix, J.C. Philippot, J. Radioanal. Chem. , 15 , p. 87-101 (1973)
8. N. Meguellati, D. Robbe, P. Marchandise, M. Astruc, Conf. Heavy Metals

in the Environment, Page Bros (Norwich) Ltd, Heidelberg, p.1090-1093 (1983)

9. G.D Robinson, Chem. Geol. , 47 , p. 97-112 (1984)

10. P.S Rendell, G.E. Batley, A.J. Cameron, Environ. Sci. & Tech. , 14 , 3, p. 314-318 (1980)

VAPORISATION OF SIMULANT FISSION PRODUCTS: IDENTIFICATION OF CHEMICAL SPECIES BY MEANS OF MATRIX ISOLATION-INFRARED SPECTROSCOPY

B R Bowsher*, A L Nichols* R A Gomme+, J S Ogden+ and N A Young+

*Chemistry Division, AEE Winfrith, Dorchester, Dorset;
+Department of Chemistry, The University, Southampton.

ABSTRACT

During a hypothetical severe accident in a light water reactor (LWR) with the subsequent failure of the emergency core cooling system, the core would overheat with the release of volatile fission products. Some proportion of these fission products would be transported through sections of the primary circuit and the containment building to the environment. Detailed assessments of such accidents would benefit from a knowledge of the chemical species released from the fuel and any subsequent chemical reactions within the primary circuit and containment building.

This paper describes the development of the matrix isolation technique to study the chemical forms of the fission product vapours released from overheated fuel. Vapour species generated from well-defined sources are condensed rapidly and isolated in a large excess of inert material (eg, argon or nitrogen). The resulting sample can then be studied at very low temperatures (~12K) by various analytical techniques, including infrared spectroscopy. In the initial studies non-radioactive simulant fission products have been volatilised from separate compounds or mixtures. This work has concentrated upon caesium, iodine and tellurium compounds mixed with uranium dioxide, and volatilised at temperatures up to 1900K in the presence and absence of water vapour. Various combinations of compounds have been examined, including Cs_2MoO_4, Te/UO_2, $Te/TeO_2/UO_2$, Cs_2UO_4/Te, CsOH/Te and CsI/B_2O_3. The spectra of the vapour species from these systems were characterised in terms of equivalent data from a series of supporting studies. Caesium oxo-anion salts were found to be stable in the vapour phase and mixed oxides were evolved from the Cs/Te/O and I_2/CsOH systems. Studies of the $CsI/B_2O_3/H_2O$ system at low temperatures indicated the formation of HI in the vapour phase.

1 INTRODUCTION

Detailed assessments of hypothetical severe reactor accidents require data defining the quantities and physicochemical forms of the radioactive species released from the overheated core (1). The extent of any release from a damaged reactor core depends upon the plant design, burn-up history of the fuel and the accident sequence, and for the purposes of this paper attention will be focussed on the three radiobiologically important elements caesium, iodine and tellurium (1,2). Transport and attenuation of these fission products in the primary circuit and containment building will depend in part upon the chemical species formed. Under accident conditions in a pressurised water reactor (PWR), they would be released from the UO_2 fuel into an immediate environment of Zircaloy-zirconium oxide/stainless steel/control rods/steam/boric acid, and thus a realistic data base for chemical speciation involves at least sixteen elements: U, Cs, I, Te, Zr, Sn, Fe, Cr, Ni, Mn, Ag, In, Cd, H, O and B.

There are two basic approaches to the general problem of establishing the chemical species formed in such a complex system. The first of these is essentially theoretical, and relies upon the use of existing thermodynamic data to predict equilibrium concentrations for the most likely chemical species. This approach appears very attractive because it allows predictions to be made over a wide range of pressure, temperature and elemental composition, and the recent study by Garisto (3) illustrates this flexibility. However, it is totally dependent on the correct assumptions being made regarding the molecules present. If the required thermodynamic data are not available, or if the system contains a significant proportion of an unsuspected or unknown species, the resulting predictions will not be reliable.

An alternative approach is to study the relevant high temperature systems in situ and to establish the identities and concentrations of the molecular species experimentally. Unfortunately, the difficulties encountered in the direct study of high temperature vapours are considerable and, despite recent advances in mass spectrometric sampling (4), most results continue to be obtained from the classical combination of vapour transport and chemical analysis of the condensate (5). However, in recent years matrix isolation has emerged as a powerful technique in the characterisation of high temperature vapours (6,7). This paper describes a series of matrix isolation experiments to study the interaction and vaporisation of Cs, I and Te compounds from UO_2 in the presence of boric acid and steam, and so aid in the assessment of hypothetical severe accidents involving PWRs. Caesium is believed to exist in the fuel with oxo-anions to form such compounds as caesium uranate (8) and molybdate (9). However, it has been proposed that the major caesium release from damaged fuel will be as caesium oxide and hydroxide (10, 11, 12, 13). The iodine release will be as caesium iodide vapour, and tellurium is predicted to evolve either as elemental tellurium, tellurium oxides (TeO, TeO_2 or Te_2O_2) or hydrogen telluride (12, 14, 15), depending on the accident conditions. Although there are a number of uncertainties associated with these predictions, they do provide a suitable basis for simulant fission product studies.

The aim of the matrix isolation-infrared (ir) experiments was to identify the most important vapour species via their characteristic ir absorptions and to build up a body of matrix spectral data which will be useful in future studies. The systems considered to be of direct relevance were Cs_2MoO_4, Cs_2UO_4/Te, CsOH/Te, CsOH/TeO_2, Te/UO_2, CsI/H_3BO_3 and I_2/H_2O. Although literature data were available for the binary oxides of uranium (16), some systems, particularly those containing tellurium and iodine, required additional background information, and a number of supporting experiments were necessary in order to assist spectral interpretation: Cs_2CO_3/TeO_2, Cs_2TeO_3, Cs_2CO_3/B_2O_3, TeO_2, TeO_3, Te/TeO_2/UO_2, CsI/B_2O_3 and HI/H_2O.

2 MATRIX ISOLATION

In a typical matrix isolation study of a high temperature system, the vapours of interest are produced in a Knudsen cell and traverse a distance of 10 to 20 cm in high vacuum before impinging on a suitable deposition surface cooled to cryogenic temperatures. A large excess (~1000X) of an inert gas is condensed simultaneously with the species of interest, solidifying and trapping these species in a rigid matrix on the cooled surface. Cryogenic temperatures are most conveniently attained using a closed-cycle refrigerator operating at ~10K, and sample deposition takes place over periods of ~1 hour using nitrogen or argon as matrix gases. Under these conditions the isolated high temperature molecules collected on the cooled surface have very extended lifetimes and can be studied at leisure. Although the vapour phase concentrations of these species may be very low, the cumulative build-up of the sample on the cooled surface means that a wide range of spectroscopic methods can be used to study the trapped species, viz, ir, uv/vis, Raman and ESR.

Infrared spectroscopy is the most useful exploratory technique, and has been used to study the characteristic vibrational transitions of the trapped species (in which the intensities of such bands are dependent on concentration). In common with more conventional forms of sampling, matrix-isolated molecules are subject to harmonic oscillator and symmetry selection rules. However, in contrast to gas phase or solution spectra, the matrix isolation-ir vibrational bands are typically single sharp lines (1 to 3 cm^{-1} wide), without any rotational structure at these low temperatures. This not only reduces the chance of accidental overlap, but also permits the extensive use of isotopic labelling to confirm the identity of the trapped molecule.

3 EXPERIMENTAL TECHNIQUE

Vapour species were generated using inductive heating up to ~1900K and deposited with either a nitrogen or argon matrix gas onto an alkali halide central window cooled by thermal contact with a cryogenic reservoir (Figure 1). The majority of samples were vaporised from silica or alumina holders located inside the inductively-heated tantalum sleeve. The caesium iodide windows used to collect the samples transmitted ir radiation down to 200 cm^{-1}, and the necessary cooling was provided by an Air Products "Displex" closed-cycle refrigerator operating at ~12K. After deposition, the central window was rotated through 90° and transmission ir spectra were then recorded using a Perkin Elmer Infrared Spectrophotometer (5000 to 200 cm^{-1}).

4 RESULTS

4.1 Cs_2MoO_4

Figure 2 shows part of the high resolution spectrum obtained after heating a sample of caesium molybdate, and isolating the vapour species in a nitrogen matrix. A detailed analysis of this complex spectrum (17) shows that molecular Cs_2MoO_4 has been isolated, and that this species has a bis-bidentate structure.

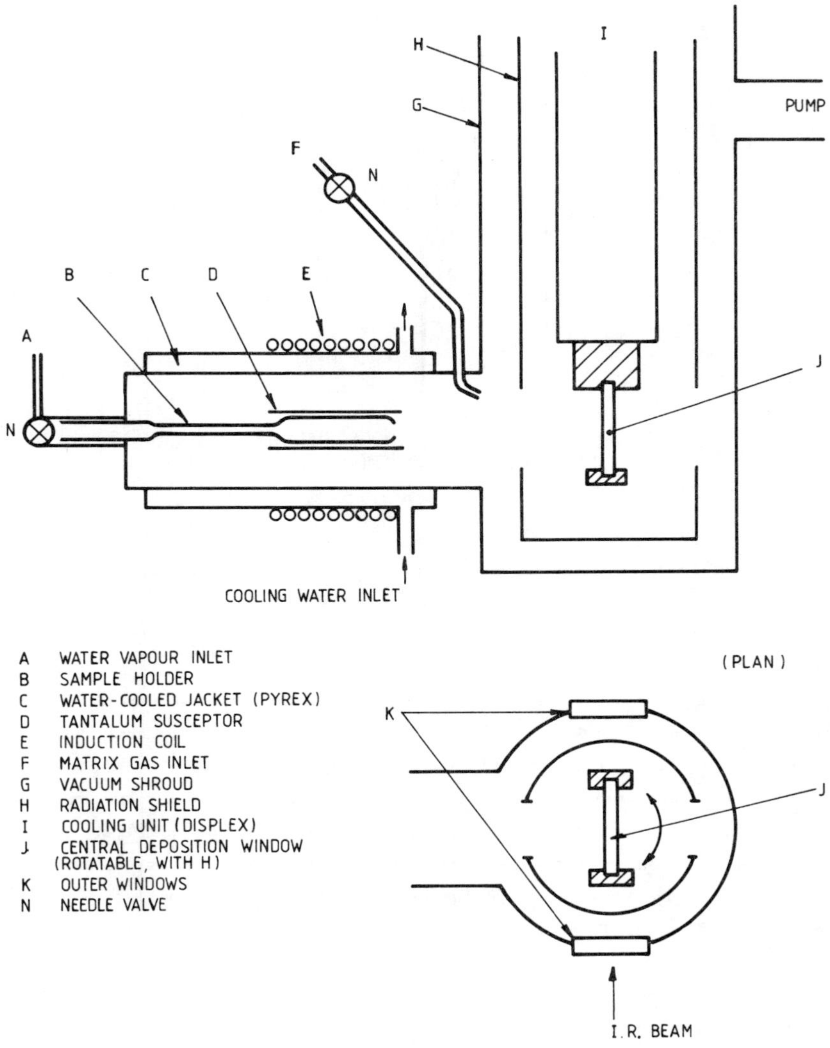

FIG.1 MATRIX ISOLATION - INFRARED SPECTROSCOPY APPARATUS

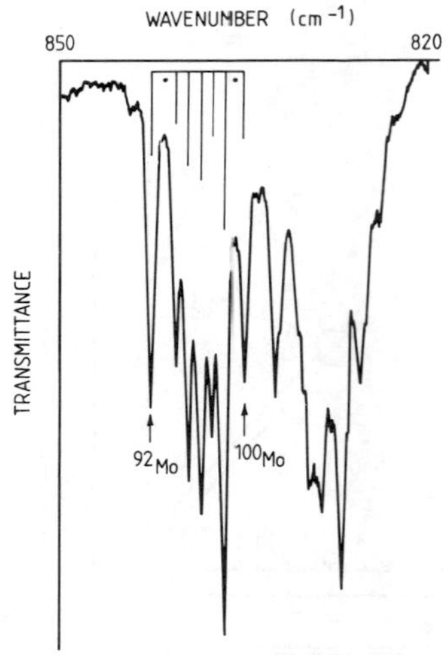

FIG. 2. HIGH RESOLUTION NITROGEN-MATRIX INFRARED SPECTRUM OBSERVED FROM THE VAPORISATION OF Cs_2MoO_4

4.2 Te/U/O Systems

Figure 3 summarises the spectral data obtained from experiments with TeO_2 and TeO_3, and mixtures of Te/UO_2 and $Te/TeO_2/UO_2$. All of the matrix spectra show characteristic absorptions of molecular TeO_2, together with a number of unassigned features.

The spectrum shown in Figure 3(e) was obtained from 5 mg TeO_2 mixed with an excess of inert Al_2O_3 powder. The excellent signal-to-noise ratio demonstrates the sensitivity of this technique, and indicates that sample weights of approximately 1 to 2 mg TeO_2 would give detectable absorptions. This implies that matrix isolation-infrared spectroscopy could be used to study the fission product chemical species released from irradiated fuel pellets since the amount of tellurium present in a 10 g pellet of highly irradiated UO_2 fuel is approximately 6 mg.

4.3 Cs/Te/U/O Systems

The spectra obtained from ternary and quaternary mixtures of these elements are shown in Figure 4. Both molecular TeO_2(A) and Cs_2TeO_3 (F and G) are evolved. Although many of the remaining features are still unexplained, these spectra provide an excellent illustration of the fingerprinting aspect of the technique, and further work is in progress to identify and characterise all of the species in this system.

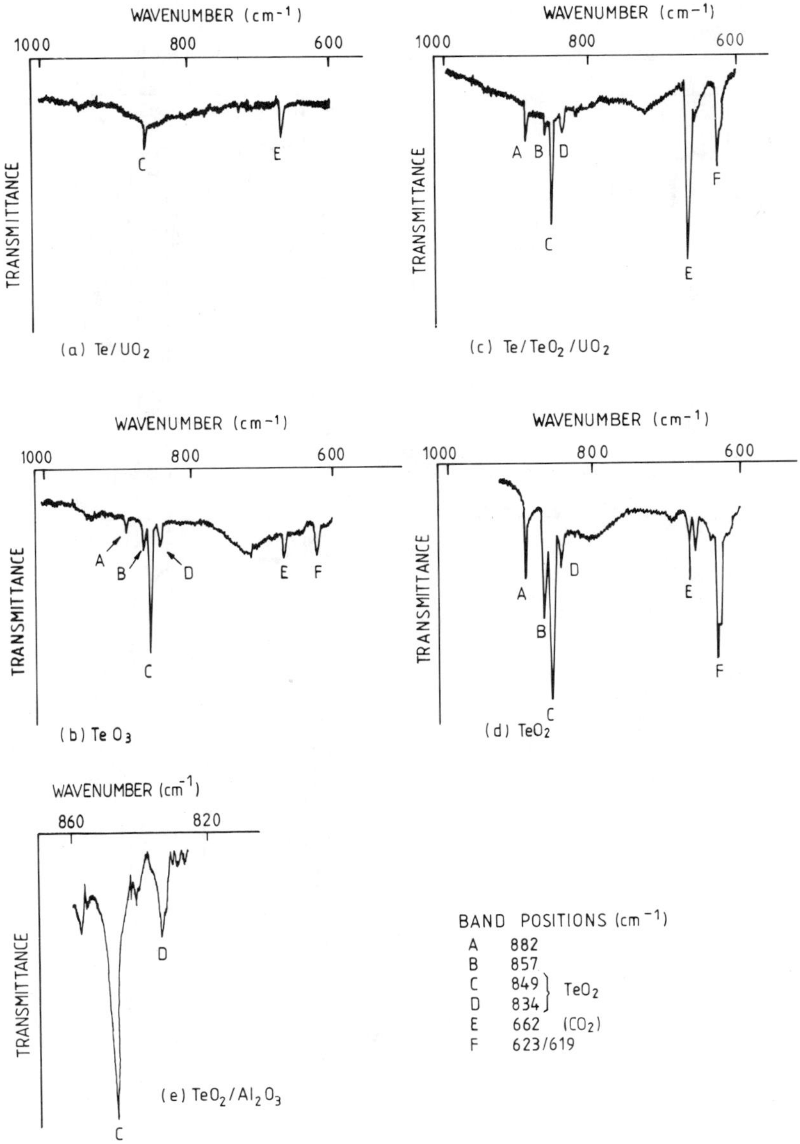

FIG.3 NITROGEN-MATRIX INFRARED SPECTRA OBSERVED FOR THE Te/U/O AND SUPPORTING SYSTEMS

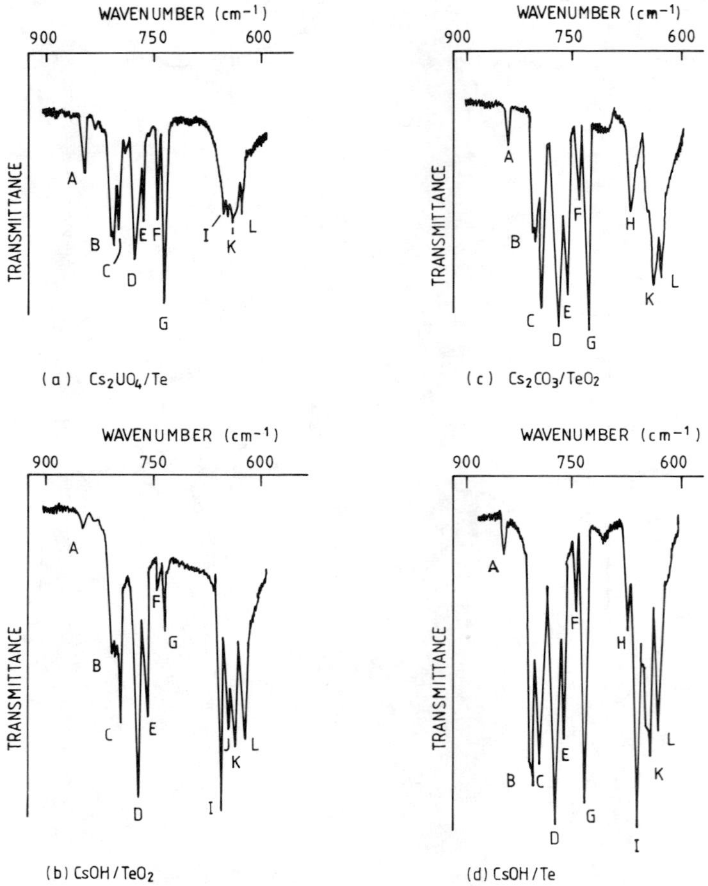

BAND POSITIONS (cm^{-1})

A	849 TeO_2	E	766	} Cs_2TeO_3	I	662	(CO_2)
B	811 / 808	F	747	} Cs_2TeO_3	J	650	
C	799	G	738	} Cs_2TeO_3	K	642	
D	777	H	674		L	632	

FIG. 4. NITROGEN-MATRIX INFRARED SPECTRA OBSERVED FOR THE Cs /Te /U /O AND SUPPORTING SYSTEMS

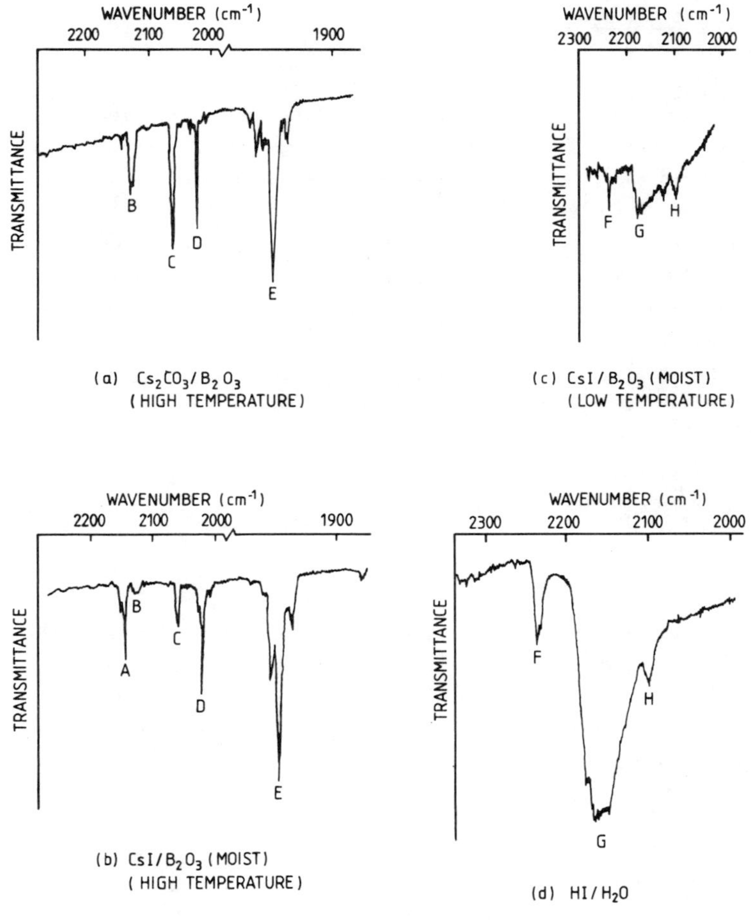

BAND POSITIONS (cm^{-1})

A	2140	CO	F	2237·2	HI
B	2125	B_2O_3	G	2160	$HI-H_2O$ COMPLEX
C	2055	B_2O_3	H	2100	$HI-H_2O$ COMPLEX
D	2015	$CsBO_2$			
E	1945	$CsBO_2$			

FIG. 5 NITROGEN-MATRIX INFRARED SPECTRA OBSERVED FOR THE Cs/I/H/O/B SYSTEMS

4.4 Cs/I/H/O/B Systems

The number of permutations within these systems is quite large, and fingerprint matrix-ir spectra have been obtained for molecular $CsIO_3$, $CsBO_2$, B_2O_3 and HBO_2. However, the possible interaction between caesium iodide and boric acid at high temperatures has emerged as an important chemical phenomenon, and Figure 5 summarises some recent results for this system.

Molecular $CsBO_2$ is a well-characterised high temperature species which can be produced by heating a mixture of Cs_2CO_3 and B_2O_3, and a spectrum of this system is shown in Figure 5(a). In addition to the $CsBO_2$ ir absorptions (18), this spectrum shows prominent bands due to molecular B_2O_3. Figure 5(b) shows that $CsBO_2$ is also produced by heating a mixture of CsI and moist B_2O_3 to 800°C and it is of some interest to discover the fate of the iodine evolved from this system.

It seems most likely that the formation of hydrogen iodide would be favoured in the absence of oxygen:

$$2CsI + B_2O_3 + H_2O \rightarrow 2CsBO_2 + 2HI$$

Figure 5(c) shows part of the matrix-ir spectrum obtained after heating a mixture of CsI and moist B_2O_3 to only 100°C and isolating the reaction products in a matrix. A sharp band (F) and broader features (G and H) are observed, which correspond very closely to the matrix-ir spectrum obtained from aqueous hydriodic acid (Figure 5(d)).

5 CONCLUSIONS

From an experimental viewpoint these matrix studies have yielded satisfactory spectra from samples weighing as little as 5 mg. Inductive heating proved to be a satisfactory method for sample vaporisation, and the ir frequency reproducibility was within 1 cm^{-1} for the isolated species.

These studies demonstrate the stability of molecular Cs_2MoO_4, Cs_2TeO_3 and $CsBO_2$ as high temperature vapours. This implies that species such as Cs_2RuO_3(19), $CsSbO_2$(20), $CsAsO_2$ (20), $CsTcO_4$ and Cs_2SeO_3 might similarly be involved in the transport of the fission products Cs, Ru, Sb, As, Tc and Se under severe reactor accident conditions, and that the existence of such molecular oxo-anion salts should be taken into account in thermodynamic modelling of these systems.

Our next objective is to apply this matrix isolation technique directly to the study of the vapour species emitted from overheated irradiated fuel. Future experimental work will concentrate on the characterisation of new species and the development of a library of fingerprint matrix-ir spectra. Improvements in apparatus design and instrumentation will be pursued so that smaller quantities of sample can be analysed.

6 REFERENCES

1. Reactor Safety Study - An Assessment of Accident Risks in US Commercial Nuclear Power Plants, WASH-1400, NUREG-75/014 (1975).
2. J.A. Gieseke, P. Cybulskis, R.S. Denning, M.R. Kuhlman, K.W. Lee and H.Chen, BMI-2104, Vols I-VI (1984).
3. F. Garisto, AECL-7782 (1982).
4. C.A. Stearns, F.J. Kohl, G.C. Fryburg and R.A. Miller, NBS Special Publication No 561, 303 (1979).
5. B.R. Bowsher, S. Dickinson and A.L. Nichols, AEE Winfrith, unpublished data (1983).
6. W. Weltner, Adv. High Temp. Chem., 2, 85 (1969).
7. A. Snelson, in Vibrational Spectroscopy of Trapped Species, edited by H.E. Hallam (John Wiley, London, 1973).
8. D.C. Fee and C.E. Johnson, J. Nucl. Mater., 78, 219 (1978);
   D.C. Fee and C.E. Johnson, J. Inorg. Nucl. Chem., 40, 1375 (1978);
   D.C. Fee and C.E. Johnson, J. Nucl. Mater., 99, 107 (1981).
9. I. Johnson, J. Phys. Chem., 79, 722 (1975).
   I. Johnson, and C.E. Johnson, Thermodynamics of Nuclear Materials, Vol I, IAEA-SM-190/43 (1975).
10. Degraded Core Analysis, Report of a UKAEA Committee Chaired by J.H. Gittus, ND-R-610(S) (1982).
11. J.A. Gieseke, P. Cybulskis, R.S. Denning, M.R. Kuhlman, K.W. Lee and H. Chen, Radionuclide Release Under Specific Accident Conditions, BMI-2104, Vol V, PWR-Large Dry Containment Design (Surry Plant Recalculations) (1984).
12. Technical Basis for Estimating Fission Product Behaviour During LWR Accidents, NUREG-0772 (1981).
13. M. Levenson and F.J. Rahn, Nucl. Tech., 53, 99 (1981).
14. S. Levine, G.D. Kaiser, W.C. Arcieri, H. Firstenberg, P.J. Fulford, P.S. Lam, R.L. Ritzman and E.R. Schmidt, ALO-1008, NUS 3808 (1982).
15. D. Cubicciotti, and J.E. Sanecki, J. Nucl. Mater., 78, 96 (1978).
16. S.D. Gabelnick, G.T. Reedy and M.G. Chasanov, J. Chem. Phys., 58, 4468 (1973); 59, 6397 (1973).
17. L. Bencivenni and K.A. Gingerich, J. Chem. Phys., 76, 53 (1982).
18. K.S. Seshadri, L.A. Nimar and D. White, J. Mol. Spectrosc., 30, 128 (1969).
19. R.A. Gomme and J.S. Ogden, University of Southampton, unpublished data (1983).
20. J.S. Ogden, and S.J. Williams, J. Chem. Soc. Dalton Trans., 825 (1982).

THE INFLUENCE OF ENVIRONMENTAL FACTORS ON THE SOLUBILITY OF PU, AM AND NP IN SOIL-WATER SYSTEMS

R.M.J.PENNDERS, M.PRINS AND M.J.FRISSEL
Laboratory for Radiation Research, RIVM,
P.O.Box 1, 3720 BA Bilthoven, The Netherlands

ABSTRACT

Speciation phenomena result in differences in solubility and therefore also in differences of bioavailability. The solubilities of ^{241}Am, $^{239+240}Pu$ and ^{137}Cs as a function of environmental conditions were observed in sediment-water and sand-water systems. Particular attention was given to the pH, E_H, CO_2 pressure and action of microorganisms. Numerous leaching experiments were carried out. The similarity in behaviour of Am and Eu on one hand, and Cs and Pu on the other hand is remarkable. A correlation was noticed between the differences of the solubility of Am, Pu and Cs and the differences of the uptake by food crops. Differences in uptake are more pronounced than differences in solubility.

1. INTRODUCTION

From a biological point of view the bioavailability of a radionuclide for specified environmental conditions such as pH, E_H (redox potential), and presence of complexing agents is one of the most characteristic properties which results from speciation. Since the early days of systematic chemical soil science research investigators have tried to correlate the bioavailability with results of extracting procedures, sometimes with, but often without success. Despite this the authors have reviewed their extraction and solubility data in the light of existing data on the uptake of Pu, Am, Np and Cs by plants from soils.

2. MATERIALS

- Sediments from the Irish sea near Sellafield (Ravenglass estuary). This material has been chosen because it has been in situ in contact with radionuclides for a rather long time and the chemical form of the radionuclides is expected to be adapted to the environmental conditions. Moreover, the concentration of ^{241}Am, $^{239+240}Pu$ and ^{137}Cs is sufficiently high to allow experiments without further additions of radionuclides. Composition: Sand (approx. 100-150 μm) 80%, (approx. 300-400 μm) 10%. Ferruginous floc + organics (< 10 μm) 8%. Fine siliceous silt (< 10 μm) 1%, shell 1%. The sand consists mainly of equidimensional grains of quartz containing ferruginious bodies, weathered feldspar, and trace amounts of iron ore, zircon, barytes, etc.

- Glauconite containing sandy deposits of eocene age (Brussel's sand). This material has been choosen because it is present in the geological formations which surround salt domes in The Netherlands which are expected to be selected for nuclear waste deposition.
  Because no samples from near the depository were available, samples were taken from a top soil near Leuven (B) were this Brussel's sand reaches the surface. Composition: Sand (50-2000 μm) 91.1%. Silt (2-50 μm) 2.9%. Clay (< 2 μm) 6.0%. pH (KCl) 5.6. Org.matter content 0%, $CaCO_3$ 0%, N 14 mg/100 g soil, P 5 mg/100 g soil. CEC 3.7 meq/100 g soil. SO_4^{2-} 0,01%.
  Radionuclides have been added in the form of nitrates, Tc as pertechnetate.

3. EXPERIMENTAL

3.1 E_H and pH experiments in culture vessels

110 gram samples of sediments or glauconite containing sand were suspended into 1100 ml of aqueous solution within a closed vessel which allowed control of pH, pCO_2 and growth of bacteria. The pH and E_h were measured continuously. The CO_2 pressure was assumed to be controlled by the composition of the gas mixture delivered to the system. Because microorganisms also produce CO_2 this is not strictly correct. However, because a continuous gas flow system was used errors were probably insiginificant. When sediments were investigated the growth of microorganisms was induced by pulse application of glucose. After each alteration of the conditions, e.g. a change of the CO_2 pressure, these conditions were maintained for 7-14 days.
Thereafter the vessel was opened and a one ml suspension sample was taken and the radionuclide concentration measured in the solid and liquid phase.

3.2 Extraction of sediments

165 gram samples of sediments were placed into small cylinders, at the bottom closed with 0.2 μm filters. To investigate a possible leaching of radionuclides in the 0.2 - 70 μm range also leaching cylinders closed with 70 μm filters were prepared. Leaching occurred with 60 - 150 l per sample. The pH of the leaching solutions was adjusted by 0.1 N NaOH or 0.1 N HCl. In one experiment citric acid (0.1 N) was added. The leaching rate was appr. 1.7 $cm.d^{-1}$.

Small differences in leaching rate and amount of leacheate occurred; the results are standardized by linear interpolation.

3.3 Salt concentration

Sea water has a salt concentration of 0.55 molair, soil water in contact with geological salt formations may reach levels of 3 to 4 molair depending on the salt composition. Extractions occurred at concentrations of 1, 0.1 and 0.03 molair. The composition of the 1 molair solution was: NH_4^+ 0.0009 $mol.l^{-1}$, Na^+ 0.945 $mol.l^{-1}$, K^+ 0.0028 $mol.l^{-1}$, Ca^{2+} 0.0231 $mol.l^{-1}$, Sa^{2+} 0.0007 $mol.l^{-1}$, Mg^{2+} 0.0210 $mol.l^{-1}$, Mn^{2+} $1.09.10^{-5}$ $mol.l^{-1}$, SO_4^{2-} 0.0094 $mol.l^{-1}$, Cl^- 1.039 $mol.l^{-1}$, HCO_3^- 0.0015 $mol.l^{-1}$, Br^- 0.0007 $mol.l^{-1}$. This is a similar ratio as has been found in water from aquifers near a salt dome at about 500 m depth, provided that ferro ions were omitted to prevent forming of iron hydroxides. The other molarities were made by dilution of the 1 molair solution. To simulate soil water in a salt free soil a 0.000166 N solution was used for which chlorides of Ca, Na and K were used in the ratio 3 : 1 : 1. The experiments were carried out as batch experiments with 2 g soil samples in 20 ml solution. Equilibration time varied from 24 to 504 h. The radioactivity was determined in the liquid phase after centrifugation.

4. RESULTS

4.1 E_H, pH experiments

The results of the E_H, pH experiments are shown in fig. 1 and 2. The observations strongly suggest that two Pu species were present, the one being PuO_2^+ and the other one Pu^{3+} and/or $PuCO_3^+$. There were no signs of irreversibility, but measurements were only made after an equilibration time of 14 days. The solubility of the Pu^{3+} species decreases with increasing pH, as can be expected from the E_H, pH diagram (fig. 3). The solubility of the PuO_2^+ increases with increasing pH. This cannot be concluded from the E_H, pH diagram and has probably to be attributed to the formation of Pu hydroxide colloides.

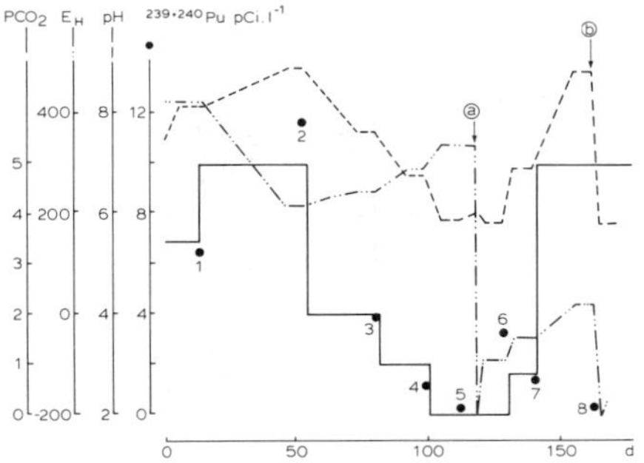

FIGURE 1 The solubility of $^{239+240}Pu$ versus time, under varying conditions of pH, E_H, pCO_2 and bacterial growth. The CO_2 pressure is controlled by the supply of different gas mixtures (pCO_2 of pure CO_2 gas is 0, pCO_2 of N_2 gas containing 0.001% CO_2 is 5). At time a a yeast extract and sufficient glucose to reach a 0.1% level was added, at b this is repeated for glucose. E_H and pH were measured continuously. After each change of the conditions there was a waiting (equilibration) time of 14 days before the Pu concentration was measured.

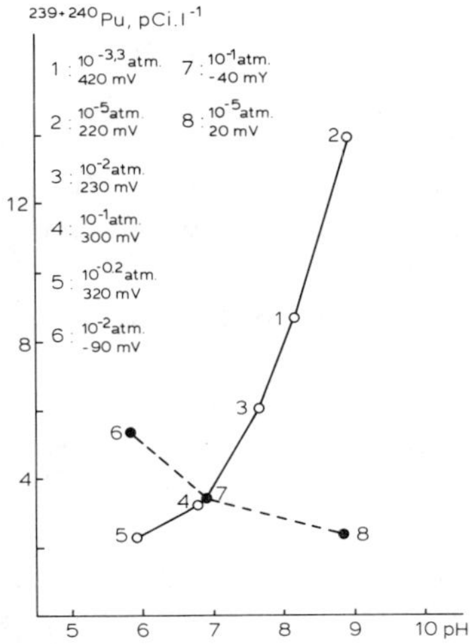

FIGURE 2 The solubility of $^{239+240}Pu$ as function of the experimental conditions. The symbols 1-8 refer to the sequence of sampling as indicated in fig. 1. The observations strongly suggest that two Pu species are present. The solubility of both species seems to be controlled by the pH and is for at least one of the species reversible. The CO_2 pressure does influence the pH, but not the solubility. A comparison of the data with E_H, pH diagrams (e.g. fig. 3) shows that the curve representing the symbols 1-5 agrees with the PuO_2^+ stability area. The curve representing the symbols 6-8 agrees with Pu^{3+} or $PuCO_3^+$ stability zone.

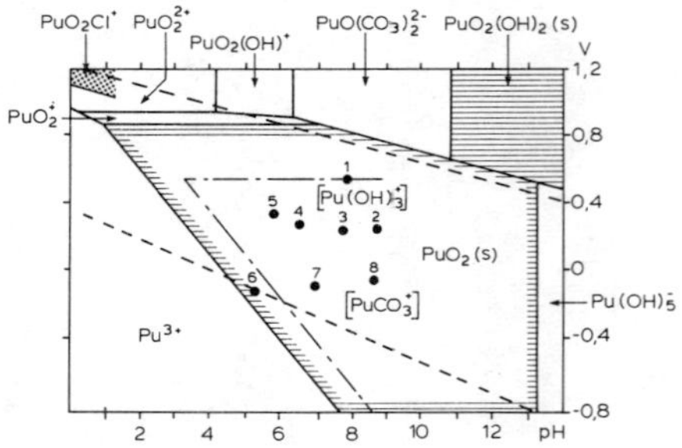

FIGURE 3 E_H, pH diagram of Pu. Data mainly from J.Paquette and R.J.Lemire. Nucl.Sci. and Engl. 79, 26-48 (1981). Dissolved species activity = 10^{-9}, 25°C. Total carbonate content 10^{-2} molal.

(S) Solid chemical form dominates, soluble form is indicated in [ ].

--- Stability limits of water.

E_H and pH experimental conditions as described in fig. 1 and 2.

—·— Stability boundary for 10^{-12} molal PuO_2 (S).

⁘ Stability area for PuO_2Cl^+ complex.

4.2 Extraction of sediments

The results of the sediment extraction experiments are shown in table 1. The role of particle mediated transport was investigated by using filters of 0.2 and 70 µm. In all cases the extraction with the 70 µm filter is equal or higher than with the 0.2 µm filter, indicating particle mediated transport indeed. This effect is, however, small. The application of citric acid, a well known complexing agent, increases the extractability considerably. Anaerobic conditions seem also to increase the extractability, but the mutual differences between the anaerobic extractions are large. This is caused by differences in the redox potential of the anaerobic samples.

An overview of the mean (aerobic) extraction fractions for Am, Eu, Cs and Pu is shown in fig. 5. The similarity in behaviour of Am and Eu on the one hand and Cs and Pu on the other hand is remarkable. For Am and Eu the similarity is understandable because of their position in the periodic chart; similarity of Cs and Pu can only be explained by the fact that both nuclides are strongly adsorbed on clay particles. The increase of the Pu extraction at a pH of 8 can be explained by solubilisation of Pu hydroxides.

TABLE 1 The precentage of radionuclides extracted as a function of the pH and some other specified conditions. Filters of 0.2 µm were used, unless specified otherwise.

pH	3	4	4	5	5	7	7	7	7	7		8	8
	aer.	aer.	aer. 70 µm	aer.	aer.	aer.	aer.	aer. citr.	an.	an.	an.	aer.	aer.
^{241}Am	35	14	20	6	4	7	5	19	11	0	3	7	1
^{239}Pu	12	38	38	27	27	30	21	34	31	27	-	29	29
^{155}Eu	57	16	25	8	2	6	9	10	15	10	9	10	0
^{154}Eu	45	16	19	9	7	8	9	17	14	8	8	6	0
^{144}Ce	65	29	31	16	23	10	23	27	34	22	8	19	14
^{137}Cs	7	23	32	10	10	21	19	24	26	17	8	14	6
^{134}Cs	-	23	30	8	10	22	17	25	27	16	7	12	7
^{106}Ru	9	21	32	16	15	22	18	30	29	25	13	18	12
^{60}Co	52	62	63	50	42	44	56	59	55	53	52	41	44
^{40}K	-	9	16	4	3	1	1	8	1	3	-	1	-

^{239}Pu = $^{239+240}Pu$
aer. = aerobe
an. = anaerobe
citr. = citrate buffer
70 µm = 70 µm filter

4.3 Salt concentration

The influence of the salt concentration on the extractability of Am, Pu and Np appeared to be dependent on the equilibration time. Typical values for $^{239+240}Pu$ and ^{241}Am extraction in a 0,1 molair salt solution after 24, 48, 144 and 504 h are 16, 5.5, 2.1 and 1.2% and 3.3, 3.3, 1.6 and 1.4% respectively. The extractability after 504 h as a function of the salt concentration is shown in table 2. The good extractability of Np compared to those ones of Am and Pu is striking. Further analysis of the data shows that the extractability of Np is hardly influenced by the salt concentration, but that for Am and Pu the salt concentration is important indeed.

This is more shown clearly if the percentages extraction are transferred into K_D values. Depending on the salt concentration there are differences of a factor 10, which means that also migration rates may differ by a factor 10.

TABLE 2 The extractability of radionuclides, in percentages, after an equilibration time of 504 h. The values between brackets are K_D values.

Salt conc. molair	^{241}Am	$^{239+240}Pu$	^{237}Np
1	5.4 (177)	2.2 (458)	54 (8.6)
0.1	1.4 (707)	1.2 (798)	49 (10.4)
0.03	1.0 (1049)	1.0 (964)	50 (10.1)
rain water	2.4 (409)	0.3 (3752)	40 (10.0)

4.4 Speciation and bioavailability

One of the most important reasons for studying speciation is that speciation is assumed to influence the uptake by organisms. Therefore, the authors have compared their results with uptake data of radionuclides by food crops. The IUR workgroup Soil-to-Plant Transfer Factors derived an equation for the uptake as a function of the environmental conditions. For the uptake of Pu, Am, Np and Cs by cerials sufficient data are available to provide factors for the influence of the pH of the soil range pH 4-8, the organic matter content and equilibration time. The results are shown in fig. 4.

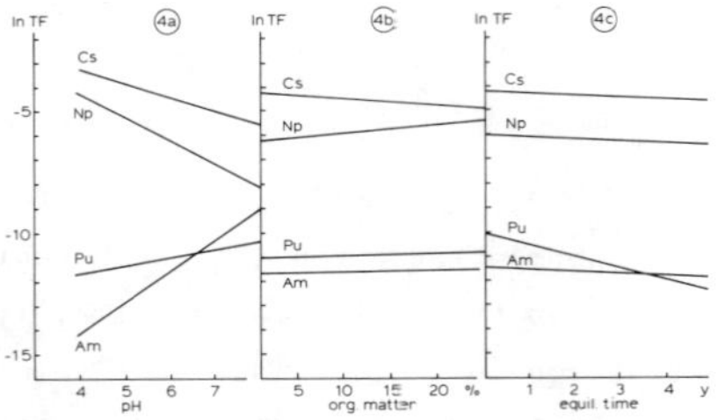

FIGURE 4 The uptake of radionuclides versus the pH (fig. 4a), the organic matter content (fig. 4b) and the equilibration time, i.e. the time elapsed since the application of the radionuclide and harvest of the product analysed (fig. 4c). Unless specified otherwise the data refer to cerials, a pH of 6, an organic matter content of 4% and an equilibration time of 2 years. The curves are statistically derived from the IUR data file of Soil-to-Plant Transfer Factors ($ln = {}^{e}ln$).

The following conclusions can be drawn:

- An increasing pH results in an increase of the uptake of Pu by cerials (fig. 4a), this is in agreement with the increased solubility of PuO_2^+ (fig. 2), it is also in agreement with the extraction pattern shown in fig. 5.

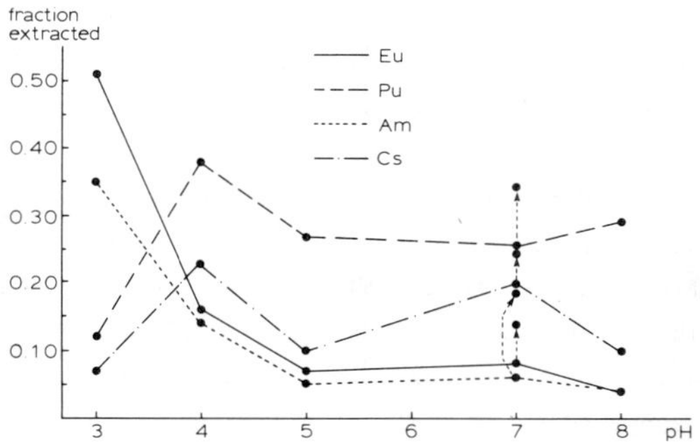

FIGURE 5 The extraction of radionuclides from Ravenglass sediments versus the pH. The vertical lines refer to extractions with a citric acid buffer solution.

- Fig. 5 shows a decrease of the Am extraction with increasing pH, this is not in agreement with the higher uptake of Am at higher pH values (fig. 4a).
- Fig. 5 shows a decrease of the Cs extraction with increasing pH (range 4-8) which is in agreement with the lower uptake of Cs at higher pH values indeed (fig. 4a).
- Fig. 2 suggests that the effect of microorganisms is limited to changes of the redox potential; an important role of complexing agents produced by microorganisms seems to be absent. When it is assumed that there exist a fair correlation between the presence of complexing agents in a soil and its organic matter content, then this is in agreement with the absence of a strong influence of the organic matter content on the uptake (fig. 4b).
- The results of the extractions with 0.03 molair salt (table 2) provide for the reciproque K_D values the rations Am : Pu : Np = 1 : 1.1 : 21. The cerial uptake experiments (fig. 4) show for the uptake the ratios Am : Pu : Np = 1 : 1.8 : 221. The trends are identical, but the differences in uptake are more pronounced than the differences in K_D.
- The extractions with different salt concentrations show a strong decrease of the extractability with equilibration time. The same phenomena is observed for the uptake by plants (fig. 4c). The uptake data refer, however, to equilibration times of years, while the extraction data refer to equilibration times of weeks. It must therefore be concluded that short term equilibration times show a correct trend, but that they cannot be extrapolated to long term data.

5. CONCLUSION

One of the most characteristic parameters controlled by speciation of radionuclides is the extractability or solubility of the nuclides. The extractability appeared, for conditions as they exist in nature, to be mainly controlled by the pH, redox potential and salt concentration. The direct effects of microorganisms play probably a minor role on the solubility, but their indirect effect due to the introduction of low redox potentials is very important. The correlation between the bioavailability and extractability was investigated by comparing uptake data by cerials with extractability data. It appeared that in almost all cases the uptake is positively correlated with the extractability. Speciation research which enlightens the dependence of the solubility on environmental conditions, is therefore a useful tool in obtaining insight in the bioavailability of radionuclides.

6. ACKNOWLEDGMENT

The authors are very grateful for the assistence of Dr.A.Knight (NRPB), who provided them with sediments from the Ravenglass estuary and of Dr.E.I.Hamilton (Institute for Marine Environmental Research), who specified the composition of the sediments.

EVALUATION OF A LARGE-VOLUME WATER SAMPLING TECHNIQUE FOR DETERMINING THE CHEMICAL SPECIATION OF RADIONUCLIDES IN GROUNDWATER

DE Robertson

Pacific Northwest Laboratory
Richland, Washington 99352

ABSTRACT

A laboratory evaluation of a large volume water sampling technique was conducted to determine the effects of various environmental parameters on the retention behavior of radionuclides processed through the sampler. The Battelle Large Volume Water Sampler (BLVWS) consists of an inline membrane prefilter assembly connected to a housing containing multiple beds of cation resin, anion resin and activated aluminum oxide. This sampler quantitatively extracts radionuclides from large volumes (up to 4000 liters) and partitions the radionuclides according to their chemical forms. The effects of cation resin type (H^+ or Na^+), pH, and additions of humic acid and EDTA on the chemical species of radionuclides in simulated groundwater were evaluated. It was shown that the use of H^+ vs. Na^+ cation resin greatly affected the uptake behavior of $^{237}Pu(IV)$, and $^{237}Pu(VI)$, ^{54}Mn, ^{59}Fe, ^{60}Co and ^{155}Eu tracers on the cation and anion resin beds. It is believed that pH reductions in the water passing through the H^+ cation resin section disturbs the chemical equilibria of the species of radionuclides present in the water causing changes from anionic to cationic forms. By adding as little as 5 ppm of humic acid or EDTA, it was possible to simulate the anionic forms of ^{54}Mn, ^{60}Co, ^{63}Ni, ^{55}Fe, ^{155}Eu and Pu observed in field sampling operations. The role of organic complexation of these radionuclides to form mobile anionic species is strongly implicated. Thus, the alternative use of Na^+ and H^+ cation resins, in combination with anion resin and aluminum oxide, appears to provide a technique for studying chemical speciation of radionuclides in groundwaters.

Work supported by the
U.S. Nuclear Regulatory Commission
under Contract DE-AC06-76RLO 1830
NRC Fin B2862

INTRODUCTION

One of the major gaps in understanding the environmental behavior of the long-lived radionuclides associated with commercial radioactive waste disposal is an inadequate knowledge of the mechanisms by which these radionuclides may be mobilized and transported if groundwater intrusion of the disposal site occurs. Radionuclide transport modeling is necessary to predict the movement of radionuclides from a waste disposal site by intruding groundwater. In order to provide greater confidence in predictive radionuclide transport modeling and to identify further areas of biogeochemical research needed for model enhancement, the present models need to be tested and evaluated at actual field sites where radionuclides have been migrating in groundwaters for many years. Such field studies require measurements of very low concentrations of various physicochemical forms of the migrating radionuclides.

Large volume water sampling techniques have been developed and utilized at our laboratory to partition and collect particulate, cationic, anionic and nonionic species of radionuclides from hundreds to thousands of liters of slightly contaminated ground and surface waters. This technique involves pumping water directly from sampling wells or surface waters through a sampling device loaded with a 0.4 μm membrane filter, followed by multiple beds of cation resin, anion resin, and activated aluminum oxide to respectively remove particulate (>0.4 μm), cationic, anionic, and nonionic forms of the radionuclides, respectively. Since it was recognized that this sampling technique would, to some degree, disturb the chemical equilibrium of the water being sampled, an evaluation of the method was carried out under controlled conditions.

This paper describes a series of laboratory experiments to evaluate the uptake and retention by the filters, resins and aluminum oxide of a number of radionuclides added to artificial groundwater. This evaluation has helped to interpret the empirical results of the field use of this sampling methodology and elucidate possible physicochemical forms of the mobile radionuclide species.

BACKGROUND

The Battelle Large Volume Water Sampler (BLVWS) was originally developed in the late 1960's and early 1970's as a tool for concentrating and removing particulate and dissolved radionuclides from large volumes (up to 6000 liters) of seawater.(1) The sampler removed particulate radionuclides on fiberglass filters, and soluble species were removed by adsorption onto 0.65 cm thick beds of activated aluminum oxide. Retention efficiencies in a seawater matrix for a single aluminum oxide bed varied for specific radionuclides, and ranged from 96% for ^{144}Ce to 2% for ^{124}Sb.(1,2) The BLVWS was utilized in a variety of oceanographic applications to study global fallout in the oceans,(1,3,4,5) beryllium-7 as a tracer of air-sea exchange and ocean mixing,(6,7) the distribution of the Columbia River plume in the N.E. Pacific Ocean,(1,8) and plutonium concentrations in the ocean at Thule, Greenland and Eniwetok.(9,10)

The BLVWS was subsequently modified for freshwater use by adding beds of cation resin and anion resin in front of the aluminum oxide to remove dissolved ionic species. This configuration of the BLVWS was utilized for studying radionuclide transport in the Columbia River,(11) radionuclide speciation in nuclear power plant effluents,[12] and low-level aqueous waste discharges.[13] Recently, the sampler has been used to study the mobile radionuclide species being transported in groundwaters from several low-level waste disposal sites.[13-18]

During field utilization of this measurement technique for groundwater studies, it has been observed that the mobile forms of nearly all migrating radionuclides (e.g. ^{55}Fe, ^{59}Fe, ^{60}Co, ^{63}Ni, ^{95}Nb, ^{99}Tc, ^{106}Ru, ^{125}Sb, ^{129}I, and $^{239-240}Pu$) are predominantly anionic or nonionic species. The only exceptions have been ^{90}Sr, ^{137}Cs, ^{241}Am, and ^{244}Cm which have been observed to migrate in trace amounts in cationic forms. In order to better understand the behavior of various radionuclide species during the field sampling process, a series of scaled-down laboratory experiments has been conducted using radionuclide tracers in known oxidation states and chemical forms. Artificial groundwater of varying pH and dissolved organic matter content (humic acid or EDTA) was spiked with the tracers, allowed to equilibrate overnight, and pumped through a scaled-down version of the BLVWS to determine the radionuclide distribution within the sampler and gain insight into the potential chemical forms of the radionuclides. The experimental design and the results of these experiments are discussed in the following sections.

EXPERIMENTAL

The retention of the radionuclides by the BLVWS was tested under two pH conditions ($\sim$6 and $\sim$8), with and without additions of high organic carbon concentrations (humic acid or EDTA). These conditions were chosen to be representative of the normal bounds of slightly contaminated groundwater plumes near low-level waste disposal sites. A scaled-down version of the BLVWS was used for the experiments in order to expedite the analyses and reduce the costs of materials. This scaled-down model was approximately 2.5 cm in diameter and contained two 2.5 cm thick beds of cation resin (200-400 mesh in either H^+ or Na^+ form) followed by dual beds of anion resin (200-400 mesh in the Cl form). The resin beds were followed by two 0.65 cm thick beds of activated aluminum oxide. Both H^+ and Na^+ forms of cation resin were evaluated because the release of the H^+ from the resin by displacement with dissolved cations in the water resulted in a substantial decrease in the pH of the water flowing out of the H^+ form cation resin beds. The field version of this sampler is larger and normally contains two 30 cm diameter by 2.5 cm thick beds of each resin and two 30 cm diameter by 0.65 cm thick aluminum oxide beds. Flow rates and volumes of samples in the laboratory tests were proportional to that of the full-scale version under the field conditions. In the field, approximately 1000 ℓ of solution is normally passed through the sampler at a flow rate of 4 to 8 ℓ/min. In the scaled-down version of the BLVWS, this was proportionally reduced to a total volume of 8 ℓ pumped at a flow rate of 0.06 ℓ/min.

To avoid the problems of unknown constituents in the test water, especially the presence of uncharacterized organic compounds, the test medias were prepared from freshly distilled water and added salts. The basic media was an artificial spring water with approximately the same composition of major constituents as measured in a shallow, oxygenated aquifer. This groundwater flows through unconsolidated sand, silt and gravel deposits composed primarily of feldspar minerals and quartz, with traces of calcite, biotite and amphibole.[14] The salts that were used to make this media are given in Table 1.

TABLE 1. Recipe for Formulation of Artificial Spring Water

Constituent	mg/liter
$NaHCO_3$	12
$MgSO_4 \cdot 7H_2O$	33.9
$CaCl_2$	1.72
$KHCO_3$	12.3
$CaCO_3$[1]	48.5
plus silica water[2]	9.6

(1) The $CaCO_3$ was first added to distilled water as a fine powder, and its dissolution was aided by sparging the solution for several hours with CO_2. After dissolution of the $CaCO_3$, the solution was sparged with air to remove excess CO_2. The other salts were then added as dilute solutions in proper sequence.

(2) The silica water was prepared by boiling high surface area chromatographic silica in distilled water for several hours and filtering through 0.4 μm membrane filters. The dissolved H_4SiO_4 concentration was measured colorimetrically and an appropriate aliquot was added to the artificial spring water to provide about 9.6 mg/ℓ as SiO_2.

The artificial spring water was prepared in batches of 50 liters and filtered through a 0.4 μm membrane filter prior to spiking with the radionuclide tracers.

The plutonium tracers used for the evaluation were ^{237}Pu prepared separately in nitric acid in both the Pu(IV) and Pu(VI) oxidation states by the method of Lovett and Nelson.[19] A mixture of carrier-free gamma-emitting radionuclides originally in the chloride form were prepared in a 0.5 N HCl solution, and contained the following tracers: ^{54}Mn, ^{59}Fe, ^{60}Co, ^{85}Sr, ^{95}Tc, ^{125}Sb, ^{137}Cs, and ^{155}Eu. Iodine-131 tracer was kept separately before spiking in a 0.4 N NaOH solution. A fresh dilute HCl solution of ^{65}Ni was prepared before each spiking.

The $^{237}Pu(IV)$, $^{237}Pu(VI)$, and the mixture of gamma-emitting radionuclides were each evaluated in separate uptake experiments with the scaled-down BLVWS. After spiking 8-liter aliquots of filtered artificial spring water with the radionuclide(s) of interest, the pH was adjusted to either approximately 6 or 8 using dilute NaOH. The water was then allowed to equilibrate overnight (∿18 to 20 hrs.), and was then pumped through a 0.4 μm membrane filter and then directly through the scaled down BLVWS. Small aliquots of the influent and effluent water were taken for trace element analysis and pH measurements, and 1-liter aliquots of the effluent water were collected and analyzed to determine if any breakthrough had occurred. After pumping the 8 liters of spiked water through the sampler, one liter of distilled water adjusted to the same pH was pumped through to rinse out any spiked water interstitially contained in the resin or aluminum oxide beds. The sampler was then disassembled and the membrane filter and each resin and aluminum oxide bed was counted on a Ge(Li) gamma-ray spectrometer to determine the retention of the tracers.

RESULTS AND DISCUSSION

The behavior of several of the radionuclides during transit through the scaled-down BLVWS was unaffected by variations in the pH of the water, or additions of humic acid or EDTA. Their retention behavior in these laboratory tests was essentially identical to that observed in the field sampling. These radionuclides included ^{85}Sr, ^{95}Tc, ^{137}Cs, and ^{131}I. Approximately 98 to 99% of the ^{85}Sr was always found on the first cation resin bed, indicating that the solution species was always Sr^{+2}. The ^{137}Cs behavior was similar, with >99% of the tracer always being retained by the first cation resin bed. This was not unexpected, since Cs^{+} was undoubtedly the only significant solution species present. No significant difference was observed in uptake of ^{85}Sr and ^{137}Cs by the H^{+} or Na^{+} forms of the cation resin. The ^{95}Tc and ^{131}I were always retained on the first anionic resin bed, 90-98% for ^{95}Tc and 94-98% for ^{131}I. However, 2-3% of the ^{131}I was usually retained on the first cation resin bed. In the water adjusted to a pH of 5.8 and 5 ppm of humic acid, approximately 6% of the ^{131}I was consistently retained on the first cation resin. This behavior could be due to a small fraction of the ^{131}I reacting chemically with the cation exchange resin matrix.

The behavior of ^{237}Pu(IV), ^{237}Pu(VI), ^{54}Mn, ^{59}Fe, ^{60}Co, ^{65}Ni, ^{125}Sb and ^{155}Eu during transit through the BLVWS was affected by the form of the cation resin (H^{+} of Na^{+}), the pH, and the addition of either humic acid or EDTA. Their individual behavior is discussed below.

$^{237}Pu^{+4}$

The uptake of ^{237}Pu(IV) was greatly affected by the form of the cation resin (see Table II). When the H^{+} form was used, 90-92% of the ^{237}Pu(IV) was retained on the first cation bed (Run 1). When the Na^{+} form was used, 79-81% of the tracer was retained on the anion resin (Run 2). Similar results were obtained for influent water having pH's of either 6 or 8. The different behavior of the ^{237}Pu(IV) between H^{+} and Na^{+} resins may be due to the lowering of the pH of the water during passage through the H^{+} form of the cation resin beds. The artificial groundwater having an initial pH of 8.05 was reduced to pH 3.5 after passing through the dual H^{+} cation resin beds. This reduction in pH could potentially convert the Pu from an anionic form to a cationic form which is retained by the cation resin. Since the Na^{+} cation resin does not substantially reduce the pH of the passing water, it appears that the predominant charge-form of the ^{237}Pu(IV) in the equilibrated artificial groundwater was actually anionic. Possible dissolved anionic plutonium species may be $Pu(OH)_4CO_3^{2-}$ and/or $Pu(OH)_5^{-}$. Either of these anionic species could

TABLE II. Partitioning of ^{237}Pu (IV) During Transit Through the Water Sampler

	Run 1 11/22/83		Run 2 1/25/84		Run 3 2/29/84		Run 4 3/8/84		Run 5 4/19/84		Run 6 4/25/84	
Cation Resin Form	H^+	H^+	Na^+	Na^+	H^+	Na^+	H^+	Na^+	H^+	Na^+	H^+	Na^+
Influent pH	8.00	8.00	8.00	8.00	8.09	8.09	6.05	6.05	7.93	8.01	6.50	6.50
Effluent pH	---	---	4.85	5.20	2.90	6.56	4.77	5.45	4.81	5.43	4.47	5.10
Humic Acid (ppm)	---	---	---	---	---	---	---	---	30	30	5.0	5.0
EDTA (ppm)	---	---	---	---	3.7	3.7	---	---	---	---	---	---
% Particulate	16.5	8.9	6.0	6.3	9.6	22	3.3	3.1	26.1	21.9	11.4	16.0
% of Soluble on:												
1st Cation	90.3	92.2	14.8	17.6	69.2	2.2	88.8	12.3	7.0	20.3	15.7	17.3
2nd Cation	8.0	6.1	4.1	3.9	27.3	1.8	3.8	12.8	<2.3	4.7	8.6	<1
1st Anion	1.4	1.3	81.0	78.6	3.6	96.0	7.3	74.3	40.5	14.7	45.3	28.2
2nd Anion	0.1	0.1	<0.3	<0.3	<0.3	<0.3	<3	<3	14.0	12.9	30.3	26.1
1st Al_2O_3	0.1	0.2	<0.3	<0.3	<0.3	<0.3	<3	<3	15.6	43.0	<1	10.0
2nd Al_2O_3	0.1	0.08	<0.3	<0.3	<0.3	<0.3	<3	<3	20.6	4.4	<1	17.7

TABLE III. Partitioning of ^{237}Pu (VI) During Transit Through the Water Sampler

	Run 1* 4/17/84		Run 2 4/18/84		Run 3 5/8/84		Run 4 5/9/84		Run 5 5/10/84	
Cation Resin Form	H^+	Na^+	H^+	Na^+	H^+	Na^+	H^+	Na^+	H^+	Na^+
Influent pH	7.85	7.84	7.95	7.98	5.90	5.85	7.85	7.94	6.03	6.01
Effluent pH	4.76	6.20	4.85	5.38	3.74	5.33	4.63	5.16	4.59	5.34
Humic Acid (ppm)	---	---	---	---	---	---	5.0	5.0	5.0	5.0
EDTA (ppm)	---	---	---	---	---	---	---	---	---	---
%Particulate	<1.6	1.3	<1	<1	0.67	1.05	3.1	3.4	10.6	8.6
% Soluble on:										
1st Cation	26.3	1.8	∼100	28.5	97.7	52.7	72.7	10.0	11.0	1.1
2nd Cation	32.7	1.3	<6	41.3	0.7	46.4	0.9	5.7	3.8	0.2
1st Anion	41.0	95.6	<6	26.7	0.6	0.6	10.1	59.4	31.5	30.6
2nd Anion	<1.6	<1	<6	<6	<0.5	<0.3	8.9	11.0	30.9	27.6
1st Al_2O_3	<1.6	<1	<6	<6	<0.5	<0.3	6.1	5.0	20.6	8.9
2nd Al_2O_3	<1.6	<1	<6	<6	<0.4	<0.3	1.2	8.9	2.3	31.6

* Inadvertently did not purge artificial spring water with air after CO_2 sparging. This water therefore contained relatively high concentrations of HCO_3^- and Na^+.

be converted to a cationic form at the low pH (3.5) encountered in the H^+ form cation resin beds.

When 3.7 ppm of EDTA was added to the water, the percentage of $^{237}Pu(IV)$ retained by the anion resin increased slightly. When 5 and 30 ppm of humic acid was added, the percentage of the tracer retained by the anion beds, when the H^+ form of the cation resin beds was used, increased from 7-24 percent to about 75 to 55 percent, respectively. A high percentage retention of Pu on the anion resin beds was also observed for Pu partitioning during field sampling, which suggests that the mobile Pu species may be an organic (fulvic or humic) anionic complex. A significant fraction of the $^{237}Pu(IV)$ was also retained on the aluminum oxide beds when humic acid was added to the water. The retention or the aluminum oxide beds may be due to Pu binding with large humic components to form molecular species which pass through the resin beds and are absorbed by the aluminum oxide. In the field water sampling, we have never observed significant uptake of Pu by the aluminum oxide beds. At field sites Pu may be binding with high molecular weight humic substances which readily absorb to the soil.

$^{237}Pu(VI)$

The behavior of $^{237}Pu(VI)$ in the scaled-down BLVWS was similar to that observed for the Pu(IV). This may suggest that perhaps the Pu(VI) was reduced to Pu(IV) during equilibration with the water or during reaction with the resin beds (see Table III). Again, the H^+ and Na^+ cation resins removed most of the Pu(VI). However, at pH 7.98 when the Na^+ form of the cation resin was used, about 27% of the Pu(VI) was retained by the anion resin bed. This suggests that the original form of the Pu(VI) in the spiked water was anionic. Possible dissolved anionic forms of Pu(VI) in the equilibrated artificial groundwater could be $PuO_2CO_3OH^-$ and/or $PuO_2CO_3(OH)_2^{2-}$. At the reduced pH in the H^+ form of the cation resin, these forms could be converted to cationic species. When 5 ppm of humic acid was added to the pH 6 influent water, the Pu(VI) partitioning changed from predominantly cationic to anionic when the H^+ form cation resin was used. This again approximated the behavior of Pu observed in field sampling, suggesting organic complexation. The humic acid again appeared to react with the Pu to form species which were significantly retained by the aluminum oxide.

^{54}Mn, ^{59}Fe, ^{60}Co, ^{65}Ni and ^{155}Eu

Thermodynamic-based speciation calculations for these radionuclides indicate that the dominate forms in the equilibrated artificial springwater would be cationic, e.g. Mn^{+2}, Fe^{+2}, Co^{+2}, Ni^{+2} and Eu^{+3}.[15] Their behavior in the BLVWS suggests this is generally the case, although about 20% of the ^{54}Mn was retained by the anion resin when either H^+ or Na^+ cation resin forms were used and at pH 6 and 8. The behavior of these radionuclides in the BLVWS is exemplified by ^{59}Fe and ^{60}Co (see Tables IV and V). Each radionuclide was usually retained primarily on the first cation resin bed, whether it was H^+ or Na^+ form. However, when Na^+ form cation resin was used and the spiked water adjusted to pH 8, about 54% of the ^{59}Fe was retained on the anion resin, suggesting an anionic species which is stable at this pH. Also, when 4.5 to 5.0 ppm of EDTA was added, there was a dramatic increase in the retention of these radionuclides on the anion resin. This effect was particularly pronounced when the Na^+ cation resin was used. Thus, it appears that a smaller percentage of

TABLE IV. Partitioning of ^{59}Fe During Transit Through the Water Sampler

	Run 1 5/30/84		Run 2 5/3/84		Run 3 6/1/84		Run 4 6/7/84		Run 5 6/12/84		Run 6 6/13/84	
Cation Resin Form	H^+	Na^+	H^+	Na^+	H^+	Na^+	H^+	Na^+	H^+	Na^+	H^+	Na^+
Influent pH	6.00	6.08	8.09	8.09	8.00	8.05	8.00	8.06	5.82	5.75	5.98	5.96
Effluent pH	3.72	5.07	4.57	5.74	4.63	5.06	3.58	5.39	3.36	5.24	3.58	5.68
Humic Acid (ppm)	---	---	---	---	5.5	5.5	---	---	5.0	5.0	---	---
EDTA (ppm)	---	---	---	---	---	---	4.5	4.5	---	---	5.0	5.0
% Particulate	18.0	20.9	23.0	28.2	14.7	11.4	1.1	0.49	8.4	10.6	1.6	2.1
% Soluble on:												
1st Cation	99.1	91.1	82.5	43.5	16.1	12.7	0.68	1.3	27.1	6.9	1.8	<0.8
2nd Cation	0.21	0.12	12.4	2.4	6.9	0.99	0.37	<0.5	7.7	0.86	0.38	0.13
1st Anion	0.68	8.6	4.4	53.6	27.7	24.1	54.0	98.7	31.5	30.6	19.5	99.5
2nd Anion	<0.9	0.08	0.36	0.22	31.2	20.6	42.2	0.02	24.4	24.3	20.6	0.20
1st Al_2O_3	<0.7	0.15	0.23	0.17	15.2	17.2	2.6	0.02	6.5	12.8	57.7	0.13
2nd Al_2O_3	<1	<0.3	0.1	0.17	2.9	24.4	0.1	<0.4	2.7	24.6	0.06	<0.5

TABLE V. Partitioning of ^{60}Co During Transit Through the Water Sampler

	Run 1 5/30/84		Run 2 5/3/84		Run 3 6/1/84		Run 4 6/7/84		Run 5 6/12/84		Run 6 6/13/84	
Cation Resin Form	H^+	Na^+	H^+	Na^+	H^+	Na^+	H^+	Na^+	H^+	Na^+	H^+	Na^+
Influent pH	6.00	6.08	8.09	8.09	8.00	8.05	8.00	8.06	5.82	5.75	5.98	5.96
Effluent pH	3.72	5.07	4.57	5.74	4.63	5.06	3.58	5.39	3.36	5.24	3.58	5.68
Humic Acid (ppm)	---	---	---	---	5.5	5.5	---	---	5.0	5.0	---	---
EDTA (ppm)	---	---	---	---	---	---	4.5	4.5	---	---	5.0	5.0
% Particulate	0.16	<0.4	<0.2	0.2	5.7	4.8	0.12	0.09	0.72	0.89	0.17	0.1
% Soluble on:												
1st Cation	99.9	99.9	99.7	99.5	97.1	91.5	93.2	0.18	99.3	99.2	98.7	0.07
2nd Cation	0.02	0.02	0.07	0.05	1.4	2.9	5.1	<0.18	0.14	0.15	0.11	0.07
1st Anion	0.03	0.03	0.13	0.39	0.68	1.9	1.6	99.7	0.38	0.23	0.99	99.7
2nd Anion	<0.1	0.02	<0.18	<0.2	0.47	1.4	0.02	0.05	0.12	0.15	0.11	0.13
1st Al_2O_3	<0.2	<0.1	<0.01	<0.01	0.28	1.1	0.04	0.01	0.05	0.03	0.01	<0.1
2nd Al_2O_3	<0.2	<0.01	<0.05	<0.04	0.10	1.1	0.03	0.07	0.04	0.20	0.09	<0.03

the metal occurred as metal-EDTA anionic complexes at the lower pH's generated by the H^+ cation resin beds. When 5 ppm of humic acid was added there was a significant conversion to anionic forms of ^{54}Mn and ^{59}Fe, but the effect was not as pronounced as that observed for EDTA. The humic acid additions had little effect on the ^{60}Co and ^{65}Ni behavior.

Thus, by adding complexing substances commonly found in most low-level waste leachates, it is possible to simulate the field observations of predominantly anionic mobile species of these radionuclides. The role of these complexing agents in mobilizing radionuclides thus takes on added significance.

^{125}Sb

The behavior of ^{125}Sb in the laboratory experiments was similar to that observed in the field. The predominant chemical form appears to be neutrally charged and is retained on the aluminum oxide. Variations in the cation resin form, pH, humic acid, and EDTA concentrations had little effect on the partitioning of ^{125}Sb. Thermodynamic calculations indicate that $HSbO_2^o$ would be the dominant species.

CONCLUSIONS

The large volume water sampling methods evaluated in this study can be a very informative tool in elucidating the chemical species of mobile radionuclides migrating in groundwaters. The laboratory assessment has shown that the partitioning of radionuclides observed during field sampling with the BLVWS can be approximated by the addition of organic complexing agents such as EDTA and humic acids to the artificial groundwater. These substances, which are commonly found in low-level waste leachates, appear to be responsible for the formation of the mobile anionic radionuclide species of ^{55}Fe, ^{60}Co, ^{63}Ni, ^{95}Zr, ^{106}Ru, and Pu in groundwaters from low-level waste disposal facilities. Thus, the alternative use of H^+ and Na^+ cation resin forms, in combination with anion resin and activated aluminum oxide or other specific adsorbants appears to provide a technique for studying the chemical speciation of radionuclides in groundwaters.

REFERENCES

1. W. B. Silker, R. W. Perkins and H. G. Rieck, Ocean Eng., 2, 49, (1971).
2. D. E. Robertson, Pacific Northwest Laboratory Annual Report for 1977, Part 2, Ecological Sciences, PNL-2500 Pt. 2, p. 7.27, Pacific Northwest Laboratory, Richland, WA., 1978.
3. W. B. Silker, Earth and Planet. Sci. Lett., 16, 131, (1972).
4. W. B. Silker, J. Geophy. Res., 77, 1061, (1972).
5. D. E. Robertson, W. O. Forster and H. G. Rieck, BNWL-715 (2), Vol. 2, Part 2, Pacific Northwest Laboratory, Richland, WA., 1968.
6. W. B. Silker, D. E. Robertson, H. G. Rieck, R. W. Perkins and J. M. Prospero, Science, 161, 879, (1968).
7. J. A. Young and W. B. Silker, BNWL-1551, Pacific Northwest Laboratory, Richland, WA., 1971.
8. R. W. Perkins, BNWL-1051, Part 2, Pacific Northwest Laboratory, Richland, WA., 1969.
9. D. E. Robertson, BNWL-2100, Part 2, Pacific Northwest Laboratory, Richland, WA., 1977.
10. W. C. Weimer, K. H. Abel and C. I. Gibson, BNWL-2100, Part 2, Pacific Northwest Laboratory, Richland, WA., 1977.
11. D. E. Robertson, W. B. Silker, J. C. Langford, M. R. Peterson and R. W. Perkins, in Radioactive Contamination of the Marine Environment, pp. 141-158, International Atomic Energy Agency, Vienna, 1973.
12. K. H. Abel, D. E. Robertson, E. A. Crecelius and W. B. Silker, in Proceedings, 4th Joint Conference on Sensing of Environmental Pollutants, pp. 339-344, American Chemical Society, 1978.
13. D. E. Robertson and R. W. Perkins, in Proceedings IAEA Symposium on Isotope Ratios as Pollutant Source and Behavior Indicators, pp. 123-133, International Atomic Energy Agency, Vienna, 1975.
14. D. E. Robertson, A. P. Toste, K. H. Abel and R. L. Brodzinski, NUREG/CR-3554, PNL-4773, Pacific Northwest Laboratory, Richland, WA., 1984.
15. J. S. Fruchter, C. E. Cowan, D. E. Robertson, D. C. Girvin, E. A. Jenne, A. P. Toste and K. H. Abel, NUREG/CR-3712, PNL-5040, Pacific Northwest Laboratory, Richland, WA., 1984.
16. D. R. Champ, J. L. Young, D. E. Robertson and K. H. Abel, "Chemical Speciation of Long-Lived Radionuclides in a Shallow Groundwater FLow System", in Press, Chalk River Nuclear Laboratories and Pacific Northwest Laboratory, 1985.
17. P. A. Eddy and J. S. Wilbur, PNL-3346, Pacific Northwest Laboratory, Richland, WA., 1980.
18. L. J. Kirby, NUREG/CR-1832, PNL-3510, Pacific Northwest Laboratory, Richland, WA., 1981.
19. M. B. Lovelt and D. M. Nelson, in Techniques for Identifying Transuranic Speciation in Oquatic Environments, pp. 27-35, International Atomic Energy Agency, Vienna, 1981.

DISCUSSION

SIBLEY:

When one uses the Battelle Large Volume Water Sampler with several beds of Al_2O_3 instead of cation or anion exchange beds, the 1st Al_2O_3 is often discolored by colloidal material that passes through the filters. What happens to the colloids in the configuration you use for sampling ground water?

ROBERTSON:

I have no direct proof, but I believe that most of the colloidal matter is able to pass through the resin beds and is retained by the aluminium oxide. I base this conclusion on the observation that when we add humic acid to artificial groundwater some of the humic material appears colloidal following filtration through 0.4 μ membrane filters. Both the dissolved and the colloidal humic material have a known color. When the water is passed through multiple beds of cation resin, anion resin and aluminium oxide the dissolved humic acid appears to be retained on the first anion bed, and the colloidal material appears to be retained by the aluminium oxide bed, since both resin and aluminium oxide become discolored.

SPECIATION: THE ROLE OF THE COMPUTERISED MODEL

J.E. CROSS, D. READ, G.L. SMITH AND D.R. WILLIAMS

Department of Applied Chemistry, University of Wales Institute of Science and Technology, P.O. Box 13, Cardiff. CF1 3XF

ABSTRACT

The computational modelling of chemical speciation is discussed and the different computer programs and thermodynamic databases available are described. The application of this approach to the study of radio-elements in radioactive waste disposal systems is illustrated by simulating the chemical behaviour of americium in a vault. The maximum solubility and speciation of americium in a concrete solution, representing the vault, is calculated for a range of pH values. Then this solution is mixed with a groundwater and the changes in pH and the chemical speciation of americium in the resultant solution are predicted.

INTRODUCTION

Computer simulation is widely used as an investigative method and has considerable application in the study of chemical speciation. There are two main areas of radioactive waste research where the use of chemical modelling is important. The first is its use in conjunction with experimental investigations. Computer modelling provides a relatively simple method for predicting the chemical behaviour of a system prior to any experimental work. This enables identification of the most important components present, often eliminating unnecessary experimental effort. In addition, once experimental work is underway modelling studies may continue to be useful by helping with the interpretation of the results.

The second important application of computer modelling is to investigate systems which are not possible to observe experimentally, either due to their complexity or to the long time-scales which would be involved. For example, in radioactive waste disposal it would not be possible to observe all the components of the waste form, vault and surrounding geosphere together in one experiment whereas this is possible in a computer simulation, enabling all the possible interactions to be considered. Also when radioactive waste is studied many of the transuranic elements are at very low concentrations which are below the limits of sensitivity of most of the traditional methods used to investigate chemical speciation; In these cases chemical modelling is essential.

In this paper the computerised approach is described and its use is illustrated by reference to some calculations for americium.

COMPUTER SIMULATION

Computer Programs

There are a large number of computer programs available for modelling chemical speciation in a wide variety of aqueous systems. The most suitable codes for adoption for modelling radioactive waste are those which were originally developed for geochemical systems. These codes have been compared and reviewed by Nordstrom[1]. They can usually incorporate solid phases in addition to aqueous species and may model redox reactions, temperature variation and in some cases adsorption; all important processes when radioelements are considered.

In general all the codes may be used to calculate the chemical equilibrium distribution of aqueous species in a solution. However, they vary considerably in their degree of sophistication, the extent of the user options available and the amount of data preparation that is required. The most versatile group of codes are able to take account of mass transfer into and out of the solution during a simulation, and are known as "reaction path" codes. For example, they may be used to model the addition of a reaction to the system, equilibration with a solid phase by its precipitation or dissolution and the mixing or titrating of two solutions. One such code is PHREEQE (pH redox equilibrium equations). It was originally developed by the U.S. Geological Survey and more recently its database has been extended to include radioelements[2]. Examples using PHREEQE are described below.

Method of Calculation

The first stage in any simulation is to fully characterise the system and set up a chemical model. The total concentrations of each chemical component are required together with the system's pH and Eh and a knowledge of the solid phases present. The computer code then uses this information together with a thermodynamic database to calculate the equilibrium distribution of aqueous species. When codes are capable of mass transfer, the resultant distribution is examined for supersaturated solid phases and precipitation is performed, followed by a recalculation of the species distribution. Similarly, when modelling the mixing of two solutions, or a solution in equilibrium with a solid phase, the equilibrium distribution of the aqueous species is recalculated after fixing the set conditions.

Thermodynamic Database

The quality of results obtained in any computer simulation is entirely dependent on the input data employed. Namely the data characterising the chemical system and the thermodynamic database. Inaccuracies in the former data may be overcome by scanning ranges of their values in the calculations; hence their relative importance and possible errors in the analyses may be estimated. The thermodynamic database comprises of formation constants for all the aqueous species and saturation indices for the solid phases possible in a system. It is essential that such a database is both complete and self-consistent. In general, the thermodynamic data available for the common groundwater components (eg. Ca, Mg, CO_3^{2-}) are quite reliable; however, the situation is less satisfactory for many of the radioelements where much of the data is still estimated. The data available for the transuranic elements is limited although the similar behaviour in each oxidation state within the group and comparisons with the lanthanides, enable quite good estimates to be made until reliable, experimentally determined constants are available.

The Nuclear Energy Agency (OECD) has undertaken to critically review all the available data for the important radioelements (Pu, U, Am, I, Tc, Np). This work is still in its early stages but on completion will become a valuable database.

Verification and Validation

Verification is necessary to ensure that the computer codes are operating correctly and to eliminate any computational "bugs". One of the simplest methods is to compare the results obtained by different codes on a common problem. There have been three important intercode comparison exercises for the geochemical codes.[1,3,4].

Validation of both the codes and their databases is a much more difficult problem. In a validation exercise the code is used to model a well characterised system and the predicted and experimentally observed results are compared. In general, there has been only limited validation of these codes, and particularly their use for radioelements, basically due to the experimental difficulties encountered when attempting to characterise a system. However, some validation is possible by comparing predicted solubilities and speciation with some relatively simple experimental results.

USE OF PHREEQE TO MODEL AMERICIUM IN A RADIOACTIVE WASTE VAULT

The computer modelling of the chemical behaviour of americium in a radioactive waste vault and its subsequent mixing with a groundwater is used as an example of the application of the computer code PHREEQE.

All the current vault designs for low and intermediate level radioactive wastes in the U.K. envisage an extensive use of concrete in their construction and waste-forms[5]. Hence as a first approximation, the aqueous phase in the vault is assumed to be a solution in equilibrium with concrete, and is termed "concrete water". A typical concrete water composition in equilibrium with both calcium carbonate and calcium sulphate at pH=11 is given in Table 1. It is known that as concrete "ages", there is a gradual decrease in the pH of the pore water from around pH=13 to 10, as the alkali salts are leached away. PHREEQE has been used to predict the effect of this change in pH on the maximum solubility of americium in the concrete solution. Throughout these calculations the solution was maintained in equilibrium with calcium carbonate, calcium sulphate and americium hydroxide. The results are shown in Figure 1 and clearly show how a decrease in pH is accompanied by a significant increase in the maximum solubility of americium. Recent experimental solubility measurements corroborate these results.[6] At high pH the calculations predict the dominant aqueous species as $Am(OH)_3^0$ but in the more neutral pH region mixed hydroxy-carbonate species of americium form and become dominant. These species are responsible for the increased solubility of americium.

TABLE 1 Chemical compositions of the concrete water and groundwater used in the PHREEQE calculations.

Component	Concentration ($mol\ dm^{-3}$) Concrete Water	Groundwater
Ca	2.10×10^{-2}	8.75×10^{-4}
Mg	2.00×10^{-6}	1.28×10^{-4}
Na	5.00×10^{-5}	7.83×10^{-4}
K	1.00×10^{-4}	9.00×10^{-5}
Al	3.00×10^{-5}	-
Si	3.00×10^{-5}	1.33×10^{-5}
Cl	1.60×10^{-3}	8.46×10^{-4}
S	1.22×10^{-2}	6.35×10^{-4}
C (inorganic)	5.85×10^{-6}	5.08×10^{-4}
N	-	3.06×10^{-4}
pH	11.0	8.7
Ionic Strength	0.042	0.004

The structural integrity of the vault has been estimated at between 380 and 3000 years[5], depending on its design. However, if it is assumed that the vault is eventually breached then the aqueous concrete phase will mix with the surrounding groundwater. A second set of calculations using PHREEQE have been made to simulate this mixing. An "aged" concrete solution (1 dm^3, pH=11) was titrated with portions of a groundwater until a total of five dm^3 had been added. The chemical composition of each solution is given in Table 1. Throughout the titration the solution was kept in equilibrium with calcite. Figure 2 shows the variation of the solution pH and the concentration of americium during the titration. The pH is seen to decrease from pH=11.0, initially, to pH=8.5 in the final solution. The americium concentration is also seen to decrease as the titration proceeded, this is the result of its successive dilution as more groundwater was added.

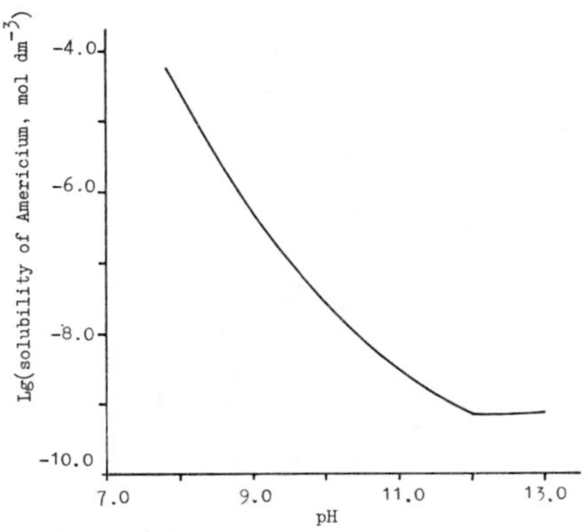

FIGURE 1 Maximum solubility of americium in a concrete water

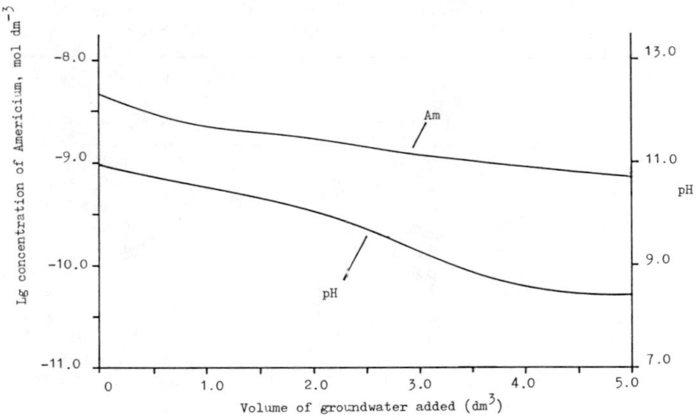

FIGURE 2 The concentration of americium and the pH of a concrete water titrated with a groundwater.

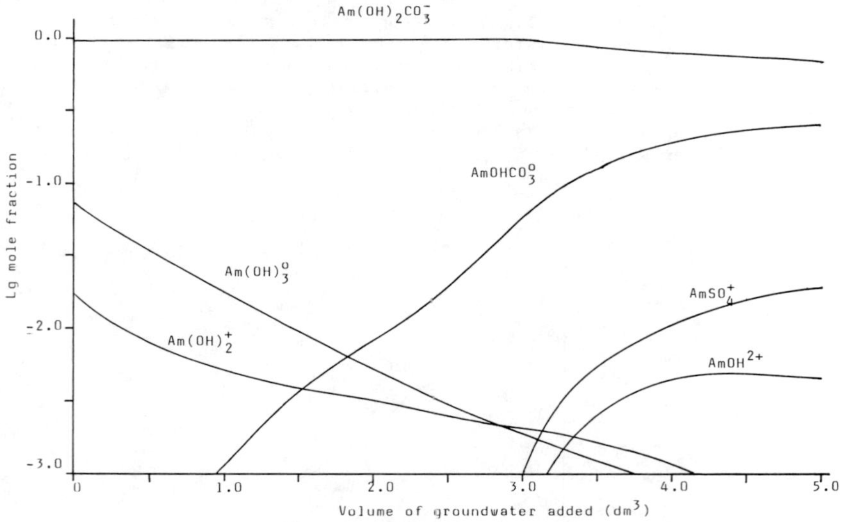

FIGURE 3 Americium speciation in a concrete water titrated with groundwater.

Figure 3 shows the changing chemical speciation of americium in the aqueous solution as the groundwater was added. In general it reflects the changing pH conditions and shows the increased importance of the $AmSO_4^+$ and $AmOHCO_3^0$ species as the pH decreased. Chemical studies of this type enable prediction of the changing bioavailability of a radioelement as it passes from the vault into the geosphere.

CONCLUSIONS

Computer modelling is a very valuable tool for the investigation of chemical speciation. Used in conjuction with experimental programmes it may be used both to give direction to the experimental work and to aid the interpretation of experimental observations. In addition it may be used to predict the chemical behaviour of systems which are not possible to observe experimentally due to their complexity or to the long time-scales involved such as those in radioactive waste disposal.

ACKNOWLEDGEMENT

We wish to thank the Department of the Environment and NIREX for their financial support.

REFERENCES

1. D.K.Nordstrom et al., in E.A.Jenne (ed.), Chemical Modelling in Aqueous Systems, ACS Symposium Series 93, (American Chemical Society, Washington D.C., 1979), pp 857-892.

2. D.L.Parkhurst et al., Report USGS/WRI-80-96, 3rd ed., (U.S. Geological Survey, 1985), 193p.

3. INTERA Environmental Consultants Inc., Report ONWI/E512-02900/CD-15 420-01G-01A, (Houston, Texas, 1982), 159p.

4. T.W.Broyd et al., A Comparison of Computer Programs which model the Equilibrium Chemistry of Aqueous Systems, presented at 3rd Plenary Meeting CEC Project MIRAGE, Brussels, 1985.

5. A.Atkinson and J.A.Hearne, Report AERE-R11465, Harwell, 1984.

6. F.Ewart, Personal Communication.

SPECIATION OF FISSION PRODUCTS IN CONTAMINATED ESTUARINE SEDIMENTS BY CHEMICAL ELUTION TECHNIQUES

D. PRIME, B. FRITH, C.I. STATHERS, and D. CHARLES.

University of Manchester, Radiological Protection Service,
Coupland III Building, Oxford Road, Manchester M13 9PL

ABSTRACT

This paper describes the use of elution ion-exchange techniques using various ionic and complexing agents in order to elucidate the species of fission products sorbed onto contaminated estuarine sediment.
The work concentrates on the fission products Cs-137, Ru-106, Zr-95, Nb-95 and Ce-144. The indications were that caesium was held mainly on inaccessible ion exchange sites; ruthenium appeared to be partially absorbed and partially held on anionic exchange sites; zirconium and niobium were sorbed chemically or physically in the form of complex hydrous oxides; cerium appeared to be in an ionic and easily complexible form on surface sites of the sediment.

INTRODUCTION

The purpose of this study was to demonstrate methods that had been used to elucidate the fission product chemical species held in contaminated sediments. Sediment samples were taken from the Ravenglass estuary in Cumbria because of their relatively high level of contamination. Previous work[1] had shown sites with high contamination levels and it was these that were chosen for investigation.

The sediments were contacted with a variety of compounds chosen for properties which included ion exchange capability, complexing ability and acidic and alkaline nature, and the degree of elution measured. Inferences could then be drawn about the chemical form in which certain fission products were held on these particular sediments.

METHODS

Fission product activity was measured using a lithium drifted germanium crystal and multichannel analyser. The activity calculations were carried out after the spectral data had been transferred to a microcomputer. The system was calibrated using mixed gamma standards and background radiation levels were determined from measurements on dry uncontaminated sand. Uncertainty limits were calculated based on counting statistics.

After initial measurement of activity the samples were divided into two halves, and each processed separately to provide a cross check. 80 cm^3 of elutant was added to each half and the sample was shaken vigorously. This was done several times daily for a period of seven days after which the sample was filtered through glass fibre paper, and dried. Weight losses were usually kept to below 4% by this method.

RESULTS

Table 1 summarises desorption results for the various fission products tested. In all cases a blank run with deionised water produced no significant desorption.

TABLE 1 Desorption Results (%) for various contact solutions

Contact Solution	% Desorption Zr-95 %	Nb-95 %	Ru-106 %	Cs-137 %	Ce-144 %
Deionised water	< 5	< 5	< 5	< 5	< 5
1 mol. $litre^{-1}$ Potassium chloride	< 5	< 5	N.T.	5	< 5
1 mol. $litre^{-1}$ Sodium sulphate	N.T.	N.T.	N.T.	N.T.	< 5
1 mol. $litre^{-1}$ Sodium phosphate	< 5	< 5	19 (0.1 mol. $litre^{-1}$)	N.T.	N.T.
1 mol. $litre^{-1}$ Calcium chloride	N.T.	N.T.	N.T.	6	N.T.
0.01 mol. $litre^{-1}$ Cerous sulphate	N.T.	N.T.	N.T.	N.T.	50
0.5 mol. $litre^{-1}$ Oxalic acid	5	46	22	N.T.	< 5
0.5 mol. $litre^{-1}$ Citric acid	63	75	16	N.T.	72
1 mol. $litre^{-1}$ Hydrochloric acid	9	< 5	12	< 5	N.T.
1 mol. $litre^{-1}$ Sodium hydroxide	< 5	< 5	22	N.T.	N.T.
12 mol. $litre^{-1}$ Hydrochloric acid	9	35	35	40	80
16 mol. $litre^{-1}$ Nitric acid	55	< 5	69	67	80

Results with less than 5% desorption were not considered significant,
N.T. means not tested.
Counting errors were less than 1% (2 S.D.) in all cases.

a) Ion Exchange Agents

Potassium chloride, sodium sulphate, sodium phosphate and calcium chloride were chosen primarily for their ion exchange capabilities. No significant desorption was obtained for zirconium, niobium or cerium. This result was not surprising in the case of zirconium and niobium since no simple cations or anions are known to be present in the marine environment[2-4]. Cerium is known to exist in cationic form in sea water[5], but the lack of exchange was not surprising because in general the more highly charged cerium cations would not be displaced by monovalent ions such as potassium or sodium.

Ruthenium was desorbed to a considerable (19%) extent by 0.1 mol. $litre^{-1}$ sodium phosphate. Since ruthenium forms no stable or soluble compounds with phosphate[6,7] an anion exchange mechanism seemed likely. This was in agreement with the findings of a number of workers that ruthenium complexes in acqueous solution tend to be anionic[7-10]. The fact that only partial desorption occurred suggested that a large fraction of the ruthenium was sorbed by a physical or chemical mechanism. A complication to this desorption was the fact that hydrolysis occurred resulting in the formation of hydroxyl ions which could have interfered with the process (solution pH 10).

Calcium chloride was slightly more efficient at desorbing caesium than potassium chloride. Calcium, being divalent, would be expected to be a better displacer of caesium ions than monovalent potassium. However, this could be offset by the greater hydrated ionic radius (Ca^{2+} 96 nm; K^+ 38 nm; Cs^+ 36 nm), which means that Cs^+ ions will not be removed from sterically hindered positions, such as collapsed interlayer sites[11]. In the event, the degree of desorption of the two ions was similar and the two effects must, therefore, have cancelled one another.

b) Isotopic Exchange Agents

Cerous sulphate was chosen as an elutant in this class since it is a readily available, easily dissolvable form of cerium salt. Since no desorption occurred of cerium with sodium sulphate it was assumed that the desorption observed was a consequence of the cerous ion. A 50% elution occurred indicating that cerium existed in an ionic form or a form which was easily dissociated to ions which were exchangeable with cerous ions. This supported the suggestion of Hirno et al[5].

c) Complexing Agents

Oxalic and citric acid have been shown to form stable complexes with all the fission products tested except for caesium. It was to be expected, therefore, that elution of complexable species would occur and generally this was shown in practice.

The lack of appreciable zirconium desorption by oxalic acid appeared initially to be anomalous. However, Blumenthal[2] has shown that zircon (ZrO_2) was completely insoluble in oxalic acid and the explanation of the observed behaviour could be that zirconium was present in the form of a zircon or zircon - related precipitate.

Oxalic and citric acids both desorbed niobium although citric acid was the more effective. Niobium has been shown to form complexes with both these agents[4]. The most likely explanation was that niobium existed as a compound, possibly a hydrous oxide, on the surface of the sediments and hence was accessible to the large molecules of oxalic and citric acid.

Ruthenium forms nitrosyl oxalate complexes but no information was found about similar citrate complexes. The ruthenium removed by these agents may have been due to a complexing action or simply anion exchange as with phosphate.

Oxalic acid forms stable complexes under certain conditions with cerium[12,13] but obviously not at the concentrations present. Cerium also complexes well with citric acid as was demonstrated in this case, thus indicating that it was present in an easily complexible form and on sites available to the citrate ion.

d) Sodium Hydroxide

Hydroxyl ions can act as ligands, and simple anions which could take part in ion exchange. However, Helfferich[14] considers that the displacing power of OH^- is low, certainly less than that of Cl^- as considered above and below. Also high alkalinity can cause alteration of sorbed species or change the redox potential enabling an easier desorption pathway than would be possible in neutral conditions.

The principle effect of sodium hydroxide in these tests was to remove a similar percentage of ruthenium as phosphate, oxalate and citrate. Ion exchange can be discounted since the chloride ions from the dilute hydrochloric acid (see above) did not have the same effect.

Hydroxyl ions have been shown to act upon nitrosyl-ruthenium complexes producing nitrosylruthenium hydroxide[8]. This degree of alkalinity would solvate both nitrate and nitro complexes and this effect could explain the observed behaviour.

e) Mineral Acids

Dilute acids have been shown to leach some metals from sediments[15]. This would not usually be due to the ion exchange capabilities of H^+. In these tests dilute hydrochloric acid had a significant effect on ruthenium and zirconium. Zirconium can exist as oxo cations in acid solution under certain conditions and it may be that such cationic forms could explain why it was effected to a small degree. Ruthenium may be bound to an iron coating and partially extractable via this acid strength[15]. The observed behaviour fitted in with this suggestion.

Concentrated acids subject the sediments to high H^+ concentrations which can attack edge sites of the aluminosilicate layers[11] and hence reduce their sorption capacity. Concentrated nitric acid is not as acidic as concentrated hydrochloric acid but is a powerful oxidising agent and a source of nitrate ions and ligands. These factors all complicate explanations of the observed behaviour.

The caesium desorption can most likely be explained by damage to absorption sites.

The amount of zirconium removed by concentrated hydrochloric acid was no different than that removed by dilute. This indicated that damage to interlayer sites and terminal functional groups on the clay caused by the high H^+ concentrations had little effect on zirconium desorption. Nitric acid, however, removed a considerable amount of zirconium. The fact that hydrous zirconia is soluble in nitric acid[2] supports the previously made suggestion that zirconium may be present in a hydrous oxide form.

Niobium does not form stable compounds with concentrated nitric acid, any nitrates immediately hydrolysing into nitric acid and the hydrous oxide. It does, however, form anionic chloride complexes[4] this could explain, therefore, the results obtained with hydrochloric acid.

The extraction of ruthenium by hydrochloric acid has been discussed above. The increase in desorption could be caused by a greater extraction of the proposed iron - ruthenium complex or even by the formation of chloronitrosylruthenium complexes in the high Cl^- concentration. Nitric acid would combine with nitrosylruthenium to form the soluble nitrate complex which could explain the high desorption effect. Alternatively, oxidation might occur to form a more soluble nitrosylruthenium complex.

Cerium appeared to be present at least partially in an easily accessible ionic form. Simple ion exchange was unlikely because of the high selectivity of cerium ions but it was possible that damage to accessible sites and complex formation could both play a part in the observed behaviour. Chloride complex formation has been shown to be slight even at these concentrations[16] and although nitrate complexes are formed at lower concentrations, a test with dilute nitric acid produced no significant desorption. It was thought, therefore, that damage to sediment sorption sites was the most likely explanation of the observations.

CONCLUSIONS

The conclusions that can be drawn from this study are:-

(i) The behaviour of zirconium and niobium in these tests could be explained if in part both were present in the form of complex hydrous oxides.

(ii) Caesium's behaviour indicated that it was most likely to be present as an ion and that these ions were in part inaccessible to large cations such as hydrated calcium.

(iii) The behaviour of ruthenium in the desorption test was more difficult to explain. In part the ruthenium appeared to be chemically sorbed onto the sediment and in part to be present in an anionic form. Either or both of these components might be nitrogenous.

(iv) The behaviour of cerium in the elutions indicated that it was present in an ionic and easily complexible form on surface sites of the sediment.

REFERENCES

1. J.A. Brown, The environmental analysis of plutonium, M.Sc. thesis, (University of Manchester Institute of Science and Technology, 1979).
2. W.B. Blumenthal, The chemical behaviour of zirconium, (Von Nostraad, Princeton, New Jersey, 1958).
3. NAS, The radiochemistry of niobium, (National Academy of Science, Washington, DC, 1961).
4. F. Fairbrother, The chemistry of niobium and tantalum, Topics in Inorganic and General Chemistry, Monograph 10. (Elsevier, Amsterdam, 1967).
5. S. Hirno, T. Koyanagi, & M. Saiki, The physico-chemical behaviour of radioactive cerium in sea water. Radioactive contamination of the marine environment, (I.A.E.A., Vienna, 1973), p. 47-55.
6. NAS, The radiochemistry of ruthenium, (National Academy of Science, Washington, DC, 1961).
7. W.P. Griffith, The chemistry of the rarer platinum metals (Os, Ru, Ir, Rh), Monographs on Chemistry (Interscience, London, 1967).
8. J.M. Fletcher, I.L. Jenkins, F.M. Lever, F.S. Martin, A.R. Powell, & R. Todd, J. Inorg. Nucl. Chem., 1, 378-401 (1955).
9. J.M. Fletcher, P.G.M. Brown, E.R. Gardner, C.J. Hardy, A.G. Wain, & J.L. Woodhead, J. Inorg. Nucl. Chem., 12, 154-73 (1960).
10. P.G.M. Brown, J. Inorg. Nucl. Chem., 13, 73-83 (1960).
11. R.E. Grim, Clay minerology, 2nd edn. (McGraw-Hill, New York, 1968).
12. G.E. Crouthmel & D.S. Martin, J. Am. Chem. Soc., 72, 1382-86 (1950).
13. P.C. Stevenson & W.E. Nervik, The radiochemistry of rare earths, (National Academy of Science, Washington, DC, 1961).
14. F. Helfferich, Ion exchange, (McGraw-Hill, New York, 1962).

15. E.K. Duursma, & D. Eisma, Neth. J. Sea, Res., 6, 265-324 (1973).
16. R.M. Diamond, K. Street, & G.T. Seabourg, J. Am. Chem. Soc., 76, 1461-69 (1954).

A METHOD FOR SPECIATION OF TRACE ELEMENTS (STABLE AND RADIOACTIVE) IN NATURAL WATERS

B. SALBU, H.E. BJØRNSTAD AND A.C. PAPPAS

Nuclear Chemistry Division, Department of Chemistry
University of Oslo, Blindern, Oslo 3, Norway

ABSTRACT

Radioactive nuclides and stable trace metals entering natural aquatic systems interact with naturally occurring particles through exchange and sorption processes. The extent of which depends not only on the elements and particles in question, but also on size distribution of particles being most pronounced for colloids having large surface areas to volume ratios. The interaction of radionuclides and trace metals with colloids changes their size and charge characteristics and thereby influences their transport, mobility and bioavailability.

In order to investigate the role of colloids as transport agents for radionuclides or trace elements, a continuous mixing and separation system, using hollow fiber cartridges has been developed. This system allows the association of traces with naturally occurring colloids of a given size distribution to be followed and information on the degree of binding, binding constants and number of sorption sites can be obtained. Based on diffusion rate measurements using the hollow fibres, the molecular weight of species diffusing through membranes can also be estimated.

INTRODUCTION

In natural waters radioactive and stable trace elements, especially multivalent metals may be present in a variety of physico-chemical forms.[1-3] These may be associated with forms ranging from simple ions or molecules via hydrolysis products and polymers to colloids, pseudocolloids and suspended particles (Figure 1).[4] The transition between categories is gradual and the kinetics and reversibility of the different transition processes involved are still open questions.

A shift towards higher dimensions may occur when sorption, complexation or aggregation of colloids takes place while desorption

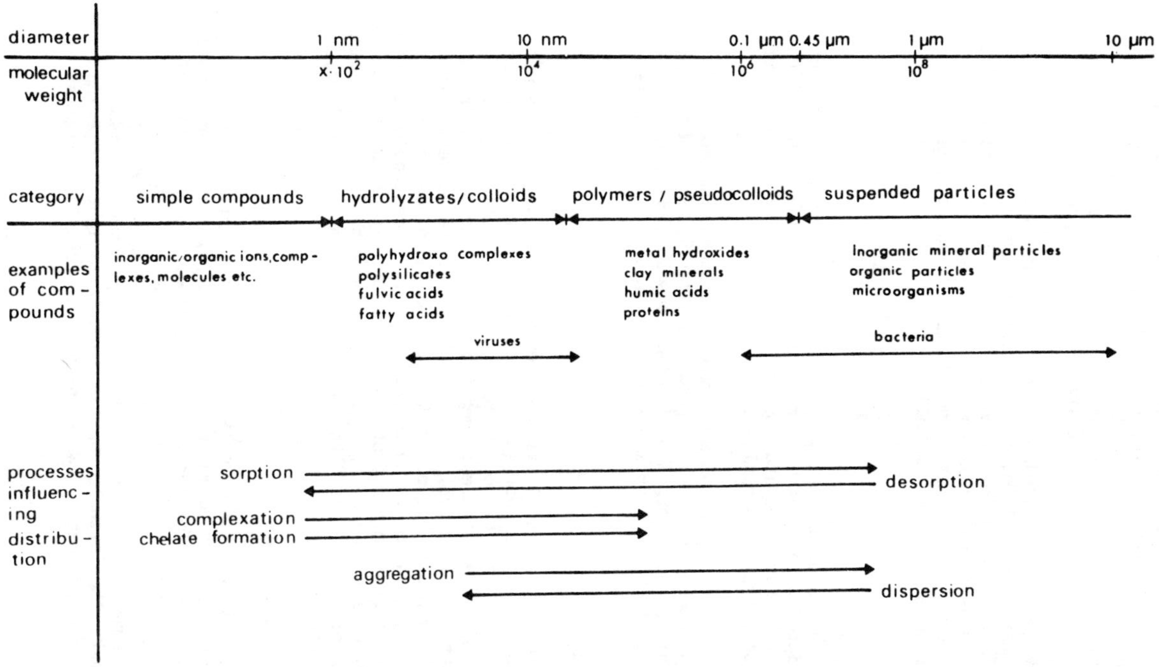

Figure 1. Association of tracc elements with compounds in differcnt size ranges. Transformation processes are indicated.

and dispersion processes may mobilize the trace metals.[4] The challenge within speciation studies is therefore to fractionate species according to size and charge and to investigate the chemical properties without a distortion of the distribution patterns (i.e. separation in situ or soon after sampling, rapid and selective fractionation techniques).

In the present work, colloidal species are fractionated using hollow fiber ultrafiltration technique.[5,6] A continuous mixing and separation system has been developed in order to investigate the role of natural colloidal species as transporting agents for traces.[4] In addition a diffusion technique for distinguishing low molecular weight species not retainable by membranes is presented.[4]

PROCEDURES

For the fractionation of naturally occurring species Amicon concentrator CH2 or CH3 equipped with hollow fiber cartridges H 1 P 100-20, H 1 P 10-8 or H 1 P 1-43 having nominal molecular weight cut-off values of 10^5, 10^4 and 10^3 respectively, is applied.[4-6]

Prior to use the cartridges are properly cleaned (suprapure HNO_3 at pH 2, destilled deionized water) and conditioned (250 ml of the sample) in order to minimize wall sorption effects.[5] Based on chemically well defined compounds the pore distribution of the membrane tubes is relatively sharp.[4]

Association studies

The continuous mixing and separation system consists of 3 compartments (Figure 2a). Low molecular test species (radioactive tracers) are transported from the test chamber to the mixing chamber containing naturally occurring colloids. The mixture is separated according to size in the molecular weight discriminator, allowing ultrafiltrate to return to the test-chamber while retainable colloids and associated test species return to the mixing chamber. The decrease of the concentration of test species (radioactivity) in the test chamber during experiment is due to a) dilution, b) wall sorption, c) the association with natural colloids. The association of $^{59}Fe^{III}$ and $^{65}Zn^{II}$ with clay and organic colloids is used for illustration.

Diffusion rate measurements

The sample to be dialyzed is circulated through the hollow fiber interiors while the dialysate (deionized water of similar pH and ionic strength) circulates at a higher rate in the fiber exteriors (opposite direction). The increase of diffusable species in the dialysate is followed by analysis of aliquots withdrawn during the experiment (Figure 3a). The method is standardized by using chemically well defined compounds. The diffusion of low molecular weight Fe-species in spiked ($^{59}Fe^{III}$) and natural water samples is used for illustration.

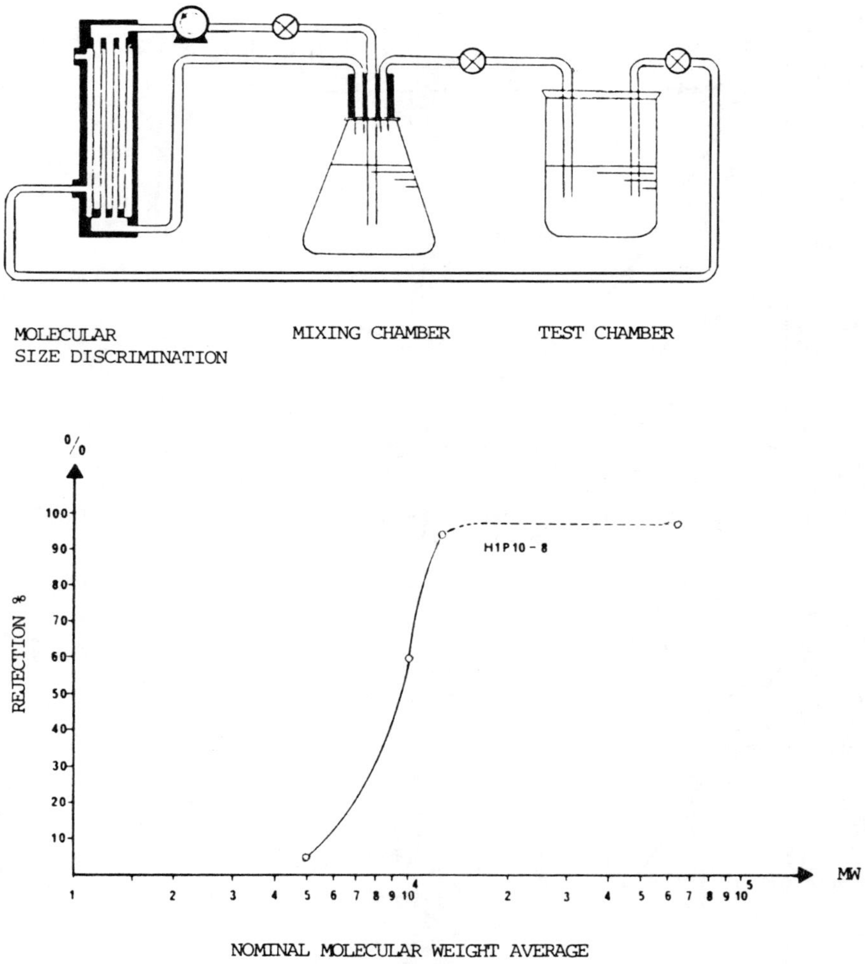

FIGURE 2a. Equipment for investigation of the association of traces with particles/colloids. (Upper): Rejection characteristics for Hollow fiber H1P10-8 (bottom).

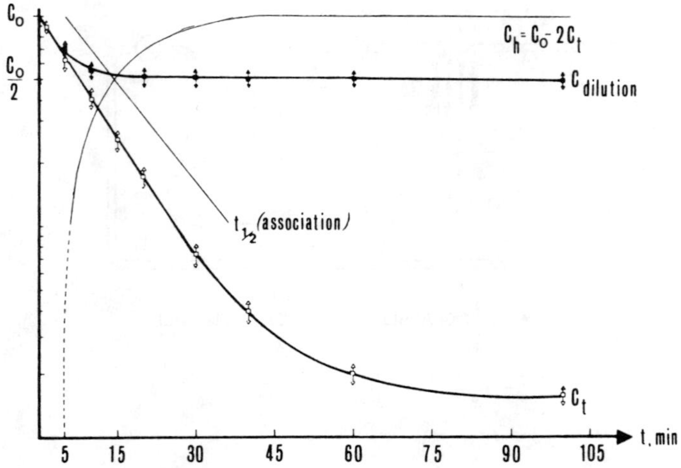

FIGURE 2b. Decrease in concentration in test chamber due to dilution and to the association with colloids.

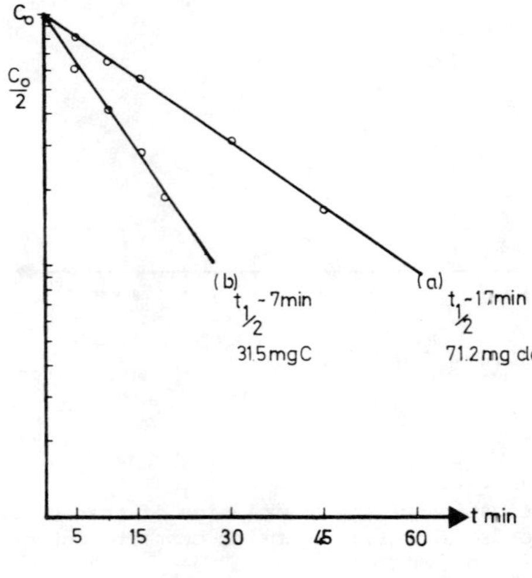

FIGURE 2c. The half-life of association of $^{65}Zn^{II}$ with clay and organic colloids.

RESULTS

Association studies

Using equal volumes in the two chambers the dilution reduces the original concentration in the test chamber, C_o, to $C_o/2$. The conditioning minimizes the wall sorption effect. Thus, the association of test species with colloids can be identified from the decrease of the concentration, C_t, in the test chamber. (Figure 2b) The slope of the curve gives information on kinetics while the level reached ($C_o-2\ C_t$) gives information on the degree of association.

Diffusion rate measurements

The diffusion of species through membranes follows first order kinetics

$$\ln (1-f) = -kt$$

where f is the relative dialysis fraction during experiment. The straight line (Figure 3b) is characterized by the half life of diffusion. As $k \propto D \propto \frac{1}{r}$ (r the radius of diffusing species) the $t_{\frac{1}{2}}$ of diffusion can be used to characterize the size of penetrating species. Based on chemically well defined compounds, a linear relationship between $t_{\frac{1}{2}}$ and the molecular weight of diffusing species is obtained (Figure 3c).

DISCUSSION

Association studies

The association of $^{59}Fe^{III}$ and $^{65}Zn^{II}$ with natural clay (illite) and organic (boggy areas) colloids is shown to follow first order kinetics and is characterized by the half life of association (Figure 2b). As illustrated for $^{65}Zn^{II}$ (Figure 2c) the association with organic colloids occurs more rapidly than with clay colloids and the rate constant, k, reflects the number of binding sites and also the binding mechanism. By varying the amount of adsorbent (i.e. colloids) sorption isotherms and association constants are obtained.[4]

Diffusion rate measurements

The $t_{\frac{1}{2}}$ of dialysis for the $^{59}Fe^{III}$ tracer added to coastal water samples is close to that of the $^{59}FeCl_3$ -standard. Thus a simple, probably hydrated form seems predominant. The $t_{\frac{1}{2}}$ of dialysis for low molecular weight Fe in organic rich waters (boggy areas) is, however, close to that of ^{59}Fe-EDTA. The $t_{\frac{1}{2}}$ for the Fe-species corresponds to that of an organic compound (UV-spectroscopy) indicating an association with organics having a molecular weight within the range 380-520 (Figure 3c).

CONCLUSION

The continuous mixing and separation equipment developed opens the possibility for investigation of physico-chemical properties of naturally occurring colloidal species. Furthermore as different reagents (e.g. acids, complexing agents) can be added so the test chamber desorption

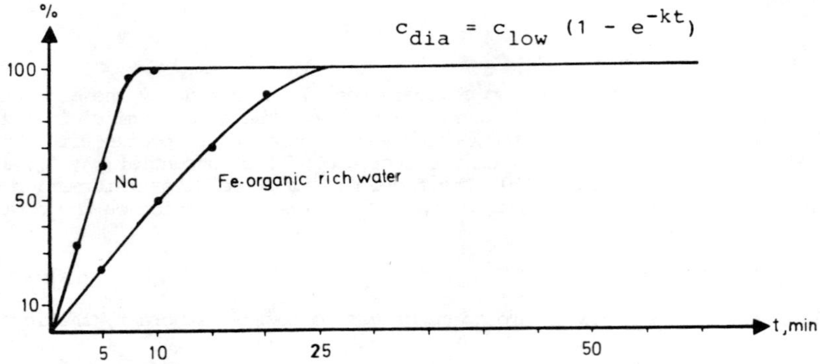

FIGURE 3a. Diffusion (% dialysis) of Na, Fe and organic component (UV) in organic rich water.

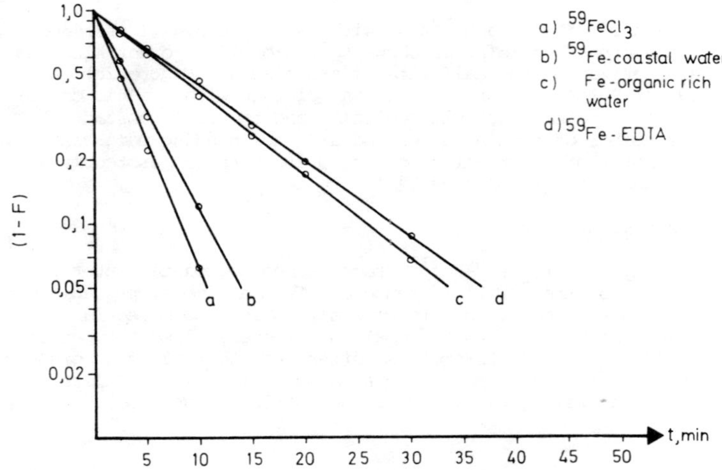

FIGURE 3b. Diffusion rates (1-f) for standards, spiked and natural water samples.

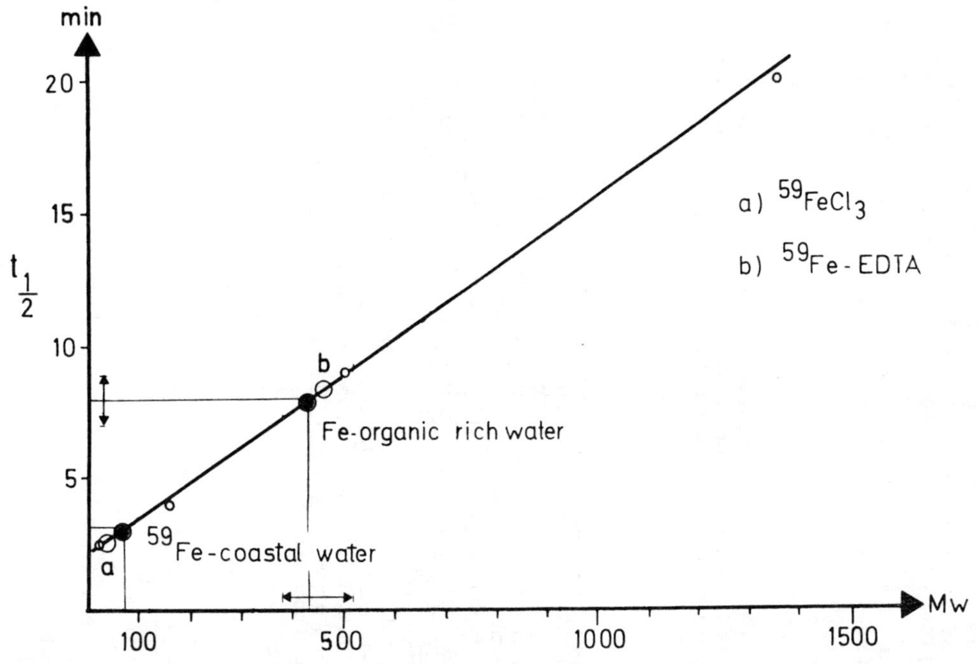

FIGURE 3c. Half-lives of diffusion versus molecular weight of diffusing species for spiked and natural water samples (O standards).

kinetics can also be followed.

The diffusion rate measurements are very useful for distinguishing between species not retainable by membranes. It should be underlined that no reagents are added and wall sorption effects are minimized by conditioning.The separation of species is based on the rate of diffusion (i.e. several aliquots) and deviation from first order kinetics is easily observed. However, for species close in $t_{\frac{1}{2}}$ of dialysis, charge and shape characteristics must be taken into account. Thus further standardization will be performed in order to increase the applicability of this technique.

ACKNOWLEDGEMENTS

B. Salbu is indepted to the Norwegian Research Council for Science and the Humanities for fellowship provided. The authors will thank the Norwegian Hydrological Committee for financial support.

REFERENCES

1. P. Beneš and E. Steinnes, Wat. Res., 8, 947 (1974).
2. P. Beneš, E.T. Gjessing and E. Steinnes, Wat. Res., 10, 711 (1976).
3. B. Salbu, H.E. Bjørnstad, N.S. Lindström, E.M. Brevik, J.P. Rambæk, J.O. Englund, K.F. Mayer, H. Hovind, P.E. Paus, B. Enger and E. Bjerkelund, Anal. Chim. Acta, 167, 161 (1985).
4. B. Salbu, Preconcentration and Fractionation Techniques in the Determination of Trace Elements in Natural Waters - their Concentration and Physico-chemical Forms. Chapts. 1, 3, 5. University of Oslo, Oslo, Norway (1984).
5. B. Salbu, H.E. Björnstad, N.S. Lindström , E. Lydersen, E.M. Brevik, J.P. Rambæk and P.E. Paus, Talanta (in press).
6. B. Salbu, H.E. Björnstad, K. Bibow, J.O. Englund, H. Hovind and J.P. Rambæk, Trace elements in fresh waters within a high mountain catchment (Under publ.).

SPECIATION OF ^{99}Tc AND ^{60}Co: CORRELATION OF LABORATORY AND FIELD OBSERVATIONS

E. A. Bondietti and C. T. Garten

Environmental Sciences Division
Oak Ridge National Laboratory
Oak Ridge, Tennessee 37831

ABSTRACT

The speciation of ^{99}Tc and ^{60}Co in groundwater and soils contaminated with leachates from retired waste-disposal trenches is discussed. The behavior of ^{99}Tc in soil is a complex interaction between soil-associated, reduced species of Tc and the very soluble TcO_4^- anion. An association between reduced Tc species and soil organic matter is evident. Humic substances solubilized with a chelating resin contained about 8% of the soil Tc, with most of it bound in a nondialyzable form except when NaOCl was present. Added $^{95m}TcO_4^-$ did not adsorb to this humic material. Density-gradient separations also confirmed the presence of a light (organic) soil fraction enriched in Tc. An anion exchange resin equilibration indicated that about 20% of the bound Tc came into solution (oxidized) at a rate of 0.015 d^{-1} under conditions favoring oxidation. The species of ^{60}Co in contaminated groundwater is discussed in the context of a review of past work.

1. INTRODUCTION

Radionuclide speciation, as it occurs in the environment, is a function of both the physical and chemical properties of the radionuclide and its interaction with the biosphere. Radiocesium was the first fission product for which the importance of this interaction was apparent. In most temperate-zone soils, ^{137}Cs is less available to plants via root uptake than ^{90}Sr. This behavior is the reverse of that observed when nutrient solutions are used to grow plants and results from the "fixation" of Cs by micaceous minerals in these soils. The ecological availability of ^{137}Cs is therefore not as predictable by its chemistry as it is by the presence of certain soil minerals. More often, however, speciation research is concerned with oxidation states and ligand complexes which affect partitioning between soil (or sediments) and solution. Soluble and insoluble complexes between organic and inorganic ligands are intensely studied on the assumption that such ligands play important roles in environmental behavior. Likewise, oxidation states are examined because they may affect the environmental fate of radioactive nuclides, particularly mobility.

As our environment becomes increasingly contaminated with man-made radioactive elements, especially those from waste-disposal sites, it is becoming increasingly important to understand how radionuclides behave after long-term interactions with ecosystems. The generic study--undertaken to develop a better understanding of how these radionuclides behave--should be supplanted as the real situation presents itself. The challenge is to understand the behavior of radionuclides in actual contamination situations and to use that information, along with controlled experiments, to improve our ability to anticipate future behavior. In particular, we can learn from actual interactions of radionuclides with other waste constituents or with ecosystems.

In this paper we present selected results of on-going and past speciation research on two radionuclides - ^{99}Tc and ^{60}Co. Both have been leaching for almost twenty years from trenches used for sludge disposal at the Oak Ridge National Laboratory (ORNL). For Tc, the existence of aged contamination in a forest ecosystem[1] offers a unique opportunity to understand the biogeochemical behavior of probably the most mobile artificial element. Co-60 is discussed because the chemical species responsible for its mobility from alkaline wastes is an excellent example of why speciation research is important.

2. STUDY SITE

The forested study area which has become contaminated with ^{99}Tc is located about 2 km south of Oak Ridge National Laboratory, near Oak Ridge, Tennessee, within an area of controlled access on the Department of Energy's Oak Ridge Reservation. The site lies nearly midway between chemical waste pits 2, 3, and 4, (approx. 120 m to the east) and chemical waste pit 5 (approx. 90 m to the west). Chemical waste pit 5 first received waste in June 1960, and received more than 300,000 curies of radioactivity before being covered and capped in 1966.[2] Trees sampled along the perimeter of pit 5 in 1962 were analyzed for gamma emitters and showed seepage away from pit 5 in a westerly direction toward the present study area.[3] Several monitoring wells are located immediately to the east of the Tc study site.

3. METHODS

The study-area soil was a silty clay loam with a mean pH of 5.5 (SD = 0.7, N = 9) as determined by using a 1:2.5 (soil:solution) equilibration with 0.01 M $CaCl_2$. The pH of shallow well water in the area is usually between 7.0 and 7.3. The mean organic-matter content of the soil, estimated by loss on combustion at 400°C, was 8.5% (dry-weight basis). The soils in the study area are poorly drained, with some areas remaining wet and muddy throughout the year because of groundwater seeps (the source of the spreading contamination in some cases).

^{99}Tc in soil, water, and vegetation samples was analyzed according to published radiochemical methods.[4] These radiochemical methods included oxidation of Tc in the sample, dissolution in 6 N H_2SO_4, and selective extraction into tributyl phosphate (TBP). Extraction recoveries were determined on representative samples to which a known amount of ^{99}Tc or ^{95m}Tc has been added. ^{99}Tc in the extract was determined by liquid-scintillation counting.

Ten-gram soil samples from different areas of the study site were extracted three times by shaking 20 min with 30 mL of the following reagents: 0.01 M $CaCl_2$, pH 6.8 (to extract water soluble pertechnetate anions); 0.1 M NaOH (to extract ^{99}Tc associated with humic substances), 0.25% NaOCl, pH 10.4 (to oxidize and render extractable the reduced forms of ^{99}Tc in the soil), and 6 N H_2SO_4 in combination with 1.3 g of potassium persulfate (to oxdize and render extractable the remaining chemical forms in the soil).

Density-gradient separations were accomplished by use of ethanol-tetrabromoethane mixtures[5] while resin extraction studies were conducted by using a chelating resin (Chelex-100, Na^+ form) and an anion-exchange resin (Dowex 1, Cl^- form). The resins were equilibrated with sieved soil (diameter, 43 μm); 0.5 g of soil and 1 g of resin were used in the anion equilibration; and 10 g resin and 20 g soil were used in the Chelex equilibration. The Tc was partially displaced (51%) from the anion resin by use of 2 M H_2SO_4 and then analyzed.

Pressurized extractions were made by using compressed air (2 MPa) and a conventional soil-moisture pressure-plate apparatus to reduce soil-water content from 34% (saturated) to usually near 24% of the dry weight.

4. RESULTS

4.1 Extractability of Tc from soil

Past studies have showed that when soils have been contaminated with ^{99}Tc in the field, either experimentally or as the result of releases from nuclear facilities, only a small fraction of the Tc is extractable with $CaCl_2$.[6] This indicates that the Tc has become immobilized in a chemical form (reduced?) different from that of the highly soluble TcO_4^- anion. But, when oxidizing extractants such as NaOCl are used, this immobile Tc is rapidly released. For soils in the Tc study site, this was also the case (Fig. 1). Between 10 and 20% of the ^{99}Tc was extracted by $CaCl_2$ (i.e., as TcO_4^-). Sodium hydroxide also solubilized ^{99}Tc by dispersing some of the soil organic matter, but weak NaOCl solution was the most effective extractant. These latter two extractions are useful only in demonstrating that solubilization can occur under different conditions and do not necessarily represent different chemical fractions.

Less drastic treatments of the soil were also undertaken to evaluate the solubility of Tc. Air-drying the soil did not affect the levels of soluble Tc found upon rewetting the soil. However, when soil was incubated at 24% water contents for two months, considerably more soluble Tc was found by pressurized extraction displacement than was observed when the soil was initially taken from the field or subsequently incubated at 15 to 20°C several months at the original 34% field-moisture content. These results are presented in Fig. 2 in the form of soil/water partition coefficients. The Tc partition coefficient at the time of sampling

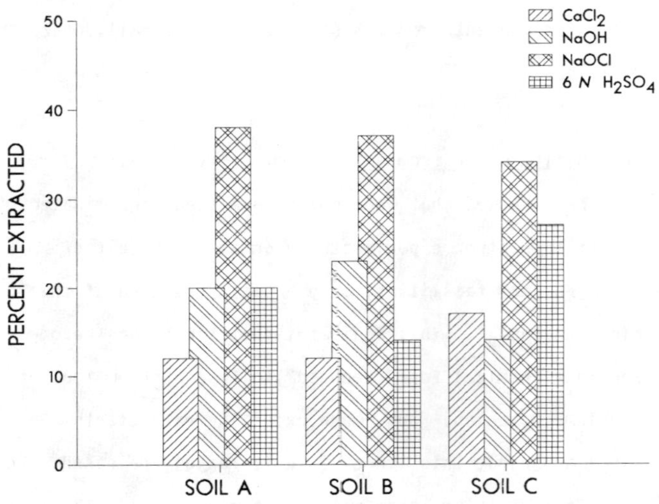

Fig. 1. Extractability of ^{99}Tc from forest soils contaminated with ^{99}Tc for the last 20 years.

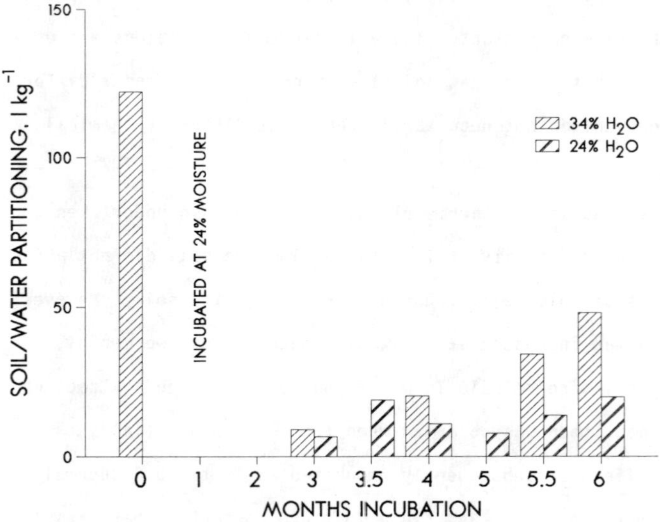

Fig. 2. Soil/water distributions of ^{99}Tc determined by pressure displacement of soil water.

was ten times higher than that found after the first 24% water incubation period of three months duration [the soil/water partition coefficient is obtained by dividing the soil Tc concentration (kg^{-1}) by the soil-water Tc concentration (L^{-1})]. At 24%-water, the Tc partition coefficient obtained from successive extractions remained between about 10 and 20 for the duration of the experiment (the soils had been brought to 34% water 24 h before extraction). However, when the soil was continuously incubated at the original field-moisture content (34%), the partition coefficient for Tc tended to increase, indicating a decline in soluble Tc.

4.2 Oxidation Rate of Tc

When soil was equilibrated for 15 d in the presence of successive changes of anion exchange resin, the extraction of $^{99}TcO_4^-$ exceeded that found by using $CaCl_2$. Figure 3 presents these results. The Y-axis intercept of a linear regression of the day-4 through -15 measurements was similar to the $CaCl_2$ value. The excess Tc found with the resin was about 20% of the bound fraction. This Tc must have formed during the extraction as the result of oxidation of the reduced Tc species in soil. From the extraction data, the rate of oxidation of the reduced pool was calculated to be 0.015 d^{-1}. This is a first-order rate constant (k) derived by evaluating the first order assumption:

$$\frac{d(\text{Tc-bound})}{dt} = k\ (\text{Tc-bound}) \qquad (1)$$

4.3 Natural humic-reduced ^{99}Tc complexes

The presence of an organically bound, reduced ^{99}Tc species was confirmed several ways. First, a fraction of the soil organic matter containing 8.6% of the soil Tc was solubilized by using the chelating resin, Chelex-100. The resin treatment had removed polyvalent cations responsible for flocculating the soil humic materials. The organic matter remaining in solution after high-speed centrifugation (9800 x g) was put in a cellulose dialysis bag. Figure 4 presents the results of dialysis experiments with this solubilized material. Compared with added $^{95m}TcO_4^-$, the indigenous ^{99}Tc showed strong retention, indicating a bound form which must not be TcO_4^-, as indicated by the behavior of $^{95m}TcO_4^-$. The addition of an oxidant, NaOCl, converted this bound ^{99}Tc to a diffusible form (data not shown).

When ultrafiltration techniques were applied to this material, 31.3% of the ^{99}Tc was found to be retained by a 0.45-µm membrane, 30.1% by a 10,000 atomic mass unit (a.m.u.) membrane, and 25.7% by a 1000 a.m.u. membrane. Ten percent passed a 500-a.m.u. membrane.

When 1 g of <43-µm sized soil was centrifuged in a 1.5 to 2.9 g cm^{-3} density gradient, 30% of the ^{99}Tc was found to be associated with soil less than 2.09 g cm^{-3} in density. This fraction contained only 5% of the soil mass, indicating a light (high-organic-matter-content) fraction enriched in ^{99}Tc. Subsequent ashing revealed a loss on ignition of 47% for this fraction, compared with only 9% for the 2.09 to 2.63 g m^{-3} fraction, which contained 90% of the mass and 63% of the Tc. Only about 10% of the soil Tc was solubilized by the gradient solution.

5. DISCUSSION

5.1 Behavior of ^{99}Tc in Soil

Organic matter in the soil appears to be an important sink for leached Tc in this forest system although its contribution to the total immobilized fraction remains unquantified. Organic-matter

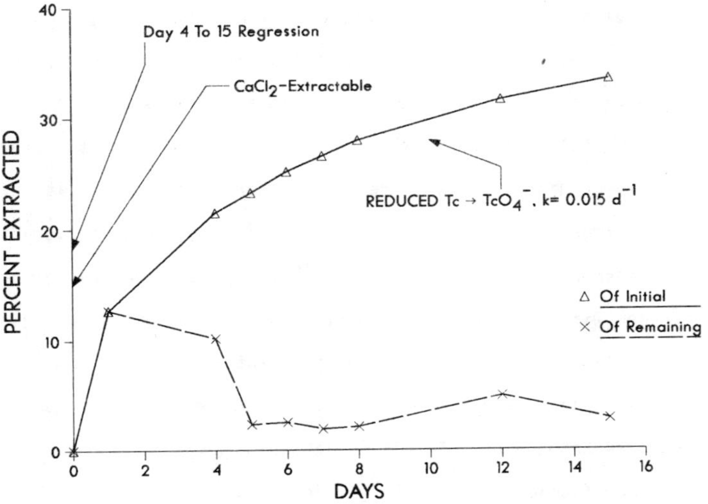

Fig. 3. Extraction behavior and apparent oxidation rate of soil ^{99}Tc equilibrated with anion exchange resin.

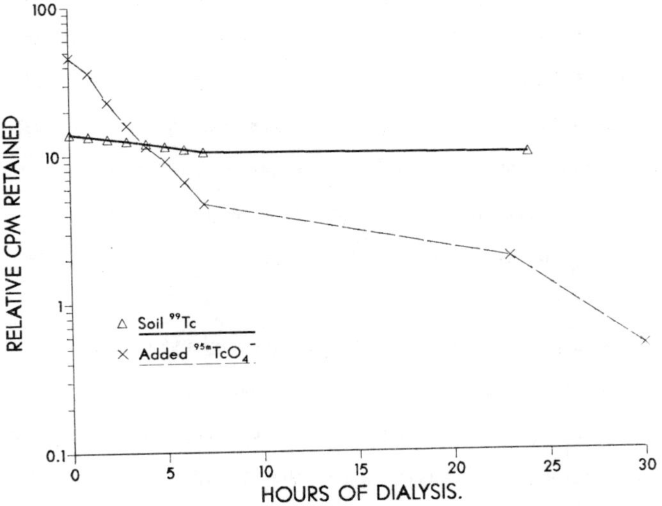

Fig. 4. Comparison of the retention of added and naturally-bound Tc by chelating resin solubilized humates from Tc-contaminated soil.

associations are apparent from the results obtained by the NaOH and chelating-resin extractions and by the density-gradient separation. These findings are consistent with other recent studies involving experimental additions of Tc.[7] A most promising result of the density-gradient work was the low leachability of the ^{99}Tc in the tetrabromoethane-ethanol mixture used for preparing the gradient. The leaching of divalent cations (^{60}Co, Cd, etc) has always been a problem with such gradients, particularly in quantifying organic associations. It is our opinion that future progress in understanding the nature of organo-technetium complexes can be acquired through use of this isolation technique.

Substantial amounts of ^{99}Tc in these soils are also water-soluble. Salt ($CaCl_2$) and pressurized extractions showed that about 10-20% of the soil Tc was in solution when incubated in the laboratory. More appeared in solution when the soil incubations were conducted at 24% water content than at the field-water content of 34%. The reduced solubility found at 34% was most likely due to the reoxidation rate of Tc being slower than the rate of water removal since 29% of the soil water was removed by each of the pressure extractions. Simple dilution of the remaining Tc upon adding makeup water can explain part of the increase in the soil/water partition coefficient.

The oxidation-rate results obtained by using the anion-resin technique can be considered an upper-limit rate for the soil:water ratio used. First, the product of oxidation, TcO_4^-, was removed from solution by the resin rather than building up in solution as it normally would. Secondly the resin was frequently replaced by sieving, which ensured that the solution remained aerated. Finally, the short period of the experiment (15 d) might reflect only the most "available" Tc that can be oxidized.

More work needs to be done on the oxidizability of soil-bound Tc. In particular, the long-term rate constant needs to be evaluated. Such a study may contribute to an understanding of the different chemical forms responsible for binding Tc to soil.

5.2 A Comment on ^{60}Co Mobility in ORNL Wastes

An estimated 57,000 GBq of ^{60}Co was disposed to Trench 7 at ORNL.[2] A very small amount of this Co rapidly leached out of trench 7 when it was first put into operation in 1962. Of considerable interest is the fact that only ^{60}Co, ^{99}Tc, and ^{3}H dominate groundwater activity.[2] Our purpose in mentioning ^{60}Co at this seminar is to illustrate how industrial processing of radionuclides can affect speciation.

In 1973 Bondietti examined the ^{60}Co behavior in a surface seep below trench 7 and concluded that the Co was anionic, that it was strongly complexed with a low-molecular weight organic compound, and that, based on a history of use as a decontamination agent, the synthetic chelating agent EDTA (ethylenediaminetetraacetic acid) was a likely suspect.[8] Later work confirmed this suspicion.[9] Most recently, S. Y. Lee showed that the oxidation state of the Co was +3.[2] The redox diagram presented in Fig. 5 indicates that the 100:1 [Co (III):Co (II)] activity line for the redox couple of the prospective unhydrolyzed EDTA complexes falls largely in the stability field for water, as contrasted to the uncomplexed equal-activity line for the uncomplexed couple. This shows that Co^{3+} is a strong oxidant in the absence of EDTA but quite stable when complexed. A series of high-pressure liquid-chromatography experiments[2] demonstrated that the anionic Co species in the

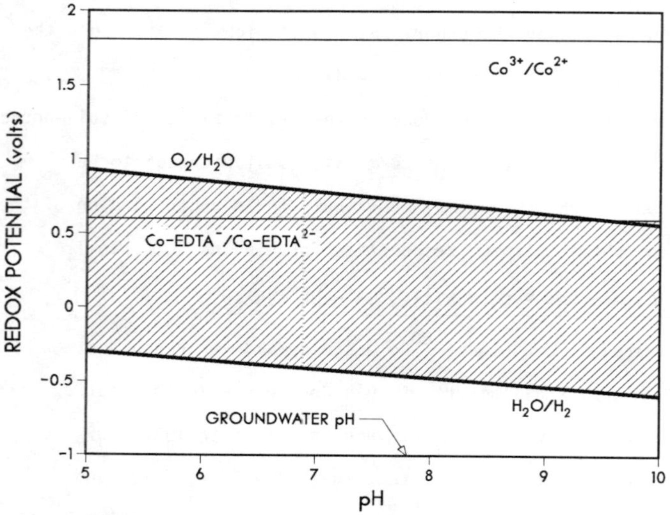

Fig. 5. Comparison of the stability fields for the uncomplexed and complexed (EDTA) Co(III)/Co(II) couple.

groundwater behaved like the mono-charged Co(III)-EDTA complex rather than the divalent Co(II)-EDTA complex. Further studies also showed that ferrous hydroxide precipitates scavenged the Co from the groundwater but that ferric hydroxide precipitates did not.[2]

This research indicated that the relatively stable Co(III)-EDTA complex was responsible for the migration of Co. The implications of this finding are important. The significance of radionuclide complexes in the environment depends on their <u>effective</u> stabilities in multielement solutions. Competition between divalent cations such as Ca^{2+} and Co^{2+} for complexing ligands depends on the respective stabilities of their complexes as well as on their respective concentrations. Also, while Co^{2+} may form an EDTA complex that is 10^6 times more stable than the Ca^{2+} complex, Co^{3+} can form a complex which is on the order of 10^{30} times more stable. This of course has a profound effect on the mobility of Co.

6. SUMMARY

This paper has presented the results of a series of experiments designed to understand the behavior of Tc in soil contaminated by Tc for a long time. The main objective of examining Tc in this soil was to evaluate the bound and soluble pools of Tc in order to help develop a better understanding of its cycling rates in the entire forest ecosystem.[1]

One of the the most significant findings was that the bound Tc is readily oxidized. This indicates that not only can groundwater contribute to the Tc found in vegetation by root uptake of water but that the Tc found in the soil and litter must also be bio-cycling. The presence of organic associations confirms previous work but the exact fraction of the soil ^{99}Tc in this pool is not known. More important, however, than the determination of the soil associations is the determination of the air-oxidizability of the bound Tc. ^{60}Co was discussed because its movement from waste-storage sites involved an interaction with a synthetic ligand which stabilized an environmentally rare oxidation state (or is it rare?). These findings also illustrate both the complexity of the problem as well as the necessity to examine environmental speciation.

Acknowledgments - Research sponsored by the Office of Health and Environment Research, U.S. Department of Energy. Oak Ridge National Laboratory is operated by Martin Marietta Energy Systems, Inc. under Contract No. DE-AC05-84OR21400 with the U.S. Department of Energy. The assistance of J. N. Brantley in these studies is deeply appreciated.

7. REFERENCES

1. C. T. Garten, C. S. Tucker, B. T. Walton (submitted, J. Environ. Radioactivity).
2. C. R. Olsen, P. D. Lowry, S. Y. Lee, I. L. Larsen, N. T. Cutshall, ORNL/TM-8839, National Technical Information Service, Springfield, VA, (1983).
3. S. I. Auerbach, J. C. Ritchie, ORNL-3492, Oak Ridge National Laboratory, Oak Ridge, TN, (1963).
4. C. R. Walker, H. S. Spring, in Radioelemental Analysis Progress, and Problems, edited by W. S. Lyon, (Ann Arbor Science Publ. Inc., Ann Arbor, MI., 1980), pp. 101-110.
5. C. W. Francis, F. S. Binkley, E. A. Bondietti, Soil Sci. Am. J., 40, 785 (1976).
6. C. T. Garten, F. O. Hoffman, E. A. Bondietti, Health Phys., 46, 647 (1984).
7. M. A. Stalmans, A. Cremers, in Scientific Seminar on the Behavior of Technetium in the Environment, Cadarache, France, October 23-26, 1984, (in press).
8. S. I. Auerbach, ORNL-4935, pp. 49, Oak Ridge National Laboratory, (1974).
9. J. L. Means, D. A. Crerar, J. O. Duguid, Science, 200, 1477, (1978).

RADIONUCLIDE SORPTION IN SOILS AND SEDIMENTS : OXIDE - ORGANIC MATTER COMPETITION

A. MAES and A. CREMERS

Laboratorium voor Colloïdale Scheikunde, K.U.Leuven
Kard. Mercierlaan 92, B-3030 LEUVEN, Belgium.

ABSTRACT

This paper deals with the speciation of europium in the solid phase and liquid phase extracts of Boom clay under in situ conditions. Using a new method for measuring the europium-humic acid complex stability constant at high pH, it is shown that carbonate complexation is poorly competitive with humic acid, europium being quantitatively present as humic acid complex, its stability constant being of the order of 10^{12}-10^{13}.

It is furthermore shown that partial oxidation of the sediment, which is strongly reducing under in situ conditions, leads to a significant increase in K_D values, whatever the liquid/solid ratio used. This effect is ascribed to the involvement of ferric oxides which are generated in the solid phase and which lead to a displacement of the metal from the humic acid sink. Some examples are presented which demonstrate that at high pH, around 9, ferric oxides may be competitive with humic acids for metal sorption and that this effect increases with pH.

1 INTRODUCTION

The chemical forms of radionuclides in the environment - geological strata, soils, sediments, groundwaters - are thought to be key factors in the understanding of bioavailability and geochemical processes. In the earlier stages, assessments were routinely based on radionuclide solid-liquid distribution coefficients but the results of such tests are of limited relevance unless they are complemented by speciation studies and refer to conditions identical to those prevailing in situ of the system of interest. Strangely enough, radionuclide K_D measurements continue to be used in a most rudimentary version, such as the equilibration of air-dried sediments with distilled water, disregarding the drastic changes which the system may have undergone. In fact, for many cases, the batch samples so studied may bear very little resemblance to the original sediment.

It is the purpose of this paper to concentrate on competitive effects of the two phases which are considered the most important in their effect on the environmental fate of many radionuclides, i.e. ferric oxides and organic matter. As a test nuclide, we have studied europium, an excellent analog for trivalent actinides. In the following, we discuss (a) the effect of dissolved humic acids (HA) on metal speciation and (b) the nuclide partitioning between oxide phases and humic acids in the solid phase, giving particular attention to the modifications in solid-liquid distribution behavior, resulting from oxidation processes.

2 LIQUID PHASE SPECIATION

The key ligands determining the geochemical forms of europium, and many other hydrolysable metals, are OH^-, CO_3^{--} and HA, the speciation being determined by the relative abundancy of these ligands and the corresponding complex formation constants. Below, we shall limit the discussion to CO_3^{--} and HA by operating under conditions where hydrolysis is not competitive. Some data are already available for carbonate[1,2] and humic acid[3,4] complexations with europium and trivalent actinides, the data being based on solvent extraction procedures.

We have recently developed a new and easy method which allows the measurement of the stability constant of the europium-HA complex, relative to the value of the carbonate complex. In essence, it is a modification of the original Schubert method[5], which is based on the measurement of the effect of increasing HA concentration on the ionic distribution coefficient between a cation exchanger and the aqueous phase. This method is limited to acid conditions[3] when dealing with hydrolysable metals. The method may be extended to slightly alkaline conditions - which are of particular interest in geochemistry - by operating in the presence of carbonate levels sufficiently high to suppress hydrolysis and ensure the exclusive formation of the $Eu(CO_3)_2^-$ complex. A solution of .1 M $NaHCO_3$ at a pH of about 9 achieves such conditions, as can readily be demonstrated by anion exchange equilibria. The method consists of measuring the change in S/L distribution coefficient of Eu with changing HA concentrations, using resinous anion exchangers. The method will be published elsewhere.

Omitting further procedural details, we shall limit the discussion to a summary of some preliminary findings. Measurements were carried out on a commercial HA (Fluka) and a sample extracted from Boom clay, using $NaHCO_3$ 0.1 M. The statistical averages for the stability constant ratio, $\log K/\beta_2$ are

Fluka HA: 0.1($\pm$0.13) (n=14)

Boom clay HA: 0.95($\pm$0.14) (n=5)

The stability constants being defined as (omitting valence signs):

$$K(\text{Eu-HA}) = \frac{|\text{Eu HA}|}{|\text{Eu}|\,|\text{HA}|} \quad \text{and} \quad \beta_2 = \frac{|\text{Eu}(CO_3)_2|}{|\text{Eu}|\,|CO_3|^2}$$

where HA is expressed in equivalent/liter. These data correspond to values of the order of 10^{12} to 10^{13} for the stability constant of the Eu-HA complex, depending on the value chosen[1,2] for β_2. This result is in excellent agreement with the value obtained by Torres and Choppin[3] for log K at pH=9: 13.25 ($\pm$1.95).

Whatever the absolute value for K (EuHA), the distribution of Eu between its HA and CO_3 forms can readily be calculated from the K/β_2 ratio using the equation

$$\frac{\text{EuHA}}{\text{Eu}(CO_3)_2} = \frac{K\,|\text{HA}|}{\beta_2\,|CO_3|^2}$$

It follows that, even at relatively high CO_3 levels, minor amounts of HA are sufficient to shift Eu into its HA complex. Taking a functional group capacity (COOH) of 3.5 meq/g[6], it would appear that, at "free" CO_3 concentrations of 10^{-3}M, HA levels of 1 ppm would be sufficient to suppress the fraction of Eu as $Eu(CO_3)_2$ complex to some 5%. (The fraction of

$Eu(CO_3)^+$, as compared to $Eu(CO_3)_2^-$ is negligible at such CO_3 levels, as can be inferred from the available stability constants[1]). Lowering the CO_3 level to 5.10^{-4}M would leave only traces of Eu as carbonate complex.

3 SOLID PHASE SPECIATION

The scavenging of radionuclides in the subphases of a mixed sedimentary system is ruled to a large extent by the nuclide "binding constants" to the various components and the composition of the mixture. Among the subphases, oxides and organic matter are the most important (for hydrolysable metals) and the sorption mechanism is one of surface complexation. The task of assessing metal partition in such a mixture is rather complex and has not been resolved satisfactorily. The various problems involved have been discussed in some detail recently[7].

Current methods for obtaining geochemical partitioning of metals in sediments are based on some form of selective extraction treatment[8,9]. These are however purely operational procedures which are not to be considered as speciation methods. Therefore, geochemical phase assignments, based on such methods are of dubious relevance because the systems are subjected to rather harsh treatments, creating conditions far remote from the in situ situation. In particular, when dealing with reducing sediments, such treatments may alter the system in generating new sorption phases in the sample. Moreover, changes in pH are liable to lead to changes in geochemical partitioning, even under conditions where the radionuclide speciation in the liquid phase is unchanged. Some of these aspects are discussed below.

3.1 Effect of air exposure

In a recent paper[7], it was shown that, in Boom clay under in situ (highly reducing) conditions and at low liquid/solid (L/S) ratios, the distribution behavior of Eu is ruled by the HA distribution, i.e. Eu is exclusively present as HA complex in both phases. At a L/S ratio of 2, the HA concentration is of the order of 10^{-3} Eq./l and K_D is about 30-40 ml/g. It was also shown that, upon partial oxidation of the clay through air exposure, the K_D(Eu) was increased by a factor of 5-6 at otherwise similar extraction conditions (pH and HA concentration of the liquid phase). The statistical average (n=10) for K_D was 202 ($\pm$ 15) at HA concentrations in the extract of 1.93 ($\pm$.13) 10^{-3}Eq/l.

The interpretation, proposed for this rather significant increase in K_D for Eu, was that hydrous ferric oxides are generated in the system (resulting from pyrite oxidation), leading to a partial displacement of Eu from the HA sink. It was estimated that the fraction of Eu in the HA phase is reduced to about 25%, such estimate being based on the EFAR concept, the Equilibrium Fractional Activity Ratio, defined as:

$$\text{EFAR}=\frac{\text{Nuclide activity / Equivalent HA in extract}}{\text{Nuclide activity / Equivalent HA in sediment}}$$

The usefulness of this parameter in geochemical phase assignments was discussed recently[10].

The extent of oxidation can conveniently be expressed in terms of soluble sulphate content. The maximum sulphate level at complete oxidation can readily be estimated from the sulfur content of Boom clay[11]: 7.75($\pm$1.6)mg/g which corresponds to 244 ($\pm$50)μMoles/g. The corresponding level of Fe_2O_3, generated in the pyrite oxidation process, would therefore amount to 9.6 ($\pm$2)mg/g clay or about 1%, a value which is to be compared with the HA concentration of the clay[11]: 22($\pm$3) mg/g. The sulphate content of the sample discussed above is 13.5 μMoles/g, which would indicate that the phase, generated in the oxidation process is an extremely active one.

The effects of oxidation are further confirmed by the behavior of Eu in partially oxidized Boom clay samples at high L/S ratios(200). The results are summarized in Figure 1, showing the effect of sulphate levels on the log K_D value for Eu in two clay samples (214 and 219 m depth).

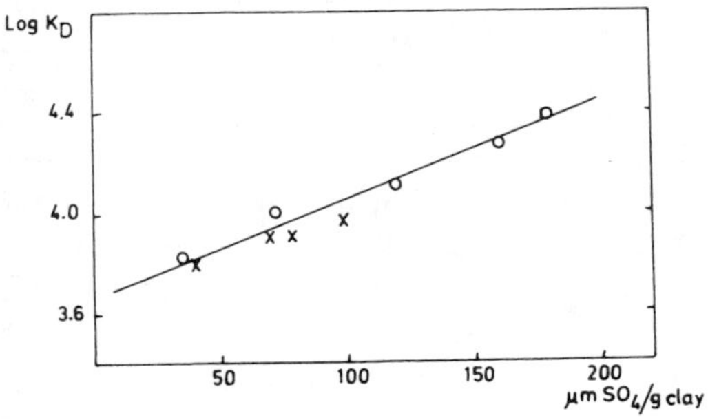

FIGURE 1 Log K_D(EU) at trace levels (< 10^{-6}M/g) versus sulphate content of Boom clay (219,214 m)

Oxidations were carried out cn 12.5 % clay slurries in distilled water at about 60°C. Samples were taken at regular time intervals and diluted fivefold with the in situ solution (essentially $NaHCO_3$ 28 meq/l, KCL 1 meq/l, $MgCl_2$ 1 meq/l, pH = 8.8-8.9). Samples were treated by N_2 bubbling and stored at 5°C. Sulphate contents were measured radiometrically (^{132}Ba) by $BaSO_4$ precipitation at pH=4 in the presence of an excess of Ag-thiourea, included to prevent ion exchange sorption[12] of barium.

It is seen that K_D values correlate quite well with sulphate content, and therefore with extent of oxidation. These data are also consistent with a series of K_D measurements at identical L/S ratios on fresh and completely oxidized samples (obtained by a two-month air exposure), using the in situ solution. The statistical averages for log K_D(Eu) at trace loadings are: fresh samples (212,214m) : 3.57($\pm$.12) (n=4)

oxidized samples (208,212,214,219m): 4.43($\pm$.10) (n=8)

It appears that the values obtained on fresh samples are in excellent agreement with predictions based on earlier data[7], relating to the effect of L/S ratios on K_D (Log K_D $\pm$ 3.4).

At the high L/S ratios used in these equilibria, we are uncertain about the speciation of Eu in the extracts. It would in any case appear that a significant fraction of Eu is present as a CO_3-complex, as evidenced by findings that the replacement of CO_3 by Cl in the in situ solution (at otherwise similar conditions) leads to K_D values of about 10^5 (as compared to $10^{3.6}$ in the presence of CO_3).

In order to validate the possible involvement of ferric oxihydroxides in the phenomena just described, it remains to be demonstrated that such phases may in fact be competitive with HA at the pH conditions of these measurements. Such an attempt is presented below.

3.2 Oxide - humic acid competition

The fractional distribution of metals between oxide and HA phases can be obtained from straightforward dilution principles. In experimental terms, known amounts of oxide and HA are equilibrated at varying pH in the presence of some known amount of radionuclide. After equilibration, radionuclide activity and HA concentrations are measured in the supernatant. using the Ag-thiourea method[13]. The fraction of nuclide, associated with the HA phase is obtained from the ratio of radionuclide activities in the equilibrium solution and the total system, both expressed per unit HA. Table I shows some examples for the case of Eu in hematite and an amorphous

$Fe(OH)_3$ system, obtained by freeze-drying from freshly precipitated $Fe(OH)_3$. A similar set of data is shown in Table II for the case of Cd.

TABLE I Effect of pH on the fraction of Eu(α) in the HA phase (Fluka) in ferric oxide - HA mixtures

Hematite				Amorphous $Fe(OH)_3$	
ox(10g/l) - HA(2g/l)		ox(10g/l) - HA(.2g/l)		ox(5g/l) - HA(.25g/l)	
pH	α	pH	α	pH	α
8	.86	7.47	.42	7.46	.98
8.44	.83	7.97	.39	8.47	.86
8.94	.78	8.54	.36	9.48	.77
9.46	.74	9.13	.33		

TABLE II Effect of pH on the fraction of Cd (α) in the HA phase (Fluka) in amorphous $Fe(OH)_3$ - HA mixtures

pH	α
7.20	.81
7.97	.71
8.91	.60
10.09	.52

These data show that the oxide phase becomes more competitive with HA at increasing pH. Such result can be explained by the fact that pH variations in the range studied, have only a minor effect on the complexing behavior in HA which is completely dissociated (pK $\pm$ 4.5). On the contrary, in the case of ferric oxides, characterized by zpc values of about 8.5, a pH change in this range can be expected to result in marked effects on sorption properties. Examination of the data in Table I indicate that the oxide phases studied are not nearly as competitive as the sink, generated by oxidation processes in the Boom clay. However, it is very likely that in this system, where the oxide phase is produced from ferro-oxidation in finely dispersed pyrite, ferric oxides may exhibit a more pronounced surface activity.

4 CONCLUSIONS

In natural sediments, the behavior of radionuclides is essentially governed by competitive effects between oxide and humic acid phases. From complexation studies of Eu in HA, it appears that in the liquid phase, carbonate is not very competitive with HA, which is able to displace Eu from its carbonate complexes at levels of the order of a few mg/l. When dealing with reducing sediments, it appears that oxidation processes, resulting from air exposure of the samples may lead to a significant increase in solid-liquid distribution coefficients. This effect is most likely due to the formation of ferric oxides which are being formed and which become increasingly competitive at higher pH.

5 ACKNOWLEDGEMENT

This work was carried out in the frame of a C.E.N./S.C.K. (Mol) programme in geological disposal of radioactive waste under contract with the C.E.C.

6 REFERENCES

1. R.Lundqvist, Acta Chem.Scand.A. 36, 741, (1982).
2. G. Bidoglio, Radiochem. Radioanal. Letters, 53, 45, (1982)
3. E.L. Bertha and G.R. Choppin, J.Inorg. Nucl. Chem., 40, 655, (1978).
4. R.A. Torres and G.R. Choppin, Radiochim. Acta, 35, 143, (1984).
5. J. Schubert, J.Phys. Coll. Chem., 52, 340, (1948).
6. F.J. Stevenson, Humus Chemistry, (Wiley Interscience, New York, 1982).
7. A. Cremers, P. Henrion and N. Monsecour in C.E.C. Int. Seminar on the Behaviour of Radionuclides in Estuaries, Renesse, The Netherlands, September 1984. (in press)
8. A. Tessier, P.G.C. Campbell and M. Bisson, Anal. Chem.,51, 844, (1979).
9. S.N. Luoma and G.W. Bryan, Sci. Total Environ., 17, 165, 1981.
10. M. Stalmans, A. Maes and A. Cremers, in C.E.C. Int. Seminar on the Behaviour of Technetium in the Environment, Cadarache, France, October 1984 (in press).
11. B. Baeyens, A. Maes, A. Cremers and P. Henrion, Radioactive Waste Management and the Fuel Cycle, (1985),(in press).
12. R. Chhabra, J. Pleysier and A. Cremers, in Proc. Int. Clay Conf. Mexico City (Applied Publ. Ltd., Wilmette, Ill., 1975, 439).
13. M. Stalmans, S. De Keijzer, A. Maes and A. Cremers, in C.E.C. Int. Seminar on the Behavior of Technetium in the Environment, Cadarache, France, October 1984 (in press).

FACTORS THAT AFFECT THE ASSOCIATION OF RADIONUCLIDES WITH SOIL PHASES

B.T. Wilkins, N. Green, S.P. Stewart, R.O. Major

National Radiological Protection Board, Chilton, Didcot, Oxon, OX11 ORQ, United Kingdom.

ABSTRACT

The use of field experiments to investigate the chemical or physical associations of some radionuclides with soil phases is limited by low levels of activity and complicated by the number of phases involved. Sequential extraction procedures provide one means of evaluating the relative importance of various phases in disposition. Although the separation steps may not be absolutely selective, these schemes can be used in a comparative manner to rationalise changes in association and disposition that can occur as soil conditions alter. In this way they can give a direction for specific laboratory studies and be of value in the prediction of the consequences of land contamination - an important aspect of radiological protection. In this paper we draw upon field and laboratory studies of the disposition of artificial radionuclides to illustrate the effects of changes in, for example, iron or organic content.

The variety of soil types that are amenable to field studies is restricted. Complementary laboratory experiments are therefore essential. Results show that the generalisations often applied to radionuclide availability are not always appropriate and that although predictions of disposition can sometimes be made on the basis of gross soil characteristics, this capability is limited and a more rigorous approach is desirable in extreme cases. The specificity of the extraction procedure is discussed and evidence is presented to support the participation of the residual phase which was previously observed in field studies of plutonium and americium.

INTRODUCTION

Sequential extraction procedures have been criticised from the geochemical viewpoint on the grounds of non-selectivity and it is true that the separation of phases is operationally defined by the method of extraction. Provided however that the results of such studies are used with caution, they can still provide valuable information on the association of trace metals and radionuclides with geochemical phases. In particular, comparative studies enable deductions to be made concerning the concerted effect of several phases and the changes in disposition that would occur as a result of credible changes in soil conditions. For these reasons, their use in radioecological investigations is becoming more frequent.

We have made use of a scheme developed by Tessier(1) in both laboratory and field experiments, the attraction of this scheme being that it attempts to simulate credible changes in environmental conditions. In the laboratory we have followed the disposition of five radionuclides over a one-year period in four soils that are very different in character but are all commonly encountered in the United Kingdom. Our intentions in the future are to include more radionuclides and chemical forms that are relevant to assessments of the long-term consequences of accidental releases from nuclear installations and waste repositories, and to take account of other soil types found in the European Community. The present data will also be used in the rationalisation of results of plant uptake experiments being undertaken at our laboratory that make use of the same four soils.

For field experiments we are able to make use of the slightly-elevated levels of certain radionuclides in the environment around the Sellafield reprocessing plant. However when deposition from atmosphere is the only source of contamination, activity concentrations in soil are sufficient for disposition to be studied only within a few kilometres of the plant. In consequence the variety of soil types that are available for such a study is limited, but adequate to provide an invaluable comparison with laboratory experiments. In addition there are areas which are occasionally inundated by the sea where the contamination is principally of marine origin. These areas can give valuable information for use in predicting the consequences of land reclamation, the use of sediment as landfill or land treatment or changes in waste management strategy. At a recent symposium we compared the disposition of several radionuclides in a pasture, a salt-marsh and in intertidal sediment, and showed that it would be inadvisable to make very general assumptions regarding radionuclide availability(2). This paper continues this theme, using the results of both laboratory and field experiments to illustrate the degree of dependence of several radionuclides upon soil character.

EXPERIMENTAL PROCEDURES

The Tessier scheme successively solubilises activity associated with the water soluble, exchangeable, carbonate, oxide, organic and residual phases. Of these, water soluble and exchangeable may be regarded as readily available, while carbonate, oxides of iron and maganese and organic material may be solubilised if soil conditions such as pH or Eh alter. Activity associated with the residual phase is unlikely to be solubilised over a considerable time-span under the conditions encountered in nature.

The extraction procedure followed that described previously(2) for both field and laboratory studies, with appropriate changes in reagent volumes. Again the procedure was modified and separate samples used for the field studies of iodine-129. Analyses of each extract made use of established methods that have been fully described elsewhere. In the laboratory experiment, radioisotopes have been chosen so as to minimise the amount of chemical processing required before measurement. In field experiments sequential analytical methods permit full use to be made of the entire extract(3), while for iodine-129 a sensitive method can be easily applied to solutions(4). Where appropriate, carriers and recovery determinants were added to each extract after separation from the solid phase and before the analyses began.

LABORATORY STUDIES

The four soils employed in this investigation correspond closely to the following classifications: Denchworth clay; Hamble loam; Fifield sand; Adventurers peat; although only soil from the plough layer (0-400 mm) has been taken for these experiments. Some of the soil characteristics are summarised in Table 1. Each soil was air-dried at room temperature and passed through a 1 mm sieve. 100 g portions were each contaminated with caesium-134, strontium-85/90, ruthenium-106, americium-241 and plutonium-241, applied as nitrates. A uniform dispersion was achieved using a combination of wrist and end-over-end-shakers, after which the moisture content was adjusted to correspond to 1000 mm of water tension. In view of the diversity of these soils, this is preferable to choosing an arbitrary moisture content. The samples were then incubated at 10°C in the dark until required for analysis. Any carbon dioxide evolved in the sample flasks was absorbed by potassium hydroxide solution and replaced by oxygen from a bleed system(5). Flasks were removed periodically and the contents divided into three portions for analysis. Gamma-ray emitting radionuclides were determined directly in each extract. For plutonium-241, an extraction procedure was required, followed by liquid scintillation counting. Results for plutonium are still not complete and will not be discussed in detail; for the other radionuclides data are available for a 6-month period but the anticipated duration of the experiment is 2 years.

Nitrate was the chosen chemical form for this first experiment on the grounds of solubility, each cation essentially being available immediately for reaction. This experiment is intended to form a baseline for comparison with future experiments involving complexed or insoluble forms. However, changes in disposition with time are still evident in some cases.

Figure 1 shows the percentage of caesium-134 found in the residual phase as a function of time, and illustrates the dependence of both rate and, more particularly, extent of transfer upon soil type. The effect of soil type upon both factors is shown by a comparisor of the percentage of the applied activity associated with the residual phase after 2 weeks and 6 months: clay, 83 and 95; loam 57 and 70; sand 35 and 60; peat 5 and 10. Furthermore, the percentages remaining readily available were 40 in the case of peat but 5-10 in the other cases. No simple correlation has been found between the percentage with the residual phase and the clay content of these soils. The minor participation of the organic phase in the peat soil could be a reflection of the general inability of caesium to form organic complexes (Table 2). However, the more significant role apparently played by this phase in loam and sand, which decreases with time, supports the contention that peroxide can attack some mineral surfaces[6]. These observations illustrate how comparative studies can be used to elucidate potential sources of error in interpretation.

The behaviour of strontium-85 is much less dependent upon soil type, disposition being dominated by the exchangeable phase. For clay, loam and sand about 85% of the applied activity remained readily available throughout the experiment. For peat the proportion is about 70%. There is no evidence of an increased participation of carbonate in the slightly alkaline loam, while the oxide phase is only of significance in peat, where it takes up about 15% of the applied activity. The organic phase plays no significant part in disposition, and these results are generally consistent with the chemical properties of strontium.

Results for americium-241 show that disposition changes little with time, although samples incubated for 6 months show a slight increase in the relative importance of the residual phase. As expected the most important phases are oxides and organic (Table 3). One objective of the laboratory study was to determine whether disposition could be adequately predicted on the basis of gross soil characteristics. These results enable us to test this hypothesis by comparing the quotient of the activities found in these phases with the quotient of the dithionate-extractable iron content and the organic carbon content, taken from Table 1. Figure 2 shows a reasonably linear trend for clay, loam and sand but this breaks down for peat (not shown) where the oxide phase is of much greater relative importance.

Data for ruthenium-106 after 6 months incubation are given in Table 4. Data for the early stages of the experiment are not given but these and the corresponding data for americium-241 enable us to comment upon the argument concerning the participation of the residual phase. In our comparison of pasture, salt-marsh and sediment[2] we reported finding substantial percentages of the americium-241 and plutonium activities in association with the residual phase. One critical opinion of the Tessier scheme is that oxidation of organic material leaves sites on mineral lattices freshly available, upon which metal ions recently released into solution may adsorb strongly. The scheme minimises this effect by the addition of ammonium acetate immediately following the treatment with 0.2 M nitric acid and peroxide. This treatment is however relatively mild compared to that normally employed in the determination of gross activities and thus a second possibility is that activity found in association with the residual fraction may be incorporated in organic matter that is more resistant to oxidation. We have two pieces of evidence to support this latter contention.

Generally, during the first 3 months of our laboratory studies of ruthenium and americium, we found the ratio of activities found in the organic and residual phases to be greater than 10 to 1, suggesting that a rapid and extensive adsorption onto mineral surfaces is not a significant problem. This experiment has involved some 30 analyses of this type for each soil, and there have been infrequent occasions when ratio was much less. We conclude that these anomalies are the result of incomplete oxidation rather than readsorption. We have also estimated the amount of organic matter remaining in the residual phase after oxidation with peroxide and 0.02 M nitric acid by determining the loss in weight of the residue after ashing at 450°C. Our findings were that, in the eight soils included in this work, at least 10% of the

Table 1. Some characteristics of soils used in laboratory experiments

	Clay	Loam	Sand	Peat
Mineral fraction < 2μm, %	79	14	8	39
pH in 0.01 M $CaCl_2$	5.7	7.1	6.1	5.9
Organic carbon, %	5.2	1.6	1.2	35
Organic content, % loss on ashing	12.8	4.2	2.9	57.6
CEC, meq/100 g	39.6	15.2	10.4	132.9
Dithionate-extractable Iron %	2.29	0.78	1.12	1.21
Water content at 1000 mm water tension, g per 100 g soil	51.6	28.7	11.4	90.3

Table 2. Disposition of caesium-134 after 6 months incubation

Phase	Percentage of total activity			
	Clay	Loam	Sand	Peat
Water-soluble	5	4	6	5
Exchangeable	< 1	2	4	36
Carbonate	< 1	< 1	< 1	11
Oxides	< 1	3	3	25
Organic	< 1	21	27	12
Residual	95	70	60	11

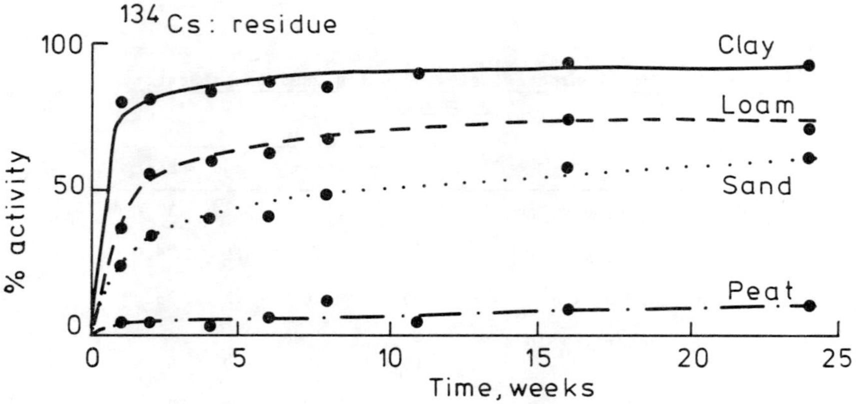

Figure 1 Incorporation of ^{134}Cs into the residual phase

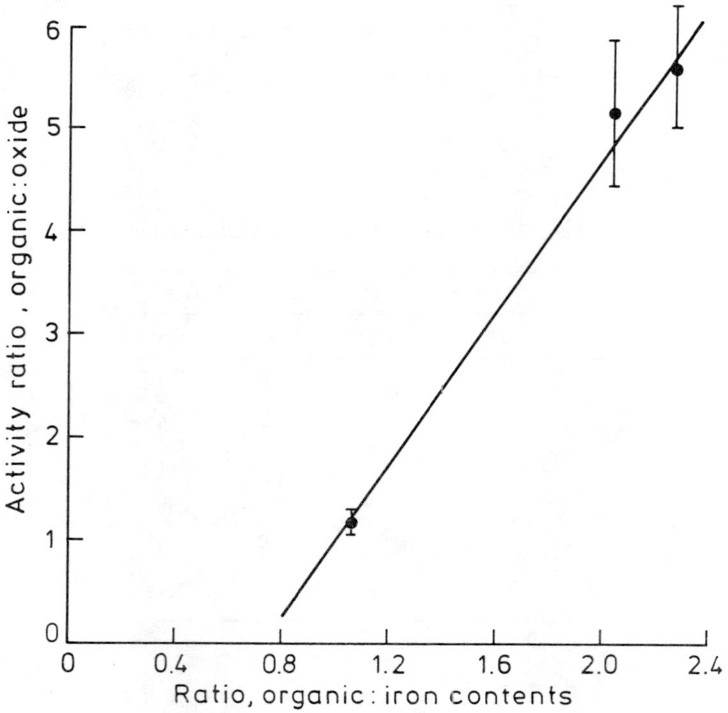

Figure 2 Correlation of ^{241}Am with soil characteristics

total organic content was found in the residual phase, the maximum being more than 30% for the Denchworth clay.

At the present time we have no new field data on the residual phase, but in view of the evidence presented here, we suggest that our earlier field results probably reflect the combined effects of long timescales and continued biological activity, factors which are largely absent in laboratory experiments.

The rate at which radionuclides are incorporated into soil phases is an important consideration in accident consequence assessments. In this experiment changes in disposition with time were most evident in the case of ruthenium-106 in the peat soil. Figure 3 shows the percentages associated with the organic phase and the sum of the water-soluble and exchangeable phases as a function of time, and indicates a relatively rapid initial rate of transfer from the readily available phases to the organic phase. The results can be fitted empirically by two-component exponential functions, as follows:

per cent of total activity in readily available phases at time t (weeks) $= 69 \exp(-0.81t) + 31 \exp(-0.091t)$

per cent of total activity in organic phase at time t (weeks) $= 75 - 36 \exp(-0.50t) - 39 \exp(-0.26t)$

These functions have no physical significance but imply that for example the initial rapid loss of activity from the readily available phases occurs with a half-time of less than 1 week. For the other soils the oxide phase becomes more important, as evidenced in Table 4. The data are not shown but changes in disposition with time are less apparent, although decreases in the proportions associated with the readily-available phases are discernible.

FIELD EXPERIMENTS

Within the limited geographical area available for an investigation of this type, we have located four significantly different soil types and again have confined our interests to surface soil, in this case 0-20 mm depth. Six cores of 100 mm diameter were collected from each area and the top 20 mm taken from each, air dried and sieved through a 1 mm mesh. The six subsamples were then blended before analysis. The areas correspond to the following soil series: Wick, Altcar, Newport, Enbourne. However, they can be more helpfully described as brown earth, alluvial, sedge peat and brown sand respectively. These soils have yet to be characterised fully but some relevant properties are shown in Table 5. Our studies of the brown earth were undertaken as part of a project on uptake by grazing cattle[7] and the results have previously been reported in full for caesium-137, strontium-90, iodine-129, americium-241 and plutonium[2]. Similar studies on the other soils are not quite complete but sufficient data are available to draw both contrasts and parallels with the laboratory experiment.

Data for strontium-90 are only available for the brown earth. These were given in reference 2 but are summarised again in Table 6. A contrast with the laboratory studies is immediately apparent: the greater participation of the oxide phase. Strontium can be effectively scavenged from aqueous solution by coprecipitation with ferric ion but with the exception of the peat soil the participation of the oxide phase in the laboratory studies is small, whereas in the brown earth about 25% of the activity is associated with this phase. The iron content of soils in the Sellafield area is generally higher than those used in the laboratory (Table 1) but that for the brown earth is very

Table 3. Disposition of americium-241 after 6 months incubation

Phase	Percentage of total activity			
	Clay	Loam	Sand	Peat
Water-soluble	2	2	2	1
Exchangeable	1	1	1	1
Carbonate	2	1	3	1
Oxides	10	15	25	5
Organic	57	70	57	85
Residual	27	11	11	6

Table 4. Disposition of ruthenium-106 after 6 months incubation

Phase	Percentage of total activity			
	Clay	Loam	Sand	Peat
Water-soluble	7	4	6	2
Exchangeable	2	1	< 1	3
Carbonate	2	8	3	1
Oxides	31	30	32	13
Organic	49	51	38	75
Residual	84	5	21	5

Table 5. Some characteristics of soils used in field experiments

	Brown earth	Alluvial	Sedge peat	Brown sand
pH in 0.01 M $CaCl_2$	5.5	5.5	5.5	5.5
Organic content, % loss on ashing	15.0	27.6	53.7	14.4
Total iron content, %	2.0	4.1	2.4	0.92

Table 6. Disposition of strontium-90 in brown earth*

	Percentage of total activity
Water-soluble	5
Exchangeable	45.1
Carbonate	20.2
Oxides	24.9
Organic	2.5
Residual	2.2

*abstracted from reference 2.

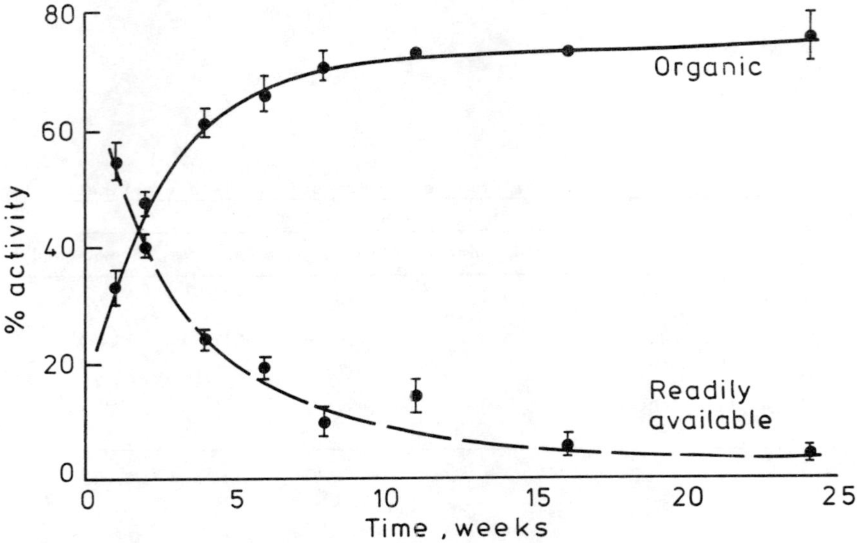

Figure 3 Ruthenium-106 in peat soil

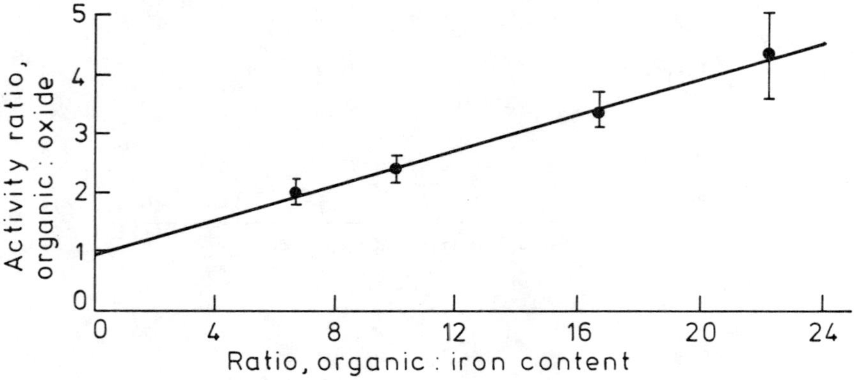

Figure 4 Disposition of ^{239}Pu : relationship with soil characteristics

similar to that for Denchworth clay. Iron content would not therefore appear to be the only controlling factor, but this conflict may be resolved as data for the other Sellafield soils become available. The importance of the oxide phase has been demonstrated in intertidal sediment and salt-marsh(2), but there the prime cause was the marine origin of the contamination. The high solubility of strontium in seawater results in an insigificant proportion of the activity being associated with the readily-available phases of sediment. This latter observation is of course at variance with assumptions generally made concerning the availability of strontium-90 in the terrestrial environment.

Results for plutonium-239 are almost complete, the exception being the residual phase, but the data available do enable us to examine the relative contributions of the oxide and organic phases in the same manner as for americium in the laboratory experiment. Figure 4 shows the quotient of the total iron content to the organic content plotted against the quotient of the plutonium-239 activities found with the two phases. Taken alone these results would suggest that relative contributions could indeed be predicted from a knowledge of gross soil characteristics, but this is a reflection of the limited variety of soil types that are available for field investigations.

Levels of iodine-129 are sufficient to undertake a similar investigation. To avoid losses due to volatility the organic extraction is carried out with hydroxylamine hydrochloride and sodium carbonate solution. In consequence the activity nominally associated with the organic phase will include essentially all that associated with the residual phase. Once again the oxide and organic phases are the most important, but here the observed disposition is not dependent upon the total iron or the organic contents. Indeed the distribution is very similar in all cases, about 60% being found with the oxide phase and a further 30% with the organic. Previous studies on sediment and salt-marsh suggest that iodine-129 is associated with a less readily reducible fraction of the oxide phase than is plutonium(2) and these results reinforce the requirement for more specific studies of this important phase.

Results for caesium-137 show disposition to be independent of soil type, with most of the activity associated with the residual phase. Measurements of activity associated with other phases were close to or below detection limits. Further rationalisation of the data for caesium-137 with those from the laboratory experiment must await a more comprehensive characterisation of the soils from the field studies.

DISCUSSION

This work reinforces strongly the need for careful implementation and cautious interpretation of sequential extraction analyses, but it also illustrates that even although the extraction steps are defined operationally, a comparative approach will provide valuable information for use in radiological protection. The necessity of undertaking both field and laboratory experiments is clearly demonstrated. Laboratory experiments can utilise relatively small sample sizes and radionuclides that are measurable without recourse to involved analytical methods. Furthermore, diverse soil types can be studied, thus providing evidence of the selectivity of the extraction procedure, and the behaviour of radioelements that are not readily measurable in the environment can be investigated. The effect of the chemical form of the contaminant is also amenable to this type of experiment. In terms of radiological protection therefore laboratory experiments can be designed to be directly relevant to assessments of the consequences of accidental releases from nuclear installations and shallow land waste repositories. The more immediate value of the experiments described here is in establishing the limitations of predicting disposition on the basis of gross soil characteristics.

Field experiments can only provide a limited range of soil types. Field results for plutonium suggest that the distribution between the oxide and organic phases could be confidently predicted on the basis of the total iron and total organic matter contents, but laboratory results for americium in the peat soil show this approach to be somewhat simplistic: when considering a wide range of soil types more rigorous correlations, for example with the complexation capacity of the organic phase(8), are likely to be more successful. Further, field results for caesium-137 support the generally-held view that caesium becomes extensively and irreversibly bound to the residual phase, or, more specifically, incorporated into the lattice of clay minerals. The laboratory study shows however that this is not so in all soil types: the absence of any correlation between fixed caesium-134 and clay content supports the view that gross characteristics cannot always be used to predict disposition.

Field studies provide us with an opportunity to test the ability of laboratory experiments to simulate behaviour in the natural environment and with an indication of the effect of factors such as biological activity which are absent in the laboratory. However, we would regard the comparative experiments described here only as a starting point in field studies. Our earlier field results for sediment and salt-marsh illustrate the usefulness of the extraction procedure. At the present time the radiological consequences of changes in the geography of the Cumbria coastal area arouse considerable interest in the United Kingdom. We expect future work to include both approaches, in which field samples are artificially amended and incubated in the laboratory before disposition is determined, with the intention of making an input to assessments of the consequences of land reclamation or the use of sediment for landfill.

REFERENCES

1. Tessier, A, Campbell, P G C and Bisson, M, Anal.Chem. 51 (7), 844 (1979).

2. Wilkins, B T, Green, N, Stewart, S P, Major, R O and Dodd, N J in Proceedings of International Seminar on the Behaviour of Radionuclides in Estuaries. Renesse, 17th-20th September 1984 (in press).

3. Green, N, in Proceedings of 4th symposium on the determination of radionuclides in environmental and biological materials. Teddington, April 1983.

4. Wilkins, B T and Stewart, S P, Int. J. Appl. Radiat. Isot. 33, 1385 (1982).

5. Clement, C R and Williams, T E, J Soil Sci 13, 82 (1962)

6. Schoer, J, Hamburg-Harburg Technical University, personal communication.

7. Sumerling, T J, Green, N and Dodd, N J, in Proceedings of 5th International Congress of International Radiation Protection Association, Berlin, May 1984.

8. Cremers, A and Henrion, P N, in Proceedings of International Seminar on the Behaviour of Radionuclides in Estuaries. Renesse, 17th-20th September 1984 (in press).

CHEMICAL SPECIATION OF RADIONUCLIDES IN CONTAMINANT PLUMES AT THE CHALK RIVER NUCLEAR LABORATORIES

D.R. CHAMP and D.E. ROBERTSON*

Environmental Research Branch, Chalk River Nuclear Laboratories,
Chalk River, Ontario KOJ 1J0
and *Battelle Pacific Northwest Laboratories

ABSTRACT

Experimental disposals of liquid and glassified wastes directly into the sands of the Perch Lake basin, Ontario, Canada, have resulted in the formation of well-defined subsurface contaminant plumes in the groundwater flow system. Using large volume water sampling techniques we have detected low concentrations of several long-lived radionuclides including isotopes of Pu, Am, Cm, Tc, I, Sr and Cs. The particulate and ionic speciation results from these studies support the conclusions of previous laboratory column studies that transport of radionuclides, particularly Cs and Pu, on particulates and/or colloids could be a significant mobilization mechanism in groundwater flow systems. We also propose, based on a comparison of the plume data with previous detailed studies on ^{60}Co that complexation reactions with natural as well as synthetic organic ligands can yield mobile anionic species of the actinides and lanthanides. Further detailed studies will be required to support this postulate.

1 INTRODUCTION

The mobility and bioavailability of radionuclides released to the environment is determined primarily by the chemical species present. Observations from actual field situations can lead us to a better understanding of the mechanisms determining the chemical form and hence the rates of groundwater transport of radionuclides. Experimental disposals of liquids and glassified wastes directly into sand in the Perch Lake basin at the Chalk River Nuclear Laboratories (CRNL, Ontario, Canada) have provided a unique opportunity to study the chemical species of migrating radionuclides following residence times in the groundwater flow system of up to 30 years.

Processes such as the formation of colloids (or pseudocolloids) and particulates, and complexation by organic or inorganic ligands to form anionic species can make the radionuclides very mobile. Results presented here from studies on radionuclides in the contaminant plumes, and laboratory columns, demonstrate the potential importance of such processes.

2 SITE DESCRIPTION AND METHODS

The contaminant plumes investigated in this study are located in the lower Perch Lake basin, a small sub-basin of the Ottawa River located about 200 km west-northwest of Ottawa, Canada. The lower basin and the four source areas for the contaminant plumes studied are shown in Figure 1.

Three of the contaminant plumes arose from liquid disposals into aeolian sand dunes; they are designated the Chemical Pit plume, the Reactor Pit plume and the A-Disposal plume. The fourth, the Glass Block plume, resulted from an experimental disposal of vitrified wastes directly into saturated sediments in the basin.

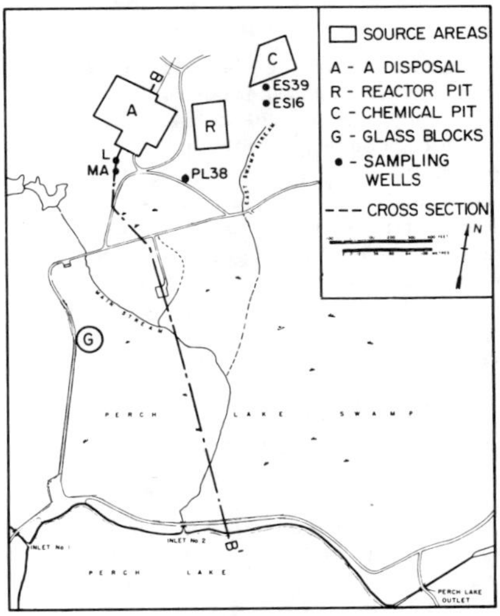

FIGURE 1 Lower Perch Lake basin showing contaminant source areas

Ground and surface waters were sampled using the Battelle Large Volume Water Sampler[1]. The sampler incorporates membrane filters to remove particulates, ion-exchange resins to remove soluble anionic and cationic species and, activated aluminum oxide to remove non-ionic species. The techniques for radiochemical and geochemical analysis of samples have been described previously[1,2].

3 RESULTS AND DISCUSSION

3.1 Colloids and Particulates

Particulates (> 0.4 µm) were evaluated for gamma emitting nuclides for all subsurface plumes and two surface waters (and Pu in three groundwaters). The percentage of each radionuclide in particulate form is shown in Table 1.

Particulate forms of most radionuclides were present; however, the percentage was very low for some elements. Isotopes of cesium were most frequently associated with particulates as evidenced by the relatively large percentage (4.2 to 86%) in particulate form at all sites. The 1-metre data from the Glass Block site supports release of particulates containing cesium from the glass blocks, presumably due to physical degradation. The absence of any cesium at 5 m downgradient of the blocks supports the removal by filtration or degradation of the particulates to a chemical form that can be sorbed. However, the cesium observed in all other sites resulted from liquid disposals such that the observed particulates may have been present in the disposed liquids or formed in the flow system. The latter is supported by the results we obtained from core columns[3]. Columns prepared from cores taken at the glass block site were spiked with ^{134}Cs, eluted with groundwater and the release of cesium monitored. Low levels of cesium were continuously released and greater than 80% was in a particulate form. The involvement of bacteria in the particulate transport process has been demonstrated[3].

A significant fraction of the Ce and Eu was present as particulates at the Glass Block site and downgradient of the Chemical Pit. This is of interest since the properties of these two elements most closely resemble the actinides. The occurrence of particulate forms of Pu was determined in groundwater from three wells, MA, ES39 and ES16L, that contained the highest concentrations of Pu. One percent of the Pu at both sites in the Chemical Pit plume was particulate, whereas only 0.01% was particulate at MA in the A-Disposal plume. Although the particulate fraction is rather small the data do support particulate transport of Pu over distances of 10's to 100's of metres in a subsurface flow system.

In column studies 20 to 45% of the plutonium released continuously, at low levels, over a period of two years was associated with particulates[4]. As with Cs, bacterial activity was implicated in particulate transport of Pu. Unlike Cs, Pu could be released from the particles by irradiation which would be consistent with disruption of organics.

3.2 Complexation Reactions

In Table 2 data are presented on the percentages of the various long-lived isotopes in various ionic forms in groundwater from the study sites. The results for ^{60}Co and ^{90}Sr are discussed first to illustrate the extremes. We previously reported[5] the results of detailed studies on ^{60}Co in the Chemical Pit plume that led us to conclude that the highly mobile anionic ^{60}Co arose from organic complexation by both synthetic and natural organic ligands. This conclusion was based upon the following observations: greater than 70% of the aqueous ^{60}Co was retained by anion exchange resins; 15%, 45% and 40% of the ^{60}Co species had apparent molecular weights of less than 500, between 500 and 1000, and greater than 1000, respectively (greater than 90% of the dissolved organics in uncontaminated groundwater in the Perch Lake basin fractionate in the 500 to 1000 molecular weight range); the anionic complexes could be disrupted by UV irradiation; tracer exchange experiments with cationic ^{57}Co showed

TABLE 1 Percent of γ-emitting isotopes, and Pu, in particulate[a] form

Isotope	Glass Block Plume 1 m	Glass Block Plume 5 m	Reactor Pit Plume	A Disposal Plume L	A Disposal Plume MA	Chemical Pit Plume ES39	Chemical Pit Plume ES16L	East Swamp Stream	Perch Lake Outlet
^{60}Co			0.08	<9	>13	0.1	0.4	2	<1
^{95}Zr						0.05	0.2		
^{106}Ru			<0.2			0.05	0.1	<0.4	
^{125}Sb			<0.3			0.1	0.2	<1	
^{134}Cs			7					8	8.3
^{137}Cs	86	N.D.[b]	>7	5	6	>56	23	>47	4.2
^{144}Ce					<1	2	29	<10	
^{154}Eu	25	<13				<1	5.7		
^{155}Eu	<26	<17				<2	<7	<5	
$^{239,240}Pu$					0.01	1	1		

a mean diameter >0.4 μm.
b N.D., not detected.

that the anionic ^{60}Co complexes were non-labile and; UV-absorbing material and ^{60}Co co-eluted from high pressure liquid chromatography columns. On the basis of UV profiles from uncontaminated and contaminated groundwater both synthetic and natural organic ligands were postulated to be involved. The absence of anionic forms of ^{90}Sr supports the thermodynamic predictions that Sr^{2+} is fully cationic and that organic complexation plays an insignificant role. Enhancement of migration of ^{90}Sr due to organic complexation has been reported[6]; however, on the basis of our data we conclude that this is unlikely to be a significant factor in Sr migration under most conditions.

Various long-lived isotopes were also present in a significant fraction as anions. Stable anionic inorganic species of ^{129}I (I^-) and ^{99}Tc (TcO_4^-) can be predicted from thermodynamics[7] and the data in Table 2 support these predictions. The observed speciation is also consistent with the formation of anionic organic complexes for which there is some support in the literature[8,9]. However, in terms of predicting the maximum potential for migration of these elements organic complexation will have little effect relative to the predicted inorganic speciation.

Redox sensitive elements, such as Sb, are of particular interest since they are generally in thermodynamic disequilibrium in groundwater[10]. The predominance of anionic species (greater than 75%) at all sites rather than the thermodynamically predicted[7] neutral species, $HSbO_2^0$ and $Sb(OH)_3^0$, can be interpreted either in terms of thermodynamic disequilibrium[11] (the formation of $Sb(OH)_6^-$) or complexation with organics.

Although organic complexation is reported[12,13] to have little effect on Cs migration the occurrence of > 60% anionic Cs at ES16L can best be explained by organic complexation since thermodynamically only uncomplexed Cs^+ and the two inorganic complexes $CsOH^0$ and $CsCl^0$ are expected in natural and polluted waters.

Anionic species accounted for greater than 97% of the Pu at all sites and, 95% and approximately 60% of the Am at two sites. Cationic species

TABLE 2 Ionic Distribution[a] of Isotopes in Percent

Isotope		Glass Block Plume			Reactor Pit Plume			A-Disposal Plume				Chemical Pit Plume				East Swamp Stream			Perch Lake Outlet		
		A	C	N	A	C	N		A	C	N		A	C	N	A	C	N	A	C	N
^{60}Co					81	18	<1	L	<15	<70	<7	ES39	82	15	3	63	34	2	59	41	<1
								MA	<27	<47	<13	ES16L	70	23	6						
^{90}Sr					<.01	100	-		<.01	100	-		<.01	100	-						
^{99}Tc												ES39	>78	<22							
												ES16L	>98	<2							
^{125}Sb					73	24	<3					ES39	87	3	9	>86	<7	6			
												ES16L	92	3	4						
^{129}I								L	>57	<43	-	ES39		N.D.[b]	-						
								MA	>60	<40	-	ES16L	>99	<1	-						
^{137}Cs	1 m	5	8	<1	<62	<27	<5	L	1	94	<1	ES39	<30	<11	<3	<15	<35	<4	<1	92	3
	5 m		N.D.					MA	<2	92	<1	ES16L	>61	<15	<4						
^{144}Ce												ES39	>80	<15	<3						
												ES16L	39	28	<3						
^{154}Eu	1 m	<15	>55	<11								ES39	>80	<13	<5						
	5 m	>50	<21	<21								ES16L	<37	>55	<5						
^{155}Eu	1 m	>18	>32	<32								ES39	>69	<25	<4	<14	>77	<4			
	5 m	>56	<12	<21								ES16L	<33	>18	<48						
^{239,240}Pu					97	3		L	98	2		ES39	97	2							
								MA	98	2		ES16L	98	1							
^{241}Am					95	5		L	62	38		ES39	2	98							
								MA	59	41		ES16L	5	95							
^{244}Cm												ES39	0	100							
												ES16L	14	86							

a the sequential numbers under headings A, C and N give the percentage distribution between anionic, cationic and neutral species, respectively.

b N.D., not detected.

accounted for more than 95% of the Am in the remaining site, the Chemical Pit plume. This site is known to receive synthetic organics and has a groundwater dissolved organic carbon concentration (DOC) approximately twice that of the other plumes and uncontaminated groundwater in the basin. Complexation by organics to yield anionic species of Pu and Am is most consistent with the observations since the inorganic form of Pu would be dominated almost entirely by OH^- and CO^{3-} complexation and $Pu(OH)_4^0$ would likely be the dominant solution species at the observed pH's[14]. In addition the tetravalent actinides would tend to form polyvalent polymeric hydroxides[15] as the thermodynamically stable polymeric species under these conditions and would possess a residual positive charge. Our data do not support hydroxy-colloids of Pu as an important groundwater species in this geochemical environment. In laboratory column experiments with Pu[4] we have shown that greater than 25% of the migrating Pu has a molecular weight of 500 to 1000 which is consistent with the formation of organic complexes. Also greater than 85% of the Pu was present as tetravalent Pu with the remainder being trivalent.

For thermodynamic reasons trivalent species of Am are expected to predominate[14] as the complexes $Am(OH)_2^+$ and $AmSO_4^+$, and less importantly Am^{3+}[7]. This prediction is consistent with the observed speciation downgradient of the Chemical Pit. Although organic complexation most readily explains the anionic Am species observed in the Reactor Pit and A-Disposal plumes, anionic colloids and pseudocolloids are a possible alternative[14].

Curium, the only other actinide detected, was present only in the Chemical Pit plume and primarily (> 80%) as a cation. This observation is consistent with the thermodynamic prediction that only trivalent Cm is stable in aqueous solutions and that the dominant species is likely to be $Cm(OH)^{2+}$. The data from ES16L indicating 14% anionic Cm do support some conversion of Cm during transport in the subsurface to a mobile anionic species.

4. CONCLUSIONS

On the basis of the complementary field and laboratory data we conclude that both colloidal or particulate transport and organic complexation can be important mechanisms for enhancing the transport of important radionuclides. Cesium is the element most generally influenced by particulate transport processes. However, the data do support the hypothesis that particulate or colloidal forms of many elements, including certain lanthanides and actinides, will be formed and transported in groundwater and surface water. The process may become important when dealing with a large source.

Further more detailed studies, similar to those reported for Co-60[7], will be necessary to confirm that the mobile anionic species are indeed organic complexes. Resolution of the chemical characteristics of the organic ligands will also be required to support predictive modelling.

5. REFERENCES

1. D. E. Robertson, C.E. Cowan, E.A. Jenne and T.R. Garland, in Speciation of Fission and Activation Products in the Environment, this Proceeding, 1985.

2. D.R. Champ, J.L. Young, D.E. Robertson and K.H. Abel, Water Poll. Res. J. Canada (1985), (in press).

3. D.R. Champ and W.F. Merritt, in Proceedings of the Second Annual Conference of the Canadian Nuclear Society, edited by F.N. McDonnell, pp 66-69 (1981).

4. D.R. Champ, W.F. Merritt and J.L. Young, in Scientific Basis for Radioactive Waste Management V, edited by Werner Zutze (Elsevier Science Publishing Co.), p 745-754, 1982.

5. R.W.D. Killey, J.O. McHugh, D.R. Champ, E.L. Cooper and J.L. Young, Environ. Sci. Technol., 18, 3, 148 (1984).

6. F.L. Himes and R. Shufeldt, COO-414-11, 1970.

7. D. Rai and R.J. Serne, PNL-2651, UC-70, Battelle, PNL, 1978.

8. H. Behrens, IAEA Symposium IAEA-SM-257/36 (1980).

9. R.E. Wildung, K.M. McFadden and T.R. Garland, J. Environ. Qual., 8, 156, (1979).

10. E.A. Jenne, in Aqueous Speciation of Dissolved Contaminants, U.S. National Bureau of Standards, pp 39-53, 1981.

11. E.A. Jenne, C.E. Cowan and D.E. Robertson, presented at the Fifth International Conference on Nuclear Methods in Environmental and Energy Research, Mayaquez, Puerto Rico (1984).

12. H. Nishita and E.H. Essington, Soil Sci., 102, 168, (1967).

13. K.C. Knoll, Report BNWL-860, (1969).

14. B. Allard, presented at OECD Workshop on Geochemistry and Radioactive Waste Management, Paris, France, (1982).

15. G.L. Johnson and L.M. Toth, Report ORNL/TM-6365, Oak Ridge, (1978).

CHANGES IN PLANT AVAILABILITY OF ^{238}Pu WITH TIME AS INDICATIONS OF CHANGES IN SPECIATION

ÅKE ERIKSSON

Department of Radioecology
Swedish University of Agricultural Sciences
S-750 07 UPPSALA, SWEDEN

ABSTRACT

A short description is given of the changes observed in the plant availability of Pu-238 in eight soils during a nine-year experimental period. It was found that during the first 4-year period with clover as test crop the plant uptake of Pu-238 was reduced with availability half-times ranging from 0.8 to 2.0 years. The reduction rate seemed proportional to the initital uptake levels except for lime rich clay soils, where the reduction rate was high regardless of the uptake level. The availability half-time was reduced from 2.0 years in the control to 0.8 years when the nuclide was placed at 10 and 20 cm depth. In 1980, when the test crop clover was replaced by spring wheat the necessary soil management operations caused intense aeration and drying in one block of the replicates. As a consequence, the Pu-uptake in that block became considerably higher than in the others. This event can be interpreted as an indirect evidence for the reversibility of that process in soil, which caused the reduction of the plant availability of Pu-238.

1 INTRODUCTION

An investigation on the behaviour of transuranics in Swedish soils and transfer to agricultural crops was started in 1976. The background for this decision was the information needed at that time in a society with an expanding nuclear industry when experimental findings showed that plutonium uptake could increase with time[1] and that complexation of plutonium in the soil might occur and increase the plant uptake considerably.[2,3]

Consequently, the principal aim of this experimental investigation was to study the trend in plant availability of the transuranics in Swedish soils and under Swedish climatic conditions. The work has been supported by the Swedish Radiation Protection authorities. (A complete report will be published elsewhere.)

2 MATERIALS AND METHODS

The experiments were carried out with lysimeters as basic experimental units. These are closed containers of stainless steel (area: 0.175 m^2, height: 1 m) for subsoil, experimental top soils, and equipment to prevent the soil particles from contaminating the crop, and with outlets for excess drainage water connected to columns with ion exchangers.

The physical and chemical characteristics of the eight soils used in the experiments are given in Table 1.

TABLE I Chemical characteristics and the content of organic matter and clay of the top soils used in the lysimeter experiments.

Soil	pH_{aq}	CEC, me/100 g dry soil	Base sat., %	Organic matter, %	Clay content, %	Dry soil per vessel, kg
1 H	5.4	57.0	73	75.4	-	10.5
2 F	5.4	7.0	53	5.6	5	36.0
3 S	6.1	11.7	27	5.6	8	49.4
4 B	5.7	13.2	65	3.9	11	50.3
5 K	5.9	12.4	70	4.8	14	40.0
6 Fo	7.3	19.5	97	5.6	15	43.0
7 L	5.7	16.4	39	3.1	39	49.1
8 St	7.5	34.1	97	3.0	57	41.0
Subsoil	5.7	-	-	0.9	4	130.0

The soils, collected from different parts of the country, were intended to cover different conditions with regard to pH-levels, cation exchange capacities and calcium and clay contents. One organic soil, 1 H, was included. The others constitute the mineral soil group dominating also among the agricultural soils. Each top soil unit was contaminated in 1976 with 4.88 MBq Pu-238 in 3M HNO_3.

The experimental crops were red clover and spring wheat, cultivated during 4 different periods:

I	1976-79	Clover.	Mainly natural precipitation.
II	1980-81	Spring wheat.	"-
III	1982	"-	"- + irrigation
IV	1983-84	Clover	"- + "-

The natural precipitation is the precipitation representative for the eastern part of Sweden. Irrigation was used to cover the water deficit between losses by evapotranspiration and precipitation.

The amount of fertilizer added to the lysimeters was intended to cover the annual needs of the crops with regard to nitrogen, phosphorus and potassium.

The clover crops were washed in 0.01 M HCl after harvest and before the preparatory and analytical work. The ears and the straw of the spring wheat were harvested and treated separately.

3 RESULTS AND DISCUSSION

3.1 Uptake of ^{238}Pu by clover in 1976-79 and 1983-84

The yield levels in the experiment corresponded to those obtained in agriculture under good field conditions. The cropping stress on the soils used with regard to the uptake of nutrients, water and contaminants has, thus, been rather normal.

The development in plant availability of ^{238}Pu can be evaluated from the content of the nuclide in the crop during the experimental periods in 1976-79 and 1983-84 (TABLE 2). Table 2 shows that each soil for the 1976-79 period can be characterized by two values: 1. Plant availability of added plutonium. 2. the half-time, Ta, for that availability. However, it should be noted that these characteristics are conditional. They also depend on variables like: 1. Speciation of the nuclide used, 2. Crop, 3. Cultivation system.

TABLE II Weighted ^{238}Pu-content, Bq/kg dry matter, in clover during Periods I and IV in experiments with 3 replicates of eight soils, Expt. A, and with two placement depths, Expt. B. Ta = availability half-time in years.

Soil	Period I					Period IV		Average
	1976	1977	1978	1979	Ta,y	1983	1984	CV, %
Expt.A								
1 H	5304	3227	3682	1143	1.6	97	71	28
2 F	38	140	65	53	1.4	30	25	34
3 S	358	318	226	128	2.0	38	45	27
4 B	908	505	142	70	0.8	56	48	24
5 K	853	638	614	279	2.0	61	54	30
6 Fo	324	183	63	46	1.0	16	19	27
7 L	66	116	94	45	1.5	17	21	29
8 St	117	86	43	32	1.5	8	7	52
Expt.B. Soil:5 K								
A pd=0-20 cm	853	638	614	279	2.0	61	54	30
B pd=10 cm	780	333	182	58	0.8	18	21	36
C pd=20 cm	383	187	49	33	0.8	14	9	22

The nuclide used, ^{238}Pu (III-VI), may have been present mainly as Pu(IV), which is more susceptible to complexation than other species in soil[3]. Among the crops, clover (1976-79 and 1983-84) represented the fodder crops, and spring wheat (1980-82) the food grain crops. The former, being a legume, has a higher capacity for uptake of minerals, and may be more deep-rooted than the latter. However, the latter may extract nutrients from different layers of the soil profile during different parts of the growing period[4].

In practice, the cultivation system depends on the crop. In leys with clover and grasses the soil surface is closed. It is not opened up by soil management operations until grains or other annual crops appear in the crop sequence. Under a more or less closed surface and in a soil with an active biological system the redox conditions may be quite different to those prevailing under grain cropping with the soil opened up in spring and with oxygen readily diffusing through the loose surface soil layer. Period I, with clover, represented a system with a closed soil surface.

In the beginning of Period I a ranking of the soils with regard to the uptake gives 1H >> 5K, 4B >> 3S, 6Fo > 2F, 7L, 8St. The availability half-time, Ta, estimated by regression analysis, differs between soils (i.e. between different redox and other conditions) in the range 0.8-2.0 years. At the end of Period IV, after 8 years, the availability of ^{238}Pu differs less and mainly between 3 groups of soils: 1H, 5K, 4B, 3S > 2F, 7L, 6Fo > 8St.

When placed at 10 and 20 cm depth the plant availability of ^{238}Pu decreases below that of the control (TABLE 2). This is due partly to a real placement effect and partly to the fact that the redox conditions prevailing in the deeper layers differs from those in the surface layer of the top soil. The data from Expt B favours an explanation based on the development of a steep gradient in the plant availability of ^{238}Pu in the upper half of the top soil layer. It is also conceivable that during the season the redox conditions may vary considerably with the contents of water and air in the soil.

3.2 Uptake of ^{238}Pu by spring wheat in 1980-82

The reduction in plant availability of ^{238}Pu proceeded at different rates in different soils and at different depths under a "closed" soil surface. Now the question arises whether the process responsible for that reduction is reversible or not if the conditions in soil are changed towards a state with aeration and oxidation.

Unintentionally, some evidence for a reversible process was found, when in 1980 the closed soil surface cultivation system was interrupted by the grain cropping during Periods II and III. The first soil management operations, including weeding and cleaning from roots, etc., were time consuming. Also, one of the blocks of replicates was more exposed to aeration and drying than the others. In this particular block of replicates a higher uptake of Pu by wheat was obtained than in the others that year. Also the uptake in the latter was higher than the level in the following year (cf. TABLE 3). The difference was marked, except in Soil 2 F, a soil with a low pH-level and perhaps stronger reducing properties than the other soils. On the other hand, the ratio between the high level in grain in 1980 and the level in grain in 1981 for Soil 6 Fo, with a high pH-value and rich in calcium carbonate, was very large, 35.

TABLE III Uptake of ^{238}Pu by spring wheat during Periods II and III as ratios for 1980 and 1982 and with the transfer factors for 1981 as the unit.

Soil	Grain				Straw			
	1980 Ratio	1981 TF_{sp}	CV%	1982 Ratio	1980 Ratio	1981 TF_{sp}	CV%	1982 Ratio
	Range	(=1.0)			Range	(=1.0)		
1 H	2.7-7.3	1.5E-6	100	1.6	3.6-10	1.5E-4	98	1.7
2 F	1.5	1.0E-6	47	1.0	0.5	9.1E-5	62	0.6
3 S	4.7-15	5.1E-7	61	1.8	3.1-4.2	5.4-5	15	1.5
4 B	2.2-14	9.3E-7	24	2.1	2.3-4.2	4.4E-5	-	1.1
5 K	4.1-13	4.1E-7	3	3.4	2.8-11	7.5E-5	-	1.3
6 Fo	1.8-11	1.1E-6	51	0.3	1.6-2.5	4.1E-5	11	0.5
7 L	1-9.1	7.0E-7	8	0.6	0.6-1.3	4.8E-5	58	0.6
8 St	1-8.0	5.0E-7	82	1.7	2.8-4.3	1.0E-5	44	0.8

In 1981 the soil management operations were carried out very cautiously to minimize soil drying and aeration. The results were a low uptake and an acceptable variation level. In 1982, irrigation was introduced to cover the deficit between precipitation and evapotranspiration. Then the plant uptake of Pu-238 increased somewhat from light soils with a loose structure and accessible to aeration but remained the same as in 1981, or decreased further, in the heavier clay soils. For the latter, irrigation may have strengthened the tendencies towards less aeration and reducing conditions so much that they outweighed the increased utilization level of the top soil layer by the crop.

The placement effects on the wheat uptake of ^{238}Pu differ between grain and straw. The effect in straw is large and similar to that in clover. However, due to the deeper plant feeding horizon during the development of the ears, placement seems to be no guarantee for a comparatively low uptake also in the grain (TABLE 4). The lower plant availability at 20 cm depth is outweighed by the more intense rooting in that soil layer during the latter part of the plant development.

TABLE IV Placement effects on the plutonium uptake by spring wheat during Periods II and III, Expt B, Soil 5 K. Uptake given as percentages of that for the control in 1980.

Placement depth, cm for ^{238}Pu	Grain 1980(I)	1981(II)	1982(III)	CV % (Av.)	Av. ratio: Straw/Grain 1980-82
0-20	100	24	81	33	124
10	76	29	29	50	17
20	61	43	114	50	7

4. CONCLUSIONS

The results obtained in the experiments indicated that the plant availability of plutonium has been greatly influenced by the soil environment. At the beginning of the experimental period the soil contents of clay and organic matter were important. Later, under cultivation of clover with a closed soil surface and restricted aeration of the soil the plant availability of ^{238}Pu was reduced. This reduction seemed to be time-dependant and should be ascribed to changes in speciation due to the reducing conditions prevailing during the growth periods.

However, after a change in cropping and in the reducing conditions it was found that this process was in fact reversible also under normal soil environments.

5 REFERENCES

1. Romney, E.M., Mork, H.M. & Larson, K.H. 1970. Persistence of plutonium in soil, plants and small mammals. Health Physics, 18, 487-491.
2. Price, K.R. 1973. Tumbleweed and cheatgrass uptake of transuranium elements applied to soil as organic acid complexes. BNWL-1755. Batelle Pacific Northwest Laboratories, Washington.
3. Bondietti, E.A., Reynolds, S.A. & Shanks, M.H. 1976. Interactions of plutonium with complexing substances in soils and natural waters. Transuranium nuclides in the environment, 273-287. IAEA, Vienna.
4. Haak, E. 1978. Studier av stråsäds rotutveckling och mineralämnesupptag. B. Matjordens och alvens bidrag till vårsädesgrödors upptag av Ca, P och K under svenska fältbetingelser. Rapporter från inst. för radiobiologi, Sveriges lantbruksuniversitet, SLU, nr 44. Diss.

URANIUM ISOTOPE DISEQUILIBRIA AS A FUNCTION OF MINERAL PHASE IN THE VICINITY OF A URANIUM BODY

R.T. LOWSON AND S.A. SHORT

Australian Atomic Energy Commission, Lucas Heights Research Laboratories, Private Mailbag, Sutherland, N.S.W. 2232, Australia

ABSTRACT

A method has been developed for the chemical separation of the principal phases in soil sampled from the vicinity of the Ranger No. 1 uranium orebody, in the Alligator Rivers region of the Northern Territory of Australia. The principal phases are identified as amorphous iron oxide, crystalline iron oxide and resistate. The distribution of ^{238}U and its decay products among the phases is determined. The isotopic distribution indicates that the phases are, to a certain extent, acting independently. The amorphous iron phase is considered to be a record of local and temporal abnormalities in the total ground water/soil system. The crystalline iron oxide phase involves a chemical control process, the kinetics of which are commensurate with or less than the half-life of ^{230}Th (7.52×10^4 y). The changes in isotope concentrations with radioactive decay in the resistate phase are interpreted in terms of a net flux of material entering or leaving the phase over the period of the decay process. In one of the models, the flux can be interpreted as a direct measure of the transport of uranium through a 'quartz-clay' crystal grain.

1. INTRODUCTION

In a closed system and over geological time scales, the half-lives of the parents and daughters of the ^{238}U decay series are such that the system eventually reaches secular equilibrium. This allows the ^{238}U decay series to be used as a geological clock. In contrast, in open systems, fractionation occurs through a variety of mechanisms, leading to the preferential loss of parents or daughters.

Recently, while leaching some uranium ores in the laboratory, Shirvington[1-3] observed that loss of ^{234}U lagged behind that of ^{238}U. Shirvington suggested that the results could be used to date conservatively the weathering process of the ores and allow an estimate of maximum migration rates of uranium from the orebody. This can have great significance for the modelling and prediction of migration of radionuclides from waste repositories.

Shirvington attempted to relate the process to the bulk mineralogy of the ore and, in particular, to the clay component. Accordingly as part of an extension to Shirvington's work, an attempt has been made to identify the nature of the 'sorbed' radionuclides and the location in the sorbed species in the rock matrix by the selective extraction of particular mineral phases.

There are a number of recommended selective phase separation schemes[4,5]. However, since there is no generalised scheme which can cover all eventualities, the adopted scheme has to be tailored to the particular requirements of the work. Several points should be taken into account when drawing up a scheme. The number of extraction steps should be kept as small as possible without losing the required detail. Inclusion of unnecessary steps overloads the subsequent analysis and may result in an inconclusive identification of the phases. Prior information may indicate the absence of some phases, while separation of other phases may not be required.

In the present context the organic phase was not considered substantial enough to justify the inclusion of an organic extraction. Previous work[6] suggested that iron oxides are preferential adsorbers of cations in mixed adsorber systems and should be treated as a separate phase. Initial work incorporated an allophane and imogolite extraction step, by boiling the residue from an amorphous alumina extraction step with 2% NaOH for exactly 2.5 minutes[7]. The step was abandoned because it was very difficult to ensure a reproducible technique and it was demonstrated that there was minimal association between the radionuclides and these materials.

The final scheme was tested for reproducibility and for the presence of mineralogical and chemical factors that could cause ambiguities in interpreting the results.

2. METHODS

2.1 Selective Phase Separation

2.1.1 Soluble salts and exchangeable ions

Two grams of crushed sample were shaken with 80 ml of unbuffered 0.1 M NH_4Cl at room temperature for 24 hours[8,9] to dissolve soluble salts and exchangeable ions.

2.1.2 Amorphous minerals and ferrihydrite

The residue from step 1 was shaken with 80 ml of Tamm's acid oxalate solution (10.92 g oxalic acid and 16.11 g ammonium oxalate in 1 l of water) at room temperature and in the dark for four hours[10-12] to dissolve the amorphous iron oxides.

2.1.3 Crystalline iron minerals

The residue from step 2 was stirred with 80 ml of Mehra and Jackson's buffered citrate-dithionite solution[13]. The extraction was repeated once more.

2.1.4 Amorphous alumina and amorphous silica compounds

The residue of step 3 was shaken with 80 ml of 5% sodium carbonate solution at room temperature for 16 hours[14,15] to dissolve the remaining amorphous inorganic compounds.

2.1.5 Resistate material

The residue of step 4 was digested with a sequence of nitric, hydrofluoric and perchloric acids. Any insoluble material remaining was fused with sodium peroxide[16].

2.2 Analysis

After each extraction, the solution was separated from the residue by centrifugation (3500 rev/min for 15-30 min) and the supernatant filtered through a 0.45 m membrane filter. The residue was washed with 15-20 ml of 0.1 M NH_4Cl, separated in the same manner and the washings were combined with the extract.

The extracts were analysed for the isotopes of uranium and thorium by adding analytical yield tracers of ^{236}U or ^{232}U and ^{232}Th, recovering the uranium or thorium by ion-exchange separation and depositing onto a stainless steel planchette to permit counting by high resolution alpha spectrometry. The uranium and thorium isotope concentrations were determined also in the gross sample by total dissolution of the sample followed by ion-exchange separation, electrodeposition and high resolution alpha spectrometry. A limited number of radium analyses were made by the radon emanation technique.

3. REGIONAL AND LOCAL GEOLOGY, AND CLIMATE

The work described here was on drill core material from the Ranger No. 1 orebody. This orebody is located in the Alligator Rivers Uranium Province, Northern Territory, Australia and forms part of the base metal and iron deposits of the Pine Creek Geosyncline. The regional geology has been described by Crohn[17] and the mine geology of the Ranger deposits by Eupene et al[18].

The orebodies and prospects of the area have a common geology and form part of a sequence of Lower Proterozoic sediments which were laid down 1900 to 1800 million years (m.y.) ago. There is evidence to suggest that the uranium underwent two periods of remobilisation. The first occurred between 850 and 710 m.y. ago and the second around 500 m.y. ago[19,20].

The Ranger deposit has a surface expression which dips sharply. Mineralisation occurs in a series of feldsparthic, quartz-mica, graphitic and chloritic schists. The surface is eroding. The depth of weathering is around 18 m but may vary from a few metres to 35 m.

4. RESULTS

The $^{234}U/^{238}U$ and $^{230}Th/^{234}U$ ratios and uranium concentrations were determined as a function of depth for a number of drill cores from the Ranger No. 1, Jabiluka and Nabarlek deposits. The results were reported by Airey et al[21,22]. The results for the Ranger No. 1 drill core S1/146 were typical of a general pattern, so four samples from this core were selected for further examination by selective phase extraction.

The isotopic data for the gross samples are listed in Table I. These values were obtained by summing the results for each phase to ensure internal consistency when discussing results for the individual phases. It was recognised that the summation could introduce cumulative errors. This was checked by carrying out an independent total assay for ^{238}U by delayed neutron activation analysis (DNA). Results from the two methods are listed in Table II. The comparison indicates the absence of accumulating errors, the absence of any trend in errors and acceptable agreement between the independently obtained values. It is concluded from this that all the material has been accounted for.

The drill core was about eight years old at the time of sampling and had been stored in a core shed at the Ranger mine site. The soil profile for this core was very similar to the generalised soil profile described by Eupene et al[18]. The position of the sample in relation to the soil profile is indicated in column 3 of Table I. A four-zone phenomenological model was developed by Airey et al[21,22] to describe the vertical transport of material through the orebody. The position of the sample in relation to this model is indicated in column 2 of Table I.

TABLE 1 Total assay for drill hole S1/146 — Ranger no. 1, as sum of phases and determined by alpha spectrometry.

Depth	Zone	Soil horiz.	$^{238}U^{(a)}$ µg/g	$^{238}U^{(a)}$ dpm/g	$^{234}U^{(a)}$ dpm/g	$^{230}Th^{(a)}$ dpm/g	$^{226}Ra^{(b)}$ dpm/g
2.7	I	B_1	47.2	34.9	42.0	71.3	65.5
4.0	I	B_1	84.3	62.3	74.0	179.1	110.6
9.1	II	B_2	35.6	26.3	27.8	47.8	43.0
13.1	III	C	352.0	220.3	233.5	383.3	248.1

Activity Ratios

Depth	$^{234}U/^{238}U$	$^{230}Th/^{234}U$	$^{226}Ra/^{230}Th$
2.7	1.2	1.70	0.49
4.0	1.19	2.42	0.59
9.1	1.06	1.72	0.90
13.1	0.89	1.64	0.67

(a) By alpha spectrometry for each phase and summing
(b) By radon emanation of the total sample and of the resistate

TABLE II Comparison of methods of analysis

Depth	DNA µg/g	DNA dmp/g	α-spectrometry µg/g	α-spectrometry dmp/g
2.7	51.3	37.9	47.2	34.9
4.0	81.1	60.1	84.3	62.3
9.1	35.6	26.3	35.6	26.3
13.1	378.0	279.5	352.0	260.3

The standing water level in the area is between 6 to 9 m below the surface during the dry season but it rises closer to the surface during the monsoon season. The $^{234}U/^{238}U$ activity ratio of the ground waters in the vicinity of the Ranger No. 1 orebody is around 0.8 and rises towards 1.2 on moving down-gradient of the orebody.

The 2.7 m and 4.0 m samples came from the B_1 soil horizon, a mottled zone underlying a ferricrete layer which is defined as zone I in the Airey model. This is the under-saturated weathered zone bounded by the surface. The samples from this zone would have been annually saturated by the fluctuating water table.

A lack of colour in the 9.1 m sample indicated that it came from the B_2 soil horizon or pallied zone. This is a saturated zone of weathered rock which corresponds to zone II of the Airey model.

The 13.1 m sample came from the C soil horizon and is saturated weathered rock below the pallied zone and corresponds with zone III of the Airey model.

The $^{234}U/^{238}U$ and $^{230}Th/^{234}U$ activity ratios in Table I are plotted in Fig. 1 as a function of depth. The ^{234}U is in slight excess near the surface and in zone I but drops to a slight deficit in zone II. The ^{230}Th is in gross excess in zone 1 but falls to near equilibrium values in zone II. The ^{226}Ra is deficient in zone I but rises to about equilibrium values within zone II.

The activity ratios and distribution of uranium and thorium between the various phases are shown in Fig. 2. The assumption here is that the reagents used to dissolve selectively a phase do not extract uranium or thorium from the remaining phases. Some pilot studies on the adsorption and desorption of ^{236}U on the 4.0 m material indicated that this is the case.

The reproducibility of the technique was checked with three separate extractions on two replicate samples. The results were in satisfactory agreement, with respect both to per cent ^{238}U and to the $^{234}U/^{238}U$ ratios.

The major part of the uranium and thorium is associated with the iron phases and this contribution increases with depth. Near the surface there is more uranium in the crystalline iron phase than in the amorphous iron phase. This is reversed with depth. Similarly, near the surface there is more

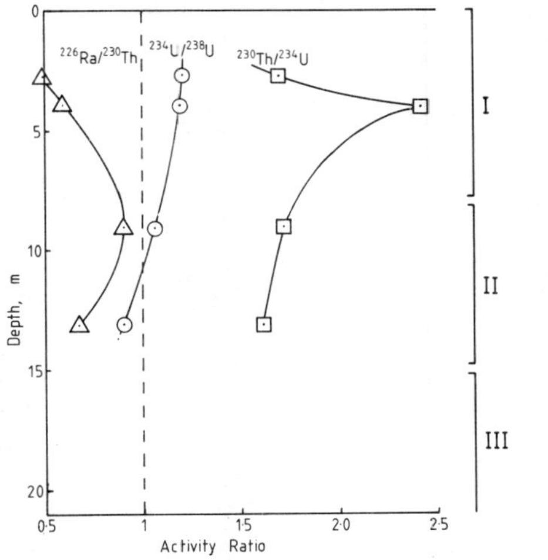

Fig. 1. Plot of activity ratios versus depth of various parent-daughter pairs for the Ranger S1/146 profile $^{234}U/^{238}U$ $^{230}Th/^{234}U$, $^{226}Ra/^{230}Th$. The zones for the Airey vertical leaching model are shown in the LHS.

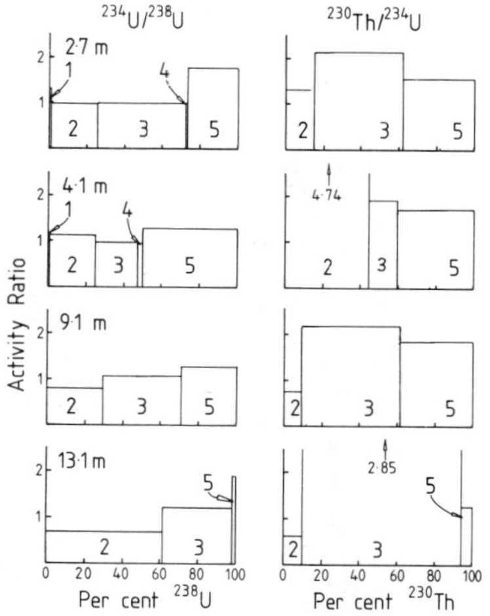

Fig. 2. Plots of activity ratio versus percent parent for $^{234}U/^{238}U$ and $^{230}Th/^{234}U$ for the five phases at different depths in the ranger profile.

thorium in the crystalline iron phase than in the amorphous iron phase but in this case the excess increases with depth.

In the near-surface sample, the uranium isotopes are close to equilibrium in the iron phases and the overall ^{234}U excess of the sample is contained in the resistate phase. There is an increasing deficiency in ^{234}U with depth in the amorphous iron phase and an overall increase in the excess of ^{234}U with depth in the resistate phase. In contrast the ^{230}Th exceeds its parent ^{234}U in the iron phases, but this excess is reduced in the resistate phase. On average, in the near-surface samples there is more ^{230}Th in the amorphous iron phase than in the crystalline iron phase; however, this trend is reversed with depth.

The radium concentration was determined only in the whole sample and the resistate. The amount of radium in phases 1 to 4 was determined by subtraction. In this case the assumption is that no radium was leached from the resistate phase or no radium was adsorbed onto the resistate phase during the extraction procedures for phases 1 to 4. The activity ratio for ^{226}Ra with its parent ^{230}Th and the distribution of radium between the phases is shown in Fig. 3.

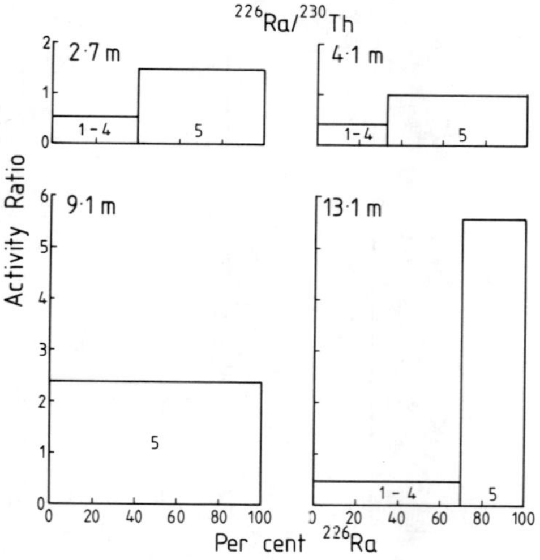

Fig. 3. Plots of $^{226}Ra/^{230}Th$ activity ratios versus per cent ^{226}Ra for the combined 1 to 4 phases and the phase at different depths to the Ranger S1/146 profile.

In the top three samples of the profile, the major portion of the radium was located in the resistate phase and, at 9.1 m, all radium was located in the resistate phase. In the deepest sample, 13.1 m, the major portion of the radium was located in the combined iron phases. The radium was in gross excess over its parent in all the resistate phases and in gross deficit in the combined iron phases.

The extracts were analysed also for aluminium, iron, manganese and silicon by inductively coupled plasma emission spectroscopy. The results for aluminium and iron are listed in Table III. The manganese and silicon concentrations were not recorded because the values were, respectively, 0.05 and 0.6% of the total respectively (with the exception of the silicon concentration in the resistate phase). The aluminium was predominantly associated (90%) with the resistate phase. The iron was distributed between the amorphous iron, crystalline iron and resistate phases, the major portion being in the crystalline phase in all cases.

TABLE III Chemical Assay for Aluminium and Iron

Depth m	Total Al %	Total Fe %	Fe in phase 2 % of total Fe	Fe in phase 3	Fe in resistate
2.7	1.41	2.81	12	77	11
4.0	1.95	1.37	14	53	33
9.1	1.56	1.17	25	39	36
13.1	6.0	3.86	16	70	14

5. DISCUSSION

Initial work using chemical and X-ray diffraction analyses and reported elsewhere[23] demonstrated that for these samples the reagents were phase selective and did not modify the residue during the extraction process.

The distribution of ^{238}U and its daughters between the phases is shown in Fig. 2 in terms of percentage amounts of isotope (effectively absolute mass) and the activity ratio with the immediate daughter. The diagram suggests that any observed fractionation of isotopes during a mild leaching process is closely associated with the mineral composition of the sample and, given the relative amounts of iron present (1 to 4 per cent), the influence of the iron oxides is far greater than expected.

An alternative approach is to examine the distribution of uranium and its daughters between the phases in terms of concentrations per phase. These values are listed in Table IV. The calculations were made for the two separate iron phases, the resistate phases and for a combined iron phase so as to include the radium results. The concentrations are expressed in terms of dpm per gram of iron for the iron phases since the actual stoichiometry of the iron phase is unknown, and dpm per gram of resistate for the resistate phase, the amount of resistate being from Table III. Since the resistate formed the major portion (in the range 95 to 99%) of the sample, the error arising from faulty estimation would have been insignificant. A number of patterns now emerge.

TABLE IV Distribution of uranium and its daughters between phases

Depth, m	2	3	2+3	R	Activity ratios 2	3	R
	(in dpm per g of Fe or phase)						
		^{238}U					
2.7	2546	743	987	9.0			
4.0	7925	1879	3142	30.8			
9.1	2625	2380	2476	7.7			
13.1	25836	3545	7692	4.7			
		^{234}U				$^{234}U/^{238}U$	
2.7	2495	735	973	16.1	0.98	0.99	1.78
4.0	9034	1803	3314	40.7	1.14	0.96	1.32
9.1	2100	2570	2386	9.7	0.80	1.08	1.27
13.1	17310	4289	6712	10.1	0.67	1.21	2.15
		^{230}Th				$^{230}Th/^{234}U$	
2.7	3269	1581	1729	25.0	1.31	2.15	1.56
4.0	42823	3463	11686	70.3	4.74	1.92	1.73
9.1	1575	5577	3942	18.0	0.75	2.17	1.85
13.1	6405	12225	11141	12.7	0.37	2.85	1.26
		^{226}Ra				$^{226}Ra/^{230}Th$	
2.7			1069	37.3			1.49
4.0			3952	72.9			1.04
9.1			0	42.3			2.35
13.3			5224	70.6			5.58

2 is the amorphous Fe phase
3 is the crystalline Fe phase
2+3 is the combined Fe phase
R is the resistate phase

(1) The concentrations of the ^{238}U and its daughters in the iron phases are two orders of magnitude greater than those in the resistate and the highest concentration is usually in the amorphous iron phase.

(2) The concentration of ^{238}U and ^{234}U in the amorphous iron phase increases with depth but there is no correlation with depth for thorium in the amorphous phase.

(3) The concentrations of ^{238}U, ^{234}U, ^{230}Th, and possibly for ^{226}Ra, have a clearly defined increase with depth in the crystalline iron phase.

(4) There is no similar trend in the concentration with depth for any of the isotopes in the resistate phase and it appears that the isotope concentration in this phase is constant and independent of depth.

(5) For the crystalline iron phase, the isotope concentrations increase down the decay chain ^{238}U ^{234}U ^{230}Th and then decrease on decay to ^{226}Ra. This occurs at all four depths.

(6) For the resistate phase, the isotope concentrations increase as ^{238}U decays to ^{226}Ra. This occurs at all four depths.

(7) The average value for the $^{234}U/^{238}U$ ratio in the amorphous phase is 0.9 and there is no correlation between the activity ratio and depth or concentration.

(8) The average value for the $^{230}Th/^{234}U$ ratio in the amorphous phase is 1.8 $\pm$ 0.8 and, although there is no correlation with depth, there is a possible correlation with ^{230}Th concentration.

(9) The $^{234}U/^{238}U$ ratio in the crystalline iron phase increases with depth; this is contrary to the trend for the $^{234}U/^{238}U$ in the total sample (see Figure 1).

(10) The average value for the $^{230}Th/^{234}U$ ratio in the crystalline iron phase is 2.4 $\pm$ 0.5. There is no trend with concentration or depth and the variability is considerably less than that found in the amorphous iron phase.

(11) The activity ratios for the resistate phase show no trend with concentration or depth.

The evidence suggests that the phases are, to a certain extent, acting independently. The amorphous iron phase is the strongest adsorber and the large and non-correlating variations in concentrations and activity ratios indicate that this phase is a record of local and temporal abnormalities in the total ground water/soil system.

There is also the possibility that the iron phases interact with the aqueous phase. The average $^{234}U/^{238}U$ ratio in the amorphous iron phase is 0.9 and in the crystalline iron phase is 1.1. This is similar to that found in ground waters at the Ranger deposit which, in the vicinity of the ore body, are about 0.8 but rise to 1.2 away from the ore body. The increase in ^{238}U, ^{234}U and ^{230}Th concentrations and the steady increase in the $^{234}U/^{238}U$ ratio with depth in the crystalline iron phase is indicative of some form of chemical control such as a steady vertical leaching process by rain-water. An alternative chemical control process would be formation of the crystalline iron phase.

In either case, the maintenance of a high $^{230}Th/^{234}U$ ratio independent of depth is indicative that the kinetics for this type of process are commensurate with or less than the half-life of ^{230}Th (7.52×10^4 y).

An unusual feature of the system is the loss of radium from the iron phases and the gain of radium in the resistate. The gain is marginal compared with the loss from the combined iron phases. The net loss from the total solid is around 50%. This suggests that radium does not stay absorbed but moves into solution. This is in general agreement with laboratory adsorption studies where it was found that only 50% of the radium was adsorbed on a high surface alumina at pH 10[24].

The behaviour of the isotope concentrations and activity ratios in the resistate is noticeably different from that observed in the iron phases. The isotope ratios fluctuate at random with concentration and depth. The isotope concentrations are reasonably constant compared to the variations which occur in the iron phases. All of the isotope concentrations are lower than those in the iron phases and there is an apparent flux of material into the resistate as the ^{238}U decays to ^{226}Ra. This is illustrated in Fig. 4 where the activity ratio for each isotope pair in the resistate phase is plotted as a function of isotopic number.

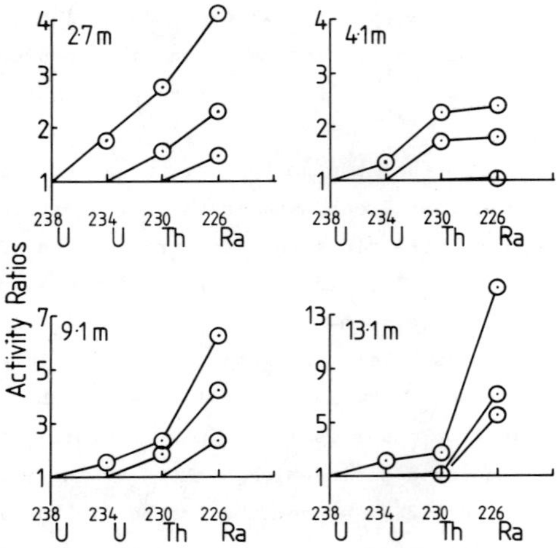

Fig. 4. Plot of various great-grandparent daughter, grandparent daughter and parent daughter activity ratios versus the isotopic number for the resistate phase at different depths in the Ranger S1/145 profile.

It is not clear whether the apparent flux of material into the resistate phase is due to an influx of daughters or an efflux of parents. Two mechanisms are proposed which unfortunately produce completely opposite solutions.

The first mechanism is termed 'alpha-recoil' and predicts an influx of daughters to the resistate. When an atom decays by alpha emission, the daughter recoils with an energy of around 70 keV which is sufficient to displace the daughter by around 10 nm in mica[25]. If the atom is within 10 nm of the surface of a phase, it could be transferred into another phase.

The alternative mechanism is termed 'selective leaching'. It is based on expected differences in leach rates from the resistate phase and predicts a greater efflux of parents over daughters. The four samples were taken from the weathered or oxidised zone. Therefore the uranium is present in the sorbed state as U^{6+}. As such it is relatively mobile as is indicated by the equilibrium between the uranium in the ground water and the iron phases, based on the similarity of the $^{234}U/^{238}U$ activity ratios. It is argued, therefore, that there is a potential for a steady loss of ^{238}U from the resistate phase. The ^{234}U is also lost, but at a marginally slower rate. This is because the ^{234}U is produced via ^{234}Th.

The dominant valency for thorium is Th^{4+}. This valency is carried into the ^{234}U for a short period of time as the ^{234}Th decays. Because U^{4+} is less mobile than U^{6+}, the loss of ^{234}U from the resistate will be slower. Eventually, the $^{234}U^{4+}$ oxidises up to $^{234}U^{6+}$ and then has the same mobility as the ^{238}U. Thus the differences in flux between ^{238}U and ^{234}U are not very large. The remaining ^{234}U decays to ^{230}Th.

Thorium is much less mobile than uranium as is evidenced by the very low levels of thorium in the ground waters and the high levels in the iron phase, which acts as a strong adsorber for thorium from the ground waters. Thus the loss of thorium from the resistate is significantly lower than that of its uranium parents. The ^{230}Th decays in turn to ^{226}Ra. As the end member of the alkaline-earth series, radium is able to exchange into alkaline-earth sites within the resistate phase. Thus radium will have a minimum tendency to travel from the resistate phase. This mechanism is favoured by the formation in the iron and resistate phases of separate particles within the heterogeneous mass of the sample.

The relative loss or gain of material by mechanisms other than decay may be determined from the measured activities. For any parent daughter pair, the activity ratio AR is given by:

$$AR = \frac{dN_d/dt}{dN_p/dt} \qquad (1)$$

where subscripts d and p refer to the daughter and parent respectively.

The activity ratio may be split up into its unit part and the residual R:

$$AR = 1 \pm R \qquad (2)$$

Substituting equation 2 into equation 1 and rearranging leads to

$$\frac{dN_d}{dt} = \frac{dN_p}{dt} \pm \frac{R.dN_p}{dt} \qquad (3)$$

The unit term represents the secular equilibrium activity and the residual term is the additional activity arising from other sources of parents or sinks of daughters; the migratory flux is then given by:

$$\text{flux} = [(dN_d/dt - dN_p/dt)] = [R.dN_p/dt] \quad (4)$$

and has been written in modulo form because of the uncertainty in defining the direction of the flux.

TABLE V Migratory fluxes between parents and daughters of the ^{238}U decay chain expressed as relative to ^{226}Ra

Depth	Isotope Pairs					
	$^{238}U:^{226}Ra$	$^{238}U:^{230}Th$	$^{238}U:^{234}U$	$^{234}U:^{226}Ra$	$^{234}U:^{230}Th$	$^{230}Th:^{226}Ra$
m	(Atoms per minute per gram of resistate phase)					
2.7	−28.3	−16.0	−7.1	−21.2	−8.9	−12.3
4.0	−42.1	−39.5	−9.9	−32.2	−29.6	−2.6
9.1	−34.6	−10.3	−2.0	−32.6	−8.3	−24.3
13.1	−65.9	−8.0	−5.4	−60.5	−2.6	−57.9

These migratory fluxes are listed in Table V as effluxes with respect to ^{226}Ra for the complete range of parent-daughter pairs. They could equally have been listed as influxes with respect to ^{238}U. The results indicate that there is an almost uniform loss of ^{238}U compared to radium from the resistate. If it is assumed that no radium is lost from the resistate, there is a steady loss of between 30 to 70 atoms per minute per gram of resistate phase (or between 10^{-6} to 10^{-5} atoms per minute per gram of ^{238}U in the resistate) which is commensurate with the decay rate of ^{238}U. Assuming that the selective leach model applies, then this value is a direct measure of the transport of uranium from the resistate phase. Alternatively, assuming that the alpha-recoil model applies, then it is a measure of net recoil into the resistate phase.

6. CONCLUSIONS

A selective phase separation method has been successfully developed for chemical separation of the principal phases in soil in the vicinity of a uranium ore body.

The principal phases were identified as amorphous iron oxide, crystalline iron oxide and resistate. Ion-exchange, amorphous alumina and silica phases were not present in significant amounts.

The distribution of ^{238}U and its decay products among the phases indicates that the phases are, to a certain extent, acting independently.

The uranium in solution is considered to be in equilibrium at least with the adsorbed uranium in the amorphous iron oxide phase and possibly with the crystalline iron oxide phase.

The amorphous iron phase is the strongest adsorber; the large and non-correlating variations in concentrations and activity ratios indicate that this phase is a record of local and temporal abnormalities in the total ground water/soil system.

The isotopic concentrations and activity ratios in the crystalline iron oxide phase have a definite structure which is indicative of some form of chemical control. The kinetics for this process are commensurate with or less than the half-life of ^{230}Th (7.52 x 10^4 y).

The change in isotope concentrations with radioactive decay in the resistate phase can be interpreted in terms of a net flux of material entering or leaving the phase over the period of the decay process. According to the model adopted, the flux can represent a direct measure of the transport of uranium through a 'quartz-clay' crystal grain.

7. ACKNOWLEDGEMENT

The work reported here forms part of a study into radionuclide migration around uranium ore bodies - analogue of radioactive waste repositories. The project is funded by the United States Nuclear Regulatory Commission as contract No. NRC-04-81-172.

8. REFERENCES

1. P.J. Shirvington, IAEA Internat. Uranium Symp. Pine Creek Geosyncline, Sydney, Australia, 1979, 509-520, June 4-8, (Pub. 1980).

2. P.J. Shirvington, IAEA/OECD-NEA Internat. Symp. Underground Disposal of Radioactive Waste, Helsinki, Finland, 2-6 July, IAEA/SM/243/129, (1979).

3. P.J. Shirvington, Geochim. Cosmochim. Acta, 47, 403-412, (1983).

4. S.K. Gupta and K.Y. Chen, Environ. Lett., 10(2), 129-158, (1975).

5. A. Tessier, P.G.C. Campbell and M. Bisson, Anal. Chem., 51(7), 844-851, (1979).

6. R.T. Lowson, Chemical Pathways Through the Environment, International Symposium, Paris, Sept. 1980.

7. I. Hashimoto and M.L. Jackson, Proc. 7th Natl. Conf. Clays and Clay Minerals, Washington, D.C., 1958, Monograph No. 5, 102-113, (Pub. 1960).

8. B.M. Tucker, Division of Soils Tech. Paper No. 23, CSIRO 1974. ISBN 0 643 00114X.

9. G.P. Gillman, R.C. Bruce, B.G. Davey, J.M. Kimble, P.L. Searle and Skjemstad, J.O., Comm. Soil Sci. Plant Nutr., (1983).

10. O. Tamm, Medd. Stat. Skogsforsoksanst., 27, 1-20, (1934). (Chem. Abs. 29:59692).

11. O. Tamm, Svenska Skogsvardsfor, Tids., No. 1-2, 231, (1934). (Chem. Abs. 29:59692).

12. U. Schwertmann, Z. Pflanzenern, Dgg., Bodenk., 105, 194-202, (1964).

13. O.P. Mehra and M.L. Jackson, Proc. 7th Natl. Conf. Clays and Clay Minerals, Washington, D.C., 1958, Monograph No. 5, 317-327, (1960).

14. E.A.C. Follett, W.J. McHardy, B.D. Mitchell and B.F.L. Smith, Clay Mineral., 6, 23-34 & 35-44, (1965).

15. B.D. Mitchell and V.C. Farmer, Clay Miner. Bull., 5, 128-144, (1962).

16. L.A. White, Scientific workshop on Environmental protection in the Alligator Rivers Region, Jabiru, N.T., May 1983, paper 53.

17. P.W. Crohn, in Economic Geology of Australia and Papua New Guinea: I Metals, ed. C.L. Knight, published by Australasian Inst. Min. Metall., Monograph Series No. 5, pp. 269-271, (1975).

18. G.S. Eupene, P.H. Fee and R.G. Colville, in Economic Geology of Australia and Papua New Guinea: I Metals, ed. C.L. Knight, published by Australasian Inst. Min. Metall., Monograph Series No. 5, pp. 308-317, (1975).

19. J.H. Hills and J.R. Richards, Miner. Deposita, 11(2), 133-154, (1976).

20. R.G. Dodson and C.E. Prichard, in Economic Geology of Australia and Papua New Guinea: I Metals, ed. C.L. Knight, published by Australasian Inst. Min. Metall., Monograph Series No. 5, pp. 281-282, (1975).

21. P.L. Airey, D. Roman, C. Golian, S.A. Short, T. Nightingale, R.T. Lowson and G.E. Calf, USNRC contract NRC-04-81-172, Annual Report AAEC/C29, (1982).

22. P.L. Airey, D. Roman, C. Golian, S.A. Short, T. Nightingale, R.T. Lowson, G.E. Calf, B.G. Davey and D. Gray, USNRC contract NRC-04-81-172, Annual Report AAEC/C40, (1983).

23. R.T. Lowson, S.A. Short, B.G. Davey and D. Gray, in preparation.

24. R.T. Lowson and J.V. Evans, Aust. J. Chem., 37, 2165-78, (1984).

25. W.H. Huang and R.M. Walker, Science, 155, 1103-1106, (1967).

PHYSICO-CHEMICAL ASSOCIATIONS OF PLUTONIUM IN CUMBRIAN SOILS

F.R. LIVENS[1,2], M.S. BAXTER[2] AND S.E. ALLEN[1]

1 : Institute of Terrestrial Ecology, Merlewood Research Station, Grange-over-Sands, Cumbria LA11 6JU

2 : Dept. of Chemistry, University of Glasgow, Glasgow G12 8QQ

ABSTRACT

The speciation of plutonium in four contrasting soil types from west Cumbria has been investigated. The soils (a sand, a gley and two brown earths) have $^{239,240}Pu$ activities in the range 8.3-6000 Bq kg^{-1} (dry weight) in their surface sections. At each site plutonium activity falls rapidly with depth, over 90% residing in the top 15 cm of the soil profile. Sequential leaching experiments show that the plutonium distributions in the different soil types are remarkably similar. Very little is readily leached and about 60% is held by organic matter. Sesquioxide phases and an intractable residual component contain the rest. Particle size analysis on a preparative scale shows enrichment of plutonium in the finer fractions, particularly in the clay (< 2 μm) material, which has a specific activity up to 50 times that in the bulk soil. Preliminary experiments on the plutonium oxidation state distribution have shown that Pu(III) or Pu(IV) is predominant. Indeed, Pu(V) and Pu(VI) are undetectable, presumably reflecting the reducing ability of the organic component of the soil.

1. INTRODUCTION

The soils in many coastal areas of west Cumbria are quite measurably contaminated with radionuclides, including plutonium, from the low-level waste discharges of the British Nuclear Fuels plc. reprocessing plant at Sellafield[1-5]. To help understand the environmental behaviour and ultimate fate of these artificial radionuclides, a detailed investigation of their physical and chemical associations in four contrasting soils from west Cumbria has been undertaken. Results, for plutonium, from both sequential leaching and particle size distribution studies are presented here, these being relevant respectively to the potentials for chemical and physical mobility of the element. The results of some experiments on the oxidation states of plutonium in soil are also discussed.

1.1 Sample Details

The samples, which were selected to give adequate activity levels and a wide range of soil properties, are a dune sand from Saltcoats (O.S. Ref SD 073966), an alluvial gley from the Esk valley (SD 113965), a brown earth from a wood near Ravenglass (SD 091959) and a brown earth from a pasture at Seascale Hall (NY 037030). Total plutonium activities down each soil profile were determined by acid digestion, ion-exchange and alpha-spectrometry[6] and are reported in Table I. In each case, over 90% of the plutonium is localised in the top 15 cm. Similar behaviour has been observed in other soils[7]. The range of plutonium activities is considerable, up to 8900 Bq kg^{-1} ($\equiv 2.4 \times 10^5$ Bq m^{-2}) although such high values are not widespread. The NRPB Generalised Derived Limit for surface soil contamination[8] is 2×10^3 Bq m^{-2}.

2. SEQUENTIAL LEACHING EXPERIMENTS

The method employed for sequential leaching is that of McLaren and Crawford[9], modified by Iu et al[10], and previously applied to radionuclide speciation studies by Cook et al[11]. Five fractions are identified:

- Exchangeable - removed by 0.01 M calcium chloride
- Inorganically adsorbed - removed by 0.5 M acetic acid
- Organically bound - removed by 0.1 M sodium pyrophosphate
- Sesquioxide bound - removed by 0.175 M ammonium oxalate at pH3
- Residual - removed by ashing and acid digestion

TABLE I Plutonium activities in soil profiles (Bq kg^{-1} ± 1 σ)

Depth (cm)	Sand		Gley		Wood		Pasture	
	$^{239,240}Pu$	^{238}Pu	$^{239,240}Pu$	^{238}Pu	$^{239,240}Pu$	^{238}Pu	$^{239,240}Pu$	^{238}Pu
0-5	74 ± 1.1	19 ± 0.4	6000 ± 60	1200 ± 26	8.3 ± 0.47	1.9 ± 0.19	22 ± 1.3	0.88 ± 0.13
5-10	21 ± 0.5	3.3 ± 0.2	850 ± 23	69 ± 6.5	2.4 ± 0.25	0.61 ± 0.13	9.0 ± 0.42	0.66 ± 0.11
10-15	3.2 ± 0.3	0.46 ± 0.07	31 ± 0.74	2.4 ± 0.20	0.93 ± 0.05	< 0.07	7.7 ± 0.20	0.77 ± 0.12
15-20	2.2 ± 0.1	0.41 ± 0.13	8.8 ± 0.67	1.1 ± 0.12	N.A.	N.A.	4.5 ± 0.19	0.38 ± 0.05
20-25	2.1 ± 0.2	0.40 ± 0.10	7.8 ± 0.25	1.1 ± 0.09	N.A.	N.A.	1.5 ± 0.10	0.11 ± 0.03
25-30	N.A.	N.A.	6.2 ± 0.20	1.2 ± 0.09	N.A.	N.A.	N.A.	N.A.

N.A. - Not Analysed.

TABLE II Sequential leaching results (% $^{239,240}Pu$ in each component ± 1 σ)

	Sand	Gley	Wood	Pasture
Exchangeable	1.0 ± 0.3	0.4 ± 0.1	1.7 ± 0.7	2.2 ± 0.3
Specific Adsorption	1.5 ± 0.2	2.5 ± 0.4	1.3 ± 0.3	4.7 ± 0.8
Organic	68.0 ± 5.5	63.5 ± 1.9	52.7 ± 3.0	58.9 ± 2.3
Sesquioxide	18.5 ± 1.3	26.4 ± 1.0	31.0 ± 2.4	17.7 ± 1.1
Residual	11.0 ± 1.3	7.2 ± 0.3	13.3 ± 1.7	16.4 ± 1.3

2.1 Sequential Leaching Results

The results of the sequential leaches are shown in Table II. Despite the variety of soil types, there is a remarkable similarity in the distribution of plutonium between the soil components.

Organic material clearly holds most (50-70%) of the plutonium whilst the remainder is associated with the sesquioxide and residual fractions.

This leaching pattern is observed in soils with a wide range of carbon contents, from 0.9% in the sand to 16% in the woodland brown earth. Sodium pyrophosphate removes only a relatively small proportion of soil organic matter, generally less than 30%[12], but this organic material has a high ash content since organically associated trace elements, mainly Fe and Al, are extracted with it[13]. For this reason, the pyrophosphate - extractable plutonium is a realistic indicator of that held by organic matter.

The sesquioxide phases, which bind 15-35% of the plutonium, are principally amorphous Fe and Mn oxy-hydroxides. These are capable of binding trace elements by ion-exchange[14] and subsequent occlusion. Reducing agents, often under acid conditions, are used to dissolve selectively the oxide precipitates and their associated trace elements[9,15,16]. Ammonium oxalate at pH3 (Tamm's reagent[17]) is one of the most selective reagents of this kind[18].

The residual fraction (7-18%) represents plutonium bound very firmly, either in the lattices of clay minerals and other silicates or irreversibly sorbed on to the insoluble, humin, component of the organic material.

3. PARTICLE SIZE FRACTIONATION EXPERIMENTS

Using a high-speed mixer, samples from each site were dispersed in distilled water for size fractionation. Fractions from 2 mm to 32 μm were isolated by wet-sieving, while the finer material was separated by sedimentation[19]. Each fraction was analysed for plutonium.

3.1 Particle Size Fractionation Results

The results of the particle size analysis of two of the soils and the plutonium activities found in the isolated fractions are reported in Table III, together with loss-on-ignition values as indicators of organic matter content[20]. The other two soils are very similar to the gley, although the activities are much lower.

There are several patterns evident in the results. Plutonium is enriched in both the coarse and fine fractions, reflecting its association with organics, since these fractions are also organic-rich. The greatest

TABLE III Particle size analysis results (Sand and Gley)

Size Class (μm)	Abundance (%)	Organic Matter (%)	^{239,240}Pu (Bq kg^{-1} ± 1 σ)
Sand:			
250-2000	2.1	5	100 ± 3.0
125-250	87.7	1	29 ± 1.1
63-125	7.8	2	110 ± 3.0
32-63	0.7	18	1100 ± 35
2-32	1.4	22	1800 ± 55
< 2	0.2	46	3100 ± 100
Bulk	-	1	54 ± 1.3
Gley:			
250-2000	1.4	46	7500 ± 240
125-250	4.2	35	7700 ± 210
63-125	20.0	12	4200 ± 150
32-63	22.8	11	2800 ± 100
20-32	26.9	10	3500 ± 110
8-20	5.5	11	4600 ± 110
2-8	6.8	13	7000 ± 260
< 2	12.2	24	10000 ± 260
Bulk	-	11	5200 ± 140

TABLE IV Results of Oxidation State Experiments

Sample	Extraction (%)	Separation (%)	Reduced ^{239,240}Pu (Bq kg^{-1})	Oxidised ^{239,240}Pu (Bq kg^{-1})
Sand	73	96	54 ± 3	< 0.5
Gley	58	70	-	-
Peroxided Gley	40	97	-	-

concentration effect occurs in the sandy soil, which, of course, has the lowest clay content. Enrichment of plutonium is generally greater in the fine fractions than in the coarse material. Clearly the fine fraction is also the most susceptible to resuspension and hence physical transport.

4. STUDIES ON PLUTONIUM OXIDATION STATES

The method of Lovett and Nelson[21] was modified in an attempt to determine the proportions of reduced (Pu(III, IV)) and oxidised (Pu(V, VI)) plutonium species present.

4.1 Method Development

A $HNO_3/H_2SO_4/K_2Cr_2O_7$ holding oxidant used to extract plutonium from suspended marine material was found to be ineffective at removing plutonium from the gley soil, dissolving only some 25% of the total. Similar difficulties were encountered during the experiments with silt samples described by Nicholson[23]. Sequential leaching of the extraction residues showed that the organically-held plutonium was insoluble.

Several other extractants were used, the most effective being a HCl/$K_2Cr_2O_7$ holding oxidant, which removes 58% of the gley soil plutonium. The plutonium component bound by organic matter is again the least soluble. The first analysis was therefore carried out on the soil with the lowest organic content, that is, the sand.

4.2 Oxidation State Analyses

Duplicate samples of the sand were extracted overnight with HCl/$K_2Cr_2O_7$ in the presence of appropriate Pu(III) and Pu(VI) tracers[24,25]. The sample was filtered and reduced plutonium isolated by co-precipitation on NdF_3. Oxidised plutonium was reduced with Fe(II) and separated by a second NdF_3 precipitation. The results are presented in Table IV.

Oxidation state analysis of the HCl - extractable portion of the plutonium in the gley soil was also attempted but showed that Pu(VI) was reduced by the sample even under chemical conditions intended to stabilise it. Peroxide treatment of the soil, to remove humic and fulvic acids[26], prevents the reduction of Pu(VI) suggesting that these are the actively reducing soil components. The results of these experiments are also shown in Table IV.

These first results suggest that the higher oxidation states of plutonium are unstable in soils, being reduced even if initially present. Similar observations were made by other workers using humic and fulvic acids[27,28,29], while Fardy and Pearson, in another context[30], postulated that hydroquinone functional groups, a constituent of soil organic matter[31], reduced Pu(IV) to Pu(III).

5. DISCUSSION

Despite the major contrasts in the physical and chemical properties of the sample soils, plutonium behaves in a consistent manner at all sites, being largely associated with organic matter and, to a lesser extent, with sesquioxide phases. While it is therefore unlikely to be easily remobilised in the short term, both organics and oxides can be altered by diagenesis so that plutonium cannot be assumed to be immobile over a longer time scale.

Physically, the plutonium is concentrated in the finest and coarsest size fractions of the soils, again reflecting its primary organic association. The fine material is obviously the most resuspensible and hence most susceptible to aeolian or aquatic transport.

Chemical conditions in the soils are such that there is no evidence of the presence of Pu(V) or Pu(VI).

6. ACKNOWLEDGEMENTS

The authors appreciate the help and advice of colleagues at both Merlewood Research Station and the University of Glasgow. The financial support of NERC is gratefully acknowledged.

REFERENCES

1. R.S. Cambray & J.D. Eakins, Nature, 300, 46, (1982).
2. P.A. Cawse, Report AERE-R9851, AERE Harwell, (1980).
3. E.I. Hamilton, Nature, 290, 690, (1981).
4. J.A. Hetherington, in Environmental Toxicity of Aquatic Radionuclides edited by M.W. Miller & J.M. Stannard (Ann Arbor, Michigan, 1976) pp. 81-106.
5. R.J. Pentreath, D.F. Jefferies & M.B. Lovett, in Proceedings of the 3rd NEA Seminar, Tokyo (OECD, Paris, 1980) pp. 203-221.
6. A.E. Lally & J.D. Eakins, in Symposium on the Determination of Radionuclides in Environmental and Biological Materials (CEGB, Sudbury House 1978).
7. P.A. Cawse, in Report AERE-PR-EMS/6 edited by M. Hainge (H.M.S.O., London 1979) p.128.
8. J.R. Simmonds, N.T. Harrison & G.S. Linsley, Generalised Derived Limits for Radioisotopes of Plutonium. National Radiological Protection Board Report NRPB-DL5, NRPB (1982).
9. R.G. McLaren & D.V. Crawford, J. Soil Sci., 24, 172, (1973).
10. K.L. Iu, I.D. Pulford & H.J. Duncan, Plant & Soil, 59, 317, (1981).

11. G.T. Cook, M.S. Baxter, H.J. Duncan, J. Toole & R. Malcolmson, Nucl. Instr. & Methods in Physics Res., 223, 517, (1984).
12. F.J. Stevenson, Humus Chemistry, Genesis, Composition, Reactions, (Wiley, 1982) p.37.
13. L.N. Alexsandrova, Sov. Soil Sci., 1960, 190, (1960).
14. R. Paterson & H. Rahman, J. Colloid Interface Sci., 98, 494, (1984).
15. D.L. Lake, P.W.W. Kirk & J.M. Lester, J. Env. Qual., 13, 175, (1984).
16. A. Tessier, P.G.C. Cambpell & M. Bisson, Anal. Chem., 51, 844, (1979).
17. O. Tamm, Medd. Stat. Skogsforsoksanst, 19, 385, (1922).
18. E.R. Landa & R.G. Gast, Clays Clay Miner., 21, 121, (1973).
19. C.B. Tanner & M.L. Jackson, Proc. Soil Sci. Soc. Am., 11, 60, (1947).
20. D.F. Ball, J. Soil Sci., 15, 84, (1964).
21. M.B. Lovett & D.M. Nelson, in Techniques for Identifying Transuranic Speciation in Aquatic Environments (IAEA, Vienna, 1981) pp. 27-35.
22. D.M. Nelson & M.B. Lovett, Nature, 276, 599, (1978).
23. S. Nicholson, Ph.D. Thesis, Manchester University, (1981).
24. J. Rydberg & L.G. Sillen, Acta. Chem. Scand., 9, 1241, (1955).
25. S.C. Foti & E.C. Freiling, Talanta, 11, 385, (1964).
26. G.W. Robinson & J.O. Jones, J. Agric. Sci., 15, 26, (1925).
27. E.A. Bondietti, S.A. Reynolds & M.H. Shanks, in Transuranic Nuclides in the Environment (IAEA, Vienna, 1976) pp. 273-287.
28. E.A. Bondietti & S.A. Reynolds, in Proceedings of an Actinide-Sediment Reactions Working Meeting at Seattle, Wash., 10-11 Feb. 1975 BNWL 2117 (1976).
29. R.A. Bulman, Struct. Bonding, 34, 39, (1978).
30. J.J. Fardy & J.M. Pearson, J. Inorg. Nucl. Chem., 36, 671, (1974).
31. F.M. Swain, Non-marine Organic Geochemistry (Cambridge, 1970) p.49.

INVESTIGATIONS OF THE CHEMICAL FORMS OF ^{239}Pu AND ^{241}Am IN ESTUARINE SEDIMENTS AND A SALT MARSH SOIL

ROBERT A BULMAN and TRACEY E JOHNSON

National Radiological Protection Board,
Chilton, Didcot, Oxon, OX11 ORQ, England

ABSTRACT

Estuarine sediments and a salt marsh soil have been fractionated by non-destructive procedures. The distribution of ^{239}Pu and ^{241}Am in these fractions has been determined by gel permeation chromatography and extraction with complexing agents.

1. INTRODUCTION

In an earlier report we have presented our investigations of the distribution of ^{239}Pu and ^{241}Am in sediments from the Ravenglass Estuary[1]. These fractionation procedures are considered to be non-destructive procedures. Here we report further studies on this material and also on soil from an adjacent salt marsh. In both cases the materials under investigation have been contaminated by marine borne radioactivity released from the nearby nuclear fuel reprocessing plant at Sellafield, formerly Windscale.

2. EXPERIMENTAL PROCEDURES

Estuarine sediments were fractionated by the previously described method[1] which employs high gradient magnetic separation (HGMS), Fig. 1. These fractions were extracted with sodium dithionite and tartrate (NaDi), a mild procedure which removes free iron oxides which act as cementing agents[2], and a dimethylsulphoxide solution (10 mg ml^{-}) of rhodotorulic acid, a ferric-complexing agent which also complexes Pu(IV) and Am(III)[3]. As a comparison a pilot study with another hydroxamic acid nitrilohydroxamic acid (NTHA) was used in aqueous solution. A similar result was obtained but the data are not presented. However, an aqueous solution of NTHA (10 mg ml^{-1}) was used to extract ^{239}Pu and ^{241}Am from salt marsh soil. These aqueous extracts were examined by gel permeation chromatography on Sephadex G-25 superfine with 2-amino-2-(hydroxymethylpropane-1,3 diol (0.1M) - sodium chloride (0.134M) buffer, p7.0, pH as the eluent.

The association of ^{239}Pu and ^{241}Am with humic substances in the salt marsh soil was determined by extraction with molten N-methylmorpholine-N-oxide (NMMNO, mp 70°C), a procedure already reported[1]. These NMMNO extracts were examined by gel permeation chromatography on glyceryl-controlled-pore glass (CG-240-200, Sigma Chemical Co. Ltd.)[1] and on a similar media (CPG-240-120) which possesses a residual negative charge not masked by glyceryl moieties. 'Blue Dextran' (Pharmacia) was used as molecular weight marker. The nature of the ionic association of ^{239}Pu and ^{241}Am with the extracted humic substances was determined by passing them over the chelating agent diethylenetriaminepentaacetic acid (DTPA) immobilized on amino-propyl-CPG-170 (Fluka)[1]. ^{239}Pu and ^{241}Am were determined as previously described[1].

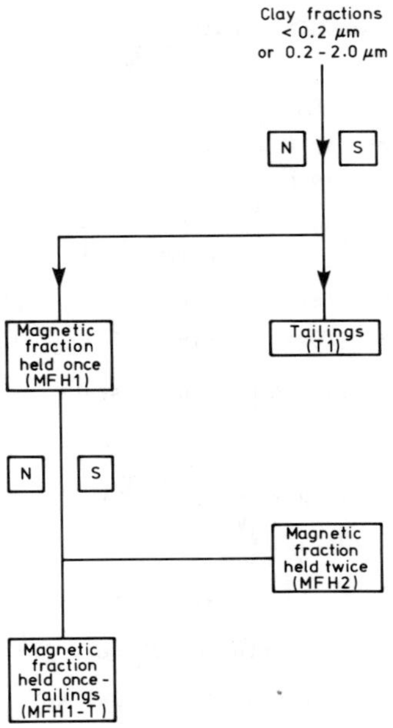

Fig. 1. Flow diagram of fractions obtained by HGMS.

3. RESULTS and DISCUSSION

From an examination of Table 1 it would appear that there are differences in the distribution of ^{239}Pu and ^{241}Am in the fractions prepared by HGMS. It is not possible to discern any pattern in the solubilization of Fe and either ^{239}Pu or ^{241}Am. Of note, however, is the apparent lack of effectiveness of the NaDi treatment at releasing ^{239}Pu and ^{241}Am from intact sediment, although there is a greater effectiveness for the various fractions. As yet there has been no attempt to characterize the minerals in these fractions. However, others have recently established that HGMS separation is a suitable qualitative procedure for fractionation of clays for subsequent infrared spectroscopy and X-ray diffraction[4].

Table 1. Percentage extraction of Fe, ^{239}Pu and ^{241}Am from estuarine sediment and some HGMS fractions

	Fe		^{239}Pu		^{241}Am	
Sample	NaDi	RA	NaDi	RA	NaDi	RA
Sediment	10.8	19.9	1.6	16.6	2.7	6.3
T_1 0.2	29.8	39.0	16.5	42.0	21.6	10.7
MFH_2 0.2	14.2	30.	9.4	38.6	10.6	26.4
T_1 0.2-2	17.6	27.5	10.1	22.7	18.2	18.6
MFH_2 0.2-2	28.0	16.7	9.6	17.3	11.6	16.5
MFH_1T 0.2-2	37.5	20.4	18.6	12.1	19.6	10.9

Extraction with NaDi and RA does not establish completely that ^{239}Pu and ^{241}Am are present as absorbed cations, for it is possible that they are present as hydrolyzed species which readily undergo solubilization.

Molten NMMNO extracts from the salt marsh soil 18% and 28% of ^{239}Pu and ^{241}Am, respectively. Elution through GG-240-200 with NMMNO:dimethylsulphoxide:water (15:15:5 by weight) as the eluent demonstrated an association between the brown material and the recovered ^{239}Pu and ^{241}Am (Fig. 2). In contrast 'Dextran Blue' eluted in fractions 5-7. As ^{239}Pu and ^{241}Am co-eluted from CPG-240-120 with the brown fractions it might be inferred that the cationic association is not weak. Nevertheless it can be inferred that the radionuclides are in cationic forms as ^{239}Pu and ^{241}Am are retained on the column bearing immobilized DTPA.

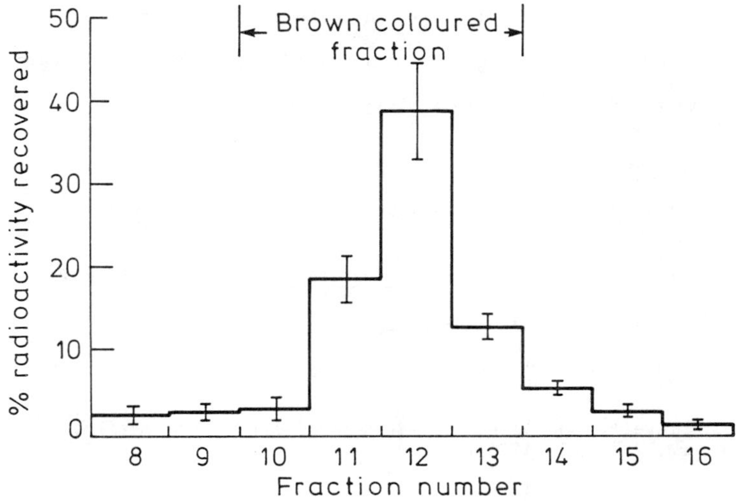

Fig. 2. Recovery of ^{239}Pu and ^{241}Am in NMMNO extracts from GG-240-220, fractions 1.2 ml.

The extraction of ^{239}Pu and ^{241}Am (14% and 53%, respectively) from the salt marsh soil (organic content 12%) by an aqueous solution of NTHA is further evidence that these transuranics should be considered as present in potentially mobile forms. The recovery of ^{239}Pu and ^{241}Am, possibly complexed by NTHA, in the low molecular weight fractions off the Sephadex G-25 superfine column substantiates that they are readily mobilized. In this context, it should be noted that hydroxamic acids, possibly of microbial origin, are present in some soils[5]. It is not known if these hydroxamic acids participate in the uptake of any elements by plants.

In conclusion, this study has demonstrated that ^{239}Pu and ^{241}Am in sediments from the Ravenglass Estuary and an adjacent salt marsh are present in forms which should be considered mobile. The extent to which ^{239}Pu and ^{241}Am in salt marsh soil are transported to plants will be the subject of a future communication. The nature of the minerals present in the HGMS subfractions will also be reported.

4. REFERENCES

1. R.A. Bulman, T.E. Johnson and A.L. Reed, Sci. Total Environ., 35, 239 (1984).

2. O.P. Mehra and M.L. Jackson, Clays Clay Min., 5, 317 (1960).

3. R.A. Bulman, Struct. Bonding, 34, 39 (1978).

4. J.D. Russell, A. Birnie and A.R. Fraser, Clay Minerals, 19, 771 (1984).

5. P.E. Powell, G.R. Cline, C.P.P. Reidand and P.J. Szaniszlo, Nature, 287, 833 (1980).

A SIMPLE MODEL FOR THE CALCULATION OF ACTINIDE SOLUBILITIES IN A REPOSITORY FOR NUCLEAR WASTES

D.C. PRYKE

Chemical Technology Division, B220, AERE Harwell, OXON, OX11 ORA

ABSTRACT

As a support to experimental measurements, a simple model has been developed to calculate the solubilities of actinides in repositories for low and intermediate level nuclear wastes. This helps to evaluate the protection provided by the near field against the release of activity.

The working of the model is briefly described and the effects of inorganic complexing agents evaluated. With cement as a matrix, predicted solubilities are all very low. They are independent of both groundwater composition and of long-term ageing of the cement.

1. INTRODUCTION

As a consequence of adopting the philosophy of redundant barriers in nuclear waste disposal[1], it is necessary to specify a package for low and intermediate level wastes whose performance is essentially satisfactory irrespective of the choice of repository site. Several sophisticated geochemical programs are presently in use to model the concentrations of radionuclides leaving a waste repository[2]. A more basic approach has been used in this study, where a model developed by Allard[3] has been adapted, and a computer program written, to predict the near field solubilities of actinides. Candidate repository materials include cement and certain polymers as immobilising matrix and cement or bentonite as the backfill. Cement is the leading choice for both duties. In general, calculations of solubilities given here compare favourably both with experimental measurements and the results of other modelling studies.

2. METHODOLOGY FOR CALCULATION OF ACTINIDE SOLUBILITIES

Assuming the groundwater/repository system to have reached equilibrium in the long-term, actinides will dissolve according to the solubility products of the various solid phases present, and distribute amongst the different oxidation states as determined by the redox potential of the system. Solids will dissolve and aqueous phases precipitate until the free energy of the system is at a minimum and the solubility product of one solid is achieved (i.e. the solution is saturated with respect to that particular solid); this identifies the solubility-limiting solid phase.

Accompanying dissolution, precipitation, and redox equilibria are a whole range of complexation reactions. The free, uncomplexed actinide cations may interact with ligands present in natural groundwater or introduced by repository materials, and this will increase solubility, possibly by orders of magnitude. The extent of complex formation is defined by the appropriate stability constants (which are determined experimentally or by extrapolation from data for similar elements) and by the concentrations of the free ligand and the uncomplexed ion. In this model it is assumed that the same constants and solubility products apply for all the actinides for a given oxidation state.

Activity coefficients are assumed unity at infinite dilution. The effects of ionic strength are corrected using the Debeye-Huckel equation in the first instance and Davies equation where insufficient data is available[4].

The oxidation states common to the actinides are (III) to (VI). The equilibria concerning hydroxyl ions and the related solid phases can be written as follows to provide a simple example of the range of species present:

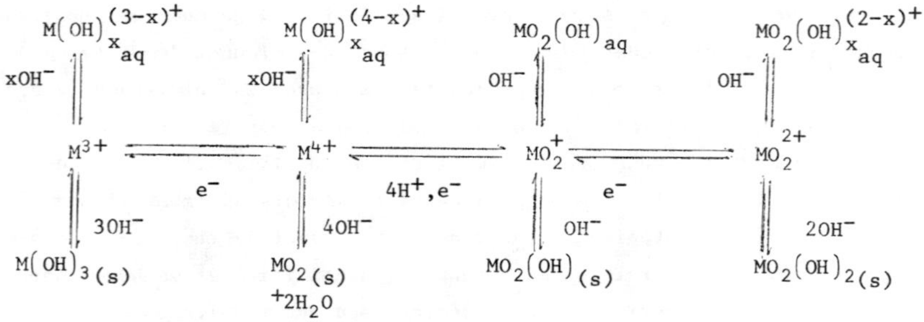

Complexing anions may be involved in further equilibria which limit their concentrations. For example, carbonate will exist extensively as bicarbonate ion and carbonic acid at intermediate and low pH, and phosphate behaves similarly. Anions may also be precipitated as sparingly soluble salts.

3. APPLICATION TO DISPOSAL CONDITIONS

On sealing a deep repository, oxygen levels are expected to fall to extremely low levels and the redox potential will be determined by geological equilibria and the corroding canister. It is expected that the $Fe(II)_{(s)}/Fe(III)_{(s)}$ couple will dominate, resulting in a reducing potential given approximately by Eh = 0.2 - 0.059 pH. This favours the four and three valent states which prove only sparingly soluble at the high pH imposed by the preferred cement matrix.

Several inorganic complexants must be considered. The carbonate anion is a strong complexant and forms soluble species with actinides in the oxidation states (III), (V) and (VI). In repository waters equilibrated with both fresh or aged cements, where calcium and pH are buffered by calcium hydroxide or a hydrated calcium silicate respectively[5], the resulting Ca^{2+} concentrations are such that carbonate levels are greatly suppressed due to the low solubility product of calcium carbonate. Other strong complexing anions are similarly affected, as shown in Table 1.

These anions are predicted to have no significant effect on solubilities. Other complexants that may be present in significant concentrations include nitrate and chloride but they interact only very weakly with actinides; sulphate is a somewhat stronger complexing agent, but at the high pH expected it will cause no significant increase in solubility. The chemistry of such systems is dominated almost totally by hydroxide species, irrespective of groundwater composition.

The predicted solubility for the major actinides with both aged and fresh cement matrices is shown in Table 2 under the reducing conditions expected.

TABLE 1 Concentrations limits for anions in cement-equilibriated waters.

Anion	Typical concentration in groundwater[6] (M)	Maximum Concentration in respository waters (M) (i) Fresh cement (pH 12.1, [Ca^{2+}] 2×10^{-2}M, I=0.1)	(ii) Aged cement (pH10.8, [Ca^{2+}] 2.5×10^{-3}M, I=0.05)
CO_3^{2-}	10^{-4}-10^{-2}*	5×10^{-7}	3×10^{-6}
F^-	10^{-6}-10^{-4}	5×10^{-5}	1×10^{-4}
PO_4^{3-}	10^{-6}-10^{-4}*	2×10^{-10}	3×10^{-9}
HPO_4^{2-}		2×10^{-5}	1×10^{-4}

* Total concentrations including protonated forms.

TABLE 2 Predicted actinide solubilities in cement waters

(1) Fresh cement Eh = -0.52V

Actinide	Solid phase	Major species	Solubility(M)
U	UO_2 (s)	$U(OH)_4$	3×10^{-9}
Pu	PuO_2(s)	$Pu(OH)_4$	3×10^{-9}
Np	NpO_2(s)	$Np(OH)_4$	3×10^{-9}
Am	$Am(OH)_3$(s)	$Am(OH)_3$	3×10^{-10}

(2) Aged cement, Eh = -0.44

Actinide	Solid phase	Major species	Solubility(M)
U	UO_2(s)	$U(OH)_4$	3×10^{-9}
Pu	PuO_2(s)	$Pu(OH)_4$	3×10^{-9}
Np	NpO_2(s)	$Np(OH)_4$	3×10^{-9}
Am	$Am(OH)_3$(s)	$Am(OH)_3$, $Am(OH)_2^+$, $AmOH(CO_3)_2^{2-}$	3×10^{-10}

4. CONCLUSIONS

Predicted solubilities for actinides in repository waters with a cement matrix or backfill are all very low and could be further reduced sorption[7]. They are apparantly unaffected by long-term ageing of the cement and the composition of groundwater entering the repository.

ACKNOWLEDGEMENTS

The author thanks the Nuclear Industry Radioactive Waste Executive for funding this study.

5. REFERENCES

1. R.H. Flowers, The Packaging of Intermediate and Low Level Radioactive Wastes, BNES Conference on Radioactive Waste Management, London, 27-29 November 1984.
2. J.E. Cross, G.L. Smith, D.R. Williams, Speciation: The Role of the Computerised Chemical Model, Speciation-85 Seminar, Oxford 16th-19th April 1985.
3. B. Allard, in Actinides in Perspective, Pergamon Press, Oxford, 1981,
4. S.L. Phillips, LBL-14313, 1982.
5. A. Atkinson and J.A. Herne, AERE-R11465, 1984.
6. A.B. Muller and L.E. Duda, The Uranium-Water System: SAND 831-0105, 1984.
7. D.C. Pryke, J.H. Rees and R.J.W. Streeton, AERE-R11391, 1984.

THE INFLUENCE OF SPECIATION ON THE GASTROINTESTINAL ABSORPTION OF ELEMENTS

John R. Cooper

National Radiological Protection Board, Chilton, Didcot, Oxon OX11 0RQ

ABSTRACT

The uptake of elements is discussed in relation to the conditions prevailing in the gastrointestinal tract. For many elements speciation markedly influences intestinal absorption but mechanisms of absorption are unclear and often the chemical species absorbed are unknown. Changes in speciation can occur in the intestine e.g. reduction or complexation by amino acids released from protein digestion. This needs to be considered when extrapolating the results of animal studies to humans were radionuclides would be consumed in mixed diets containing a variety of potential complexing agents.

1. INTRODUCTION

Early work on the nutritional requirements of animals established that the bioavailability of elements depended upon dietary composition. For example, Mellanby[1] showed that the absorption of calcium by puppies was markedly influenced by the type and amount of grain in the diet. It has also been known for many years that considerable differences exist in the absorption of various chemical forms of copper from rat intestine[2]. These studies, together with others made on different elements led to the proposal that "the availability of mineral elements to the animal is affected by the chemical form in which it is ingested "[3].

The radiological protection interest in the influence of speciation on gut absorption of elements stems from concern about the movement of radionuclides through the foodchain to man. Many early studies on the gut absorption of radionuclides produced by the nuclear fuel cycle used inorganic forms of the elements and the relevance of such work to the uptake from foodstuffs, where the chemical species may be largely unknown, is unclear. The purpose of this paper is to look at how speciation can influence the gastrointestinal absorption of fission products, activation products and actinides. The first part will look at the structure of the gastrointestinal tract in relation to uptake and the second at data on specific elements. The discussion will be limited to uptake by adult monogastric mammals

2. STRUCTURE OF THE MAMMALIAN GASTROINTESTINAL TRACT

Passage through the gastro-intestinal tract begins when food is briefly mixed with salivary juice in the mouth and then passed down the oesophagus into the stomach. The food partially neutralises the acidic gastric secretions and the pH of the bulk gastric contents rises from about 1 to between 3 and 5[4]. Although the stomach is not generally considered to be an absorptive organ, some chemical forms of a few elements can be absorbed. (Table 1) Gastric secretions may be important in facilitating the absorption of other compounds. A glycoprotein secreted by the gastric muscosa binds vitamin B12 and is necessary for its subsequent absorption by the small intestinal mucosa[5].

Food is resident in the stomach, depending upon its composition, for times varying between 1 and 3 hours[6]. The partially digested food or chyme then passes into the duodenum where it is neutralised by the alkaline secretions of the pancreas and Brunners glands. The small intestine is, for most chemical species, the major absorptive organ. Its structure is uniquely adapted to its function by folds, villi and microvilli which increase its surface area 600 times relative to a simple cylinder[7].

The luminal surface of the intestine is covered by an epithelium one cell thick. This, on the villus shaft, is made up of two main cell types (1) the mucus secreting goblet cells and (2) the absorptive cells which carry out the main processes of digestion and absorption. The epithelial cells are bonded together by tight junctions. These are normally impermeable to molecules. However, when the intestinal contents are hypertonic, as may occur following the ingestion of a carbohydrate-rich diet[8], then the tight junctions may become "leaky" and allow the passage of large molecules[9]. The effect is transient and the mucosa returns to normal within 2.5 hours[10]. Nevertheless, for the majority of chemical species in the healthy adult entry into the bloodstream is determined by the ability to traverse the absorptive cell.

Between the villi are the crypts of Lieberkhun. Cells divide in the crypts, migrate up the sides of the villi and are sloughed off at the tip. This cell extrusion zone could be a site of entry of foreign substances into the body[9]. The time taken for cells to traverse the length of the villus is about 6 days in humans[11] and approximately 17 million cells a day are shed into the intestine[12]. Thus any material entering the absorptive cell but not expelled into the bloodstream will eventually be lost from the body.

To enter the absorptive cell molecules must pass through the microvillus membrane. This is the seat of considerable hydrolytic enzymatic activity and is of central importance in the digestion of foodstuffs[13]. There is little known about the molecular aspects of absorption across the microvillus membrane. Its lipoidal character acts as a barrier to the absorption of water-soluble molecules and, broadly, two mechanisms have been proposed to explain the movement of nutritional substances across the membrane:- (1) aqueous pores exist, allowing the passage of certain specific molecules and (2) carriers in the membrane selectively bind molecules and shuttle them across. The uptake of molecules may occur against a concentration gradient by utilising metabolic energy.

Material may also enter the cell by pinocytosis - engulfment at the cell surface into small vesicles which are internalised. This may occur in the normal epithelial absorptive cell[14] and definitely does so in the Peyers patches[15]. These are aggregated masses of lymphoid tissues in the wall of the small intestine and are most common in the ileum. They do not have a normal intestinal epithelium but bear on their surface the so-called "M" cells[15]. These transport, by pinocytosis, protein antigens as well as infectious micro-organisms across the mucosa and deliver them to underlying mononuclear cells[16]. It is not known at the moment whether the M cells will translocate colloids of a non-protein nature.

The food residue passes from the small intestine into the colon where water is reabsorbed. Some other chemical species may also be taken up, for example, tellurium following oral administration of sodium tellurite[17].

3. UPTAKE OF CHEMICAL SPECIES

The following section includes data on elements for which quantifiable effects of speciation on absorption have been reported. However, the list of elements considered is not exhaustive. Iron is discussed first because there is more data on its uptake than for any other element.

3.1 Iron

Dietary iron can be broadly divided into two categories- heam and non-heam. Heam, containing iron in a porphyrin ring, is found in heamoglobin and myoglobin and accounts for over 30% of the iron in animal tissue. Possibly because of this a unique mechanism has evolved for its uptake. The intact iron-porphyrin complex is taken up by the mucosal cell and the iron is released by a specific enzyme[18]. The bioavailability of heam iron is relatively unaffected by other dietary constituents and is greater than that of non-heam iron. Heam iron represents 5-10% of the iron ingested in a western diet but accounts for over 30% of that absorbed[19].

The absorption of non-heam iron is influenced by components in the diet. Many of these, as discussed below, may be relevant to the uptake of metals of radiobiological interest such as the actinides. Ionic iron can exist in two oxidation states, ferric (Fe^{3+}) and ferrous (Fe^{2+}). Both forms of iron are probably taken up by the mucosa, although studies with soluble Fe^{3+} and Fe^{2+} complexes suggest that the uptake of the latter is faster[20]. Ferric ions have only limited solubility (10^{-18} M) at the pH of the small intestine whereas ferrous ions are more soluble (10^{-1} M). The availability of iron clearly depends upon its solubility and therefore dietary constituents that reduce ferric to ferrous ions or which complex iron and maintain it in a soluble form may enhance absorption. Ascorbic acid (vitamin C) is one of the most important and probably acts in two ways, firstly by reducing ferric to ferrous ions in the acid stomach conditions and secondly by complexation[21]. Other organic acids including lactic, citric, malic and tartaric (but probably not oxalic) acids will also promote absorption[22] by complexing iron. For this to occur, however, the complex must either be able to donate its iron to receptors in the microvillus membrane or pass intact into the absorptive cell.

Compounds that form insoluble iron salts will inhibit iron absorption. Examples are the tannates in tea, which are amongst the most potent inhibitors of iron absorption known[23], and polyphenols of coffee[19] and other plant tissue[24]. The non-digestible dietary fibres such as lignins, pectin, cellulose, hemicellulose and gum will also bind iron and thus inhibit uptake[25]. It should be mentioned that animal studies investigating metal uptake by administration of large quantities of dietary fibre may yield equivocal results. Excess dietary fibre could reduce actual uptake by decreasing intestinal transit time or by an alteration of epithelial cell biochemistry[26].

Work on the bioavailability of iron from various diets has led to the important concept of intraluminal iron pools. That is all of the non-heam iron in a meal, with a few exceptions, exchanges with a common pool from which absorption takes place[27]. The exceptions include insoluble compounds such as ferric orthophosphate, ferric oxide and also ferritin which only partially exchanges[28]. The absorption of iron is then determined by competition between the various promoting and inhibiting compounds present in the intestinal lumen.

3.2 Transuranics

The important transuranics are neptunium, plutonium and americium. The former two elements can exist in more than one oxidation state but americium commonly occurs in the trivalent form. Little is known about the effects of speciation on americium uptake but interesting observations have been made on the others.

The gut uptake of plutonium administered in the tetravalent form has been studied extensively. Some parallels exist with Fe^{3+} in that in the absence of complexing agents it probably forms poorly absorbed colloids in the small intestine. Complexing agents such as citrate[29] or diethylenetriaminetetraacetic acid[30] may facilitate absorption by holding the metal ions in solution.

Phytate (myoinostiol hexakisphosphate) may be an important dietary complexing agent for plutonium. It occurs in a variety of foodstuffs including grain, peas and potatoes[31] and forms stable soluble, complexes with plutonium[32]. It has a profound effect on plutonium uptake in some species. Rats absorb 0.13% of the plutonium from plutonium phytate whereas rabbits absorb only 0.01%[33]. This difference is related to the presence of the enzyme phytase in the small intestinal mucosa[34]. The enzyme is localised in the microvillus membrane and it is thought to break down phytate releasing the plutonium in an ionic form immediately adjacent to the absorptive surface. Thus some absorption can take place before being prevented by the competing reaction of hydrolysis to form an insoluble colloid.

Tetravalent plutonium may be reduced or oxidised under conditions that could be encountered in food and drink. Trivalent plutonium could be formed in the stomach by the action of dietary reducing agents such as ferrous ions or ascorbic acid (Table 2). The gastrointestinal uptake of Pu^{3+} is thought to be higher than Pu^{4+}.[29] The oxidation of Pu(IV) to Pu(VI) in chlorinated drinking water was shown to occur by Larsen and Oldham[35] who suggested that an increased gut uptake would result. However, in both fasted and non-fasted animals the gut uptake of Pu(VI) was indistinguishable from that of Pu(IV)[36]. This was probably because the Pu(VI) was reduced in the stomach as described above.

Reducing conditions in the intestine may also influence the absorption of neptunium. The pentavalent neptunyl ion (NpO_2^+) is stable at neutral pH and hydrolysis to form insoluble colloids would not be expected to interfere with its uptake. Administration of milligram amounts of Np^{237}(V) to rats resulted in an uptake of up to 3%[37]. However, in the same study only 0.04% of a 5 pg dose of Np^{239}(V) was absorbed. The authors explain this by suggesting that small amounts of Np(V) are reduced in the intestine to the poorly absorbed Np(IV). Consistent with this other workers have observed this mass effect in rats[38] and baboons[39] but not, interestingly, in hamsters[38].

3.3 Other metals

For many metals similar dietary factors affect absorption. Zinc, calcium, magnesium and aluminium, when present in molar excess will all form insoluble salts at neutral pH with phytate[40]. Some metals in trace amounts such as cadmium, copper, lead and zinc will also co-precipitate with calcium phytate[41]. These reactions if occurring *in vivo* would reduce absorption. Similarly, dietary fibre may bind a variety metals and render them unavailable[42].

Potentiators of uptake have been proposed for some metals. The effectiveness of human breast milk in the treatment of acrodermatitis enteropathica has been attributed to a particularly readily absorbed form of zinc. Hurley and Lönnerdal[43] proposed zinc citrate whilst another group of workers[44] suggested zinc picolinate as the readily absorbable complex. More recent work implicates histidine, released by the digestion of proteins, as an important ligand for zinc in the lumen of the intestine[45]. Those authors suggested from kinetic experiments that the zinc histidine complex was absorbed by the same mechanism as the amino acid. Copper may also be absorbed as amino acid complexes. Kirchessner and Grassman[46] found that copper added to animal diets as amino acid complexes produced greater liver accummulation than did an equivalent amount of copper sulphate. Significantly, the naturally occurring L amino acids were more effective than the D isomers. However, not all metals that form complexes with amino acids are absorbed in this form. L-Histidine complexes manganese but it appears that this metal is not absorbed as this low molecular weight complex[47].

Some metal complexes have essential biological functions and specific uptake mechanisms exist in the intestine. Heam absorption has been discussed and reference has been made earlier to the absorption of cobalt as vitamin B^{12} (section 2). The absorption of cobalt in this form, up to 70%, is much higher than that of cobalt chloride (1-16%[48]). Cobalt in marine organisms could be present as vitamin B^{12}. Inaba et al.[49-50] reported that the retention of cobalt in rats from orally administered fish or marine algae, 35% and 10% respectively, was substantially higher than when it was administered as the chloride ($\sim$ 1%).

The incorporation of other metals into foodstuffs can markedly influence absorption. Inorganic chromium is poorly absorbed from the gut. Regardless of oxidation state, dose or dietary chromium status its uptake lies between 1 and 3%[3]. However, many higher plants synthesise an anionic chromium complex of high (60%) availability to rat[51]. A different chromium complex, the so-called glucose tolerance factor, is produced by yeast and similarly exhibits a high gut uptake (30%)[51]. The mechanism of this enhancement is unknown. For mercury, conversion to methyl mercury by, for example, marine organisms[52] is important. Human and animal studies indicate that elemental mercury is virtually unabsorbed, inorganic salts exhibit 8-15% absorption whilst the uptake of methyl mercury may be complete[53]. However, not all cases of incorporation into foodstuffs result in increased availability. The gastrointestinal uptake of technetium from labelled plant tissue is lower than from pertechnate[54-55].

3.4 Non-metals

Chemical form has very little influence on the gut uptake of some non-metals of radiobiological importance such as sulphur and iodine. Commonly occurring forms of both of these elements are well absorbed[56] with the exception of elemental sulphur[57] and lignin-sulphonate[58]. Selenium chemically resembles sulphur and selenium analogues of cysteine and methionine occur naturally[59]. It is generally accepted that incorporation of this element into amino acids enhances absorption relative to the inorganic selenite. However, differences occur in this respect between animal species. In rats the intestinal absorption of selenite and seleno-methionine are similar (ca 95%)[60] whereas in humans the absorption from selenium-labelled meat (76%) may be higher than from selenite (35%)[61]. Like sulphur, elemental selenium is very poorly absorbed (~5%).

4. CONCLUSIONS

For many elements speciation markedly influences intestinal absorption. Therefore, a knowledge of the chemical forms of radionuclides in foodstuffs is essential to our understanding of their movement through the foodchain to man. However, mechanisms of absorption are unclear and often the chemical species absorbed are unknown. Complications arise because changes in speciation can occur in the intestine. Examples are 1) reduction and 2) complexation by amino acids released by the digestion of proteins. This needs to be considered when extrapolating the results of animal studies to humans where radionuclides would be consumed in mixed diets containing a variety of potential complexing agents.

Table 1

Examples of Chemical Forms Absorbed by Stomach

Form administered	Ref
Methyl mercury	Sasser et al.[62]
Niobium oxalate	Eisele and Mraz[63]
Cupric nitrate	Van Campen and Mitchell[64]

Table 2

Action of dietary reducing agents on Pu(IV)[65]

Reducing agent	Concentration (M)	% Reduction
ascorbate	0.0004	85.3 ± 7.3 (4)
"	0.00004	14.4 ± 3.4 (2)
Fe^{2+}	0.007	70.2 ± 1.7 (2)
Fe^{2+}	0.0007	0 (2)

Reducing agent, Pu^{239} (10 Bq in 0.1 M HCl) were mixed in 0.1 M HCl (2.0 ml) for 1 hour at 37°C and then extracted with thenoyl triflouroacetone[66] to determine the amount of Pu(IV) remaining.

The results are expressed as a mean ± standard deviation with the number of determinations in brackets.

For comparison the concentration of ascorbate in fruits and vegetables can be up to 0.003 M[67].

5. REFERENCES

1. E. Mellanby, Spec. Rep. Ser. Med. Res. Coun. Lond. No. 93 (1925).

2. M. D. Schultze, C. A. Elvehjem and F. B. Hart, J. Biol. Chem. 115, 453 (1936).

3. E. J. Underwood, Trace Elements in Human and Animal Nutrition 4th Edn. (Academic Press. London, 1977).

4. A. H. James and G. W. Pickering, Clin. Sci. 8, 181 (1949).

5. B. Seetharan and D. H. Alpers, Ann. Rev. Nutr. 2, 243, (1982).

6. I. Eve, Hlth. Phys. 12, 131 (1965).

7. T. H. Wilson, Intestinal Absorption (W. B. Saunders & Co., Philadelphia 1962).

8. M. L. G. Gardner, Biological Reviews, 59, 289 (1984).

9. P. G. Wheeler, I. S. Menzies and B. Creamer, Clin. Sci. and Mol. Med. 54, 495 (1978).

10. I. S. Menzies, Biochem. J., 126 (1972).

11. W. C. MacDonald, J. S. Trier and N. B. Everett, Gastroenterol., 46, 45 (1964).

12. F. Moog, Scientific American, November, 116 (1981).

13. A. J. Kenny and S. Maroux, Physiol. Rev 62, 91 (1982).

14. G. Volkheimer, F. H. Schulz, A. Lindenau and U. Beitz, Gut, 10, 32 (1969).

15. R. L. Owen, Gastroenterol., 72, 440 (1977).

16. J. L. Wolf, D. H. Rubin, R. Finberg, R. S. Kauffman, A. H. Sharpe, J. S. Trier and B. N. Fields, Science, 212, 471 (1981).

17. P. L. Wright and M. C. Bell, Am. J. Physiol., 211, 6 (1966).

18. L. R. Weintraub, M. B. Weinstein, H. Huser, S. Ratal, J. Clin. Invest. 47. 531 (1968).

19. J. D. Cook, Food Technology, Oct., 126 (1983).

20. W. A. Muir, U. Hopfer and M. King, J. Biol. Chem. 259, 4896 (1984).

21. F. M. Clydesdale in Nutritional Biochemistry of Iron (C. Kries, ed.) Am. Chem. Soc. Symp. Ser. No. 203, 55 (1982).

22. M. Gillooly, T. H. Bothwell, J. D. Torrance, A. P. MacPhail, D. P. Derman, W. R. Bezwoda, W. Mills, R. W. Charlton and F. Mayet, Brit. J. Nutr. 49, 331 (1983).

23. P. B. Disler, S. R. Lynch, R. W. Charlton, J. D. Torrance and T. H. Bothwell, Gut, 16, 193 (1975).

24. M. Gillooly, T. H. Bothwell, R. W. Charlton, J. D. Torrance, W. R. Bezwoda, A. P. MacPhail, D. P. Derman, L. Novelli, P. Morrall and F. Mayet, Brit. J. Nutr. 51, 37 (1984).

25. J. G. Reinhold in The Nutritional Bioavailability of Iron (C. Kries ed.) Am. Chem. Soc. Symp. Ser. No. 203, 143 (1982).

26. D. T. Gordon, C. Besch-Williford and M. R. Ellersieck, J. Nutr. 113, 2545 (1983).

27. R. W. Charlton and T. H. Bothwell, Ann. Rev. Med. 34, 55 (1983).

28. T. H. Bothwell, R. W. Charlton, J. D. Cook and C. A. Finch, Iron Metabolism in Man (Blackwell, Oxford, 1979).

29. M. H. Weeks, J Katz, W. D. Oakley, J. E. Ballou, L. A. George, K. K. Bustad, R. C. Thompson and H. A. Komberg, Radiat. Res. 4, 339 (1956).

30. M. F. Sullivan, L. S. Gorham and B. M. Miller, Radiat. Res. 94, 89 (1983).

31. D. J. Cosgrove, Inositol Phosphates (Elsevier, Oxford, 1980).

32. J. R. Cooper and J. D. Harrison, Hlth. Phys 46, 693 (1984).

33. J. R. Cooper and J. D. Harrison, Hlth Phys. 43, 915 (1982).

34. J. R. Cooper and H. S. Gowing, Brit. J. Nutr. 50, 673 (1983).

35. R. P. Larsen and R. D. Oldham, Science, 201, 1008 (1978).

36. J. W. Stather, J. D. Harrison, H. Smith, P. Rodwell and A. J. David, Hlth. Phys. 39, 334 (1980).

37. M. F. Sullivan, B. M. Miller and J. L. Ryan, Radiat. Res. 94, 199 (1983).

38. J. D. Harrison, D. S. Popplewell and A. J. David, Int. J. Rad. Biol. 46, 269 (1984).

39. H. Metivier, R. Masse and J. Lafuma, Radioprotection 18, 13 (1983).

40. M. Cheryan, CRC Critical Reviews in Food Science and Nutrition. 13, 297 (1980).

41. A. Wise and D. J. Gilburt, Nutrition Res. 3, 321 (1983).

42. J. G. Reinhold, B. Faradji, P. Abadi and F. Ismail-Beigi, J. Nutr. 106, 493 (1976).

43. L. S. Hurley and B. Lönnerdal, Nutr. Rev. 40, 65 (1982).

44. G. W. Evans and P. E. Johnson, Pediat. Res. 14 876 (1980).

45. R. A. Wapnir, D. E. Khani, M. A. Bayne and F. Lifshitz, J. Nutr. 113, 1346 (1983).

46. M. Kirchgessner and E. Grassmann, in Trace Element Metabolism in Animals (C. F. Mills Ed. Livingstone, Edinburgh, 1970) p277.

47. J. A. Garcia-Aranda, R. A. Wapnir and F. Lifshitz, J. Nutr. 113, 260 (1983).

48. International Commission on Radiological Protection, Publication 30, part 1 (1978).

49. J. Inaba, Y. Nishimura and R. Ichikawa, Hlth. Phys. 39, 611 (1980).

50. J. Inaba, Y. Nishimura, K. I. Kimura and R. Ichikawa, Hlth. Phys. 43, 247 (1982).

51. G. H. Starich and C. Blincoe, Science Tot. Environ. 28, 443 (1983).

52. N. Nelson, T. C. Byerly, A. C. Kolbye, L. T. Kurland, R. E. Shapiro, S. I. Shibko, W. H. Stickel, J. E. Thompson, L. A. Van den Berg and A. Weissler, Environ. Res. 4, 1 (1971).

53. G. W. Monier-Williams, Trace Elements in Food (John Wiley Inc. New York, 1968).

54. C. T. Garten, C. Myttenaere, C. M. Vandecasteele, R. Kirchmann and R. van Bruwaene in Proceedings of the Scientific Seminar on the Behaviour of Technetium in the Environment, Cadarache, France (1984) to be published.

55. M. F. Sullivan, T. R. Garland, D. A. Cataldo and R. G. Schreckhise in Biological Implications of Radionuclides Released from Nuclear Indsutries vol. 1, (IAEA, Vienna, 1979) p447.

56. International Commission on Radiological Protection, Publication 32, Part 2 (1978).

57. W. H. Johnson, R. D. Goodrich and J. C. Meiske, J. Anim. Sci. 32, 778 (1971).

58. R. Bouchard and H. R. Conrad, J. Dairy Sci. 56, 1435 (1973).

59. G. F. Combs and S. B. Combs, Ann. Rev. Nutr. 4, 257 (1984).

60. C. D. Thomson and R. D. H. Stewart, Brit. J. Nutr. 30, 139 (1973).

61. M. J. Christensen, M. Janghorbani, F. H. Steinke, N. Istfan and V. R. Young, Brit. J. Nutr. 50, 43 (1983).

62. L. B. Sasser, G. E. Jarboe, B. K. Walter and B. J. Kelman, Proc. Soc. Exp. Biol. Med. 157, 57 (1978).

63. G. R. Eisele and F. R. Mraz Hlth. Phys. 40, 235 (1981).

64. D. R. Van Campen and E. A. Mitchell, J. Nutr. 86. 120 (1965).

65. J.R. Cooper and S. Baker unpublished observations.

66. A.M. Poskanzer and B.M. Foreman, Jnr. J. Inorg. Nucl. Chem. 16, 323, (1961).

67. A.A. Paul and D.A.T. Southgate, The Composition of Foods 4th Edn. (H.M.S.O. London, 1978).

DISCUSSION

CREMERS:

Do you have an idea about the redox regime in the intestinal tract?

COOPER:

Little is known about redox potential in the intestine but it is generally assumed that conditions in the small intestine are reducing and it has been stated that "most foods have a standard reduction potential of 400 mV or slightly less" (Clydesdale, F.M. in Nutritional Biochemistry of Iron (C. Kries, ed) Am. Chem. Soc. Symp. Ser. No. 203,55 (1982)).

BONOTTO:

What would be the influence of gastrointestinal microflora on speciation?

COOPER:

Microflora only occur commonly in the colon and as only a small number of chemical species are absorbed in this region any effect of microflora will be limited.

VALENCY FIVE, SIMILARITIES BETWEEN PLUTONIUM AND NEPTUNIUM IN GASTROINTESTINAL UPTAKE*

H. METIVIER*, C. MADIC**, J. BOURGES** AND R. MASSE*
*CEA-IPSN-DPS-SPE-Section de Toxicologie et Cancérologie Expérimentale - BP 12 - 91680 Bruyères-le-Chatel - France.
**CEA-IRDI-DERDCA-DGR-SEP-Section des Transuraniens - BP 6 - 92260 Fontenay aux Roses - France

ABSTRACT

The gastrointestinal uptake of neptunium has been shown to be dependent on mass ingested in rats and in baboons. Among possible explanations, Sullivan has suggested a gastrointestinal reduction of the neptunyl ion (NpO_2^+) to Np^{4+} ion, which is less readily absorbed.

With plutonium, we have shown that for a high ingested mass (5.10^4 $\mu g.kg^{-1}$) of the 5 valency state which is stable and administered without added oxidants, the gastrointestinal uptake is high and close to Np values obtained in the same conditions (1.10^{-2}). When the mass ingested is decreased to 10 $\mu g.kg^{-1}$, the G.I. uptake decreases to 1.10^{-4}.

The two curves, G.I. transfer versus ingested mass, obtained with the two actinides in baboons were compared. Similarities and differences were explained by physiochemistry of the elements.

1 INTRODUCTION

The gastrointestinal uptake of neptunium has been shown to be dependent on ingested mass in mice[1], rats and baboons[2 3]. This increased efficiency of neptunium transfer as a function of mass administered was puzzling, since plutonium is known to behave in the opposite manner[1]. Amongst possible explanations Sullivan has suggested a gastrointestinal reduction of NpO_2+ to Np^{4+} which is less readily absorbed. This hypothesis was confirmed by addition of an oxidant to low mass of neptunium administered to rats[4] and baboons[5]. If reduction of Np(V) to Np(IV) is a good explanation, we can expect similar trends with pentavalent plutonium. The purpose of this study is to compare the mass effect for the two pentavalent actinides.

*This work was partially funded by contract No.BIO-D-565-84 F from the Commission of European Communities.

MATERIALS AND METHODS

2.1 Preparation of the Pu(V) stock solution

Due to the potential instability of Pu(V) *vs* disproportionation in aqueous solution precautions have to be taken for the preparation of the Pu(V) solution used for this study. The principle of the preparation procedure is based on the electrochemical reduction on a platinum screen of a Pu(VI) solution, previously adjusted to pH 3.7, in an electrochemical cell.

Two different types of Pu V solution were prepared using $^{239+240}Pu$ and ^{238}Pu isotopes; the different steps of the preparation are as follows

Pu(VI) solution

After dissolution of plutonium dioxide in nitrate medium containing silver (II) oxide (Azo) barium plutonate was precipitated and then dissolved in nitric acid solution. After dissolution the pH was adjusted to 3.7 and the ionic strength to 1.0:

Pu(V) solution

The plutonium(VI) solution, adjusted to pH 3.7, was put in to the cathodic compartment of an electrochemical cell. The electrochemical reduction Pu(VI) + e → Pu(V) was obtained by imposing a potential E = + 0.790 V/ENH to the platinum screen. The quantity of electricity consumed corresponds to a one electron reaction. The Pu(V) obtained was pink violet in colour.

2.2 Stability of the Pu(V) stock solution

The stability of the Pu(V) solution was studied by recording its visible absorption spectrum. After one and half days of ageing essentially all the plutonium was still on the Pu(V) form.

2.3 Preparation of the Np(V) stock solution

The solutions of 237 and 239 Np(V) were prepared in nitric acid solution by solvant extraction of Np(IV) by trylaurylammonium nitrate then back extraction using α bromocapric acid[7]. ^{239}Np is used for the low mass experiments and as tracer for high mass ones.

2.4 Measurement of transfer values

Baboons, male and female, *papio papio* used in this experiment ranged from 6.5 to 13 kg. Radionuclide solution (1 ml) was administered by gastric intubation to animals anaesthetized with 2 mg/kg of ketamine chlorhydrate 2 hours after feeding. Animals were kept in metabolism cages throughout the experiment then sacrificed 4 days after nuclide ingestion. Total gastrointestinal absorption and body retention were calculated by measuring activity in the main organs and urine.

3 RESULTS

The tissue retention results for the two actinides are given in the figure

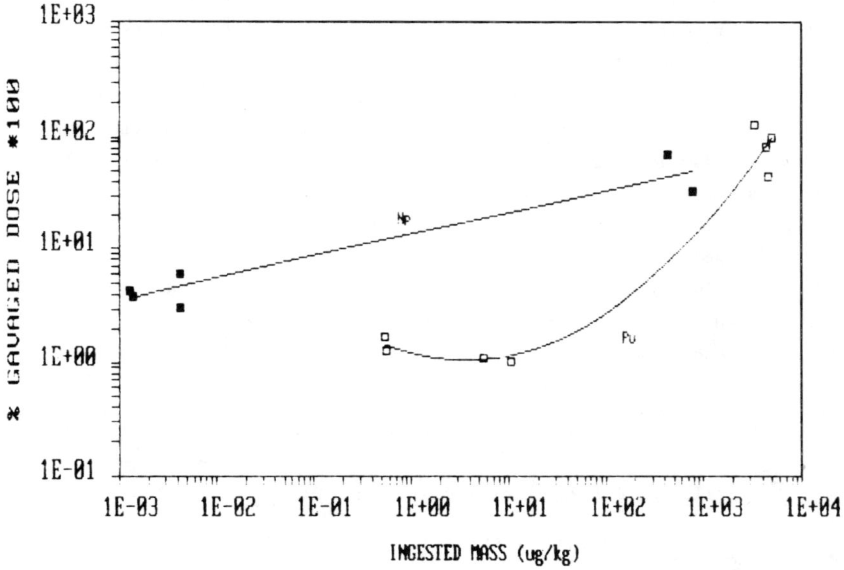

Figure: Effect of ingested mass on actinide body retention.

When the mass of Np decreased by a factor 10^5 to 10^6, the retention decreased by a factor 10. With Pu a factor 100 was observed for a mass decrease of 10^3.

4. DISCUSSION

The data presented in the figure can be discussed as follows. For high ingested mass (> 400 μg/kg) the values of percent gavaged dose absorbed for neptunium and plutonium are close and approach 1%. This high value seems characteristic of the V oxidation state of these elements. The curves obtained for Np and Pu are different but the trends observed for both elements are certainly due to the same chemical behaviour. The decrease of percent gavaged dose absorbed when the ingested mass decreases cannot be due to disproportionation reactions of Np(V) or Pu(V) because the kinetics of these reactions tend to increase with M(V) concentration.

Radiolytic reduction of M(V) to M(IV) also cannot account for the observed behaviour because of the small specific activity of the neptunium solution and the quasi constant activity of the different Pu solutions used ($^{239+240}Pu$ and ^{238}Pu isotopes were used for high and low ingested mass values resoectively). Finally reduction of Np(V) and Pu(V) by organic material present in the body of the baboons is certainly responsible for the shape of the curves. As the M(V)/M(IV) formal potential is higher for Pu than for the Np system (Pu = 1.17 V/ENH, Np = 0.74 V/EHN) the effect of reduction is more apparent for Pu. For 10 µg/kg Pu ingested mass, the percent of gavaged dose absorbed is close to that obtained for Pu(IV). The apparent increase in absorbtion for Pu ingested masses lower than $\simeq$ 5 µg/kg may be correlated to a decrease in the proportion of Pu(IV) colloidal polymer. It would be tempting to predict that data for 10^{-3} to 10^{-2} µg/kg Pu ingested dose should be identical to the neptunium data. Such a prediction is based on the similarities in the hydrolytic behaviour of Pu(IV) and Np(IV).

5. REFERENCES

1. M F Sullivan, B M Miller and J L Ryan, Radiat. Res., 94, 199, (1983)
2. H Métivier, R Masse and J Lafuma, Radioprotection, 18, 13, (1983).
3. J D Harrison, D S Popplewell and A J David, Int. J. Radiat. Biol., 46, 269, (1984).
4. M F Sullivan, P S Ruemmler and J L Ryan, Radiat. Res., 100, 519, (1984).
5. H Métivier, J Bourges, P Fritsch, D Nolibé and R Masse, Submitted to Radiat. Res., (1985).
6. C Madic, G M Begun, D E Hobart and R L Hahn, Inorg. Chem., 23, 1914, (1984).
7. C Madic and G Koehly, Nuclear Technol., 41, 323, (1978).

6. ACKNOWLEDGEMENTS

The technical assistance of M Discour, C Duserre and G Rateau during the course of these experiments was sincerely appreciated.

EXPERIMENTAL APPROACH OF MECHANISMS INVOLVED IN THE TRANSFER OF INGESTED NEPTUNIUM‡

P. FRITSCH*, M. BEAUVALLET*, B. JOUNIAUX**, H. METIVIER*, R. MASSE*
*CEA-IPSN-DPS-SPE-Section de Toxicologie et Cancérologie Expérimentale - Laboratoire de Toxicologie des Transuraniens - BP 12 - 91680 Bruyères-le-Chatel - France.
**CEA-DAM-DI-DETN-Service Controle Qualité - BP 12 - 91680 Bruyères-le Chatel - France.

ABSTRACT

We have studied in rats the in vivo transfer through jejunal segments of different soluble forms of Np (5.10^{-12} to 9.10^{-4} M) in NaCl 0.9% at pH 7.

At 5.10^{-12} M the jejunum is permeable to Np(V) nitrate and as much as 2.7% of the activity in the lumen is transferred per 10 cm per hour. As Np(V) is not complexed, the addition of DTPA Ca (100 μM) does not significantly modify this value. On the contrary 1 mM citrate decreases the transfer to 0.8%. Np(IV) is readily transformed to Np(V) when the pH is raised to 7, the pH of the working solution. In these conditions, citrate (1 mM) and DTPA Ca (100 μM) stabilize Np(IV) and the observed transfers, 0.3% and 0.9% respectively, are lower than those of Np(V) in similar solutions. Whatever the Np(V) solution used, tissue distribution of transferred Np is similar. Tissue distribution of complexed Np(IV) is quite different and varies as a function of the chelator used.

Using Np(V) nitrate at different concentrations, the transfer decreases with decreasing concentration down to 10^{-5} M. This could correspond to a limited flux of Np through the jejunal epithelium estimated at 4 nM/10 cm/hour.

Finally, histology demonstrates that after gavage in vivo with Np at masses that increase the f_1 at least 10 times, no alteration of the intestinal barrier can be established.

1 INTRODUCTION

The gastrointestinal transfer of Neptunium administered in the pentavalent state increases with the mass ingested[1,2,3].

‡This work was partially funded by contract No.BIO-D-565-84 F from the Commission of European Communities.

Two main hypothesis could account for this phenomenon : for high Np concentration either i) a toxic effect is induced at the level of gastro intestinal epithelium, as reported for other tissues[4], or ii) the rate of reduction of Np(V) to Np(IV), supposedly less transferable, is altered within the gastrointestinal contents[5]. Moreover, it is now well known that fasting has a similar effect since it increases Np transfer[3,5,6]. In order to determine experimental data that could be useful for a better understanding of factors increasing Np transfer, the purpose of this work was to develop new assays and histological techniques for measuring the transfer of Np through ligated or perfused jejunal segments.

2 HISTOLOGY

Light and electron microscopy studies were performed in the different areas of rat intestine : duodenum, jejunum, ileum, caecum and colon.

At least 4 animals were analysed, 6 hours after the ingestion of $^{237}Np(V)$ (2 mg/kg), or after fasting for 1, 2 or 3 days.

In all these animals, no significant alteration of the intestinal barrier was observed as compared to controls. Tissue modifications only occurred in fsted animals in which a decresed cellularity of crypts as a function of the fasting time was the main alteration observed.

3 Np TRANSFER THROUGH JEJUNUM AS A FUNCTION OF Np(V) CONCENTRATION

3.1 Ligated jejunum

30 to 40 cm length segments from the ligament of Treitz were instilled with 1 ml saline (pH 7) containing different concnetrations of $^{237}Np(V)$ nitrate using ^{239}Np as a tracer. Table 1 shows the amount of Np transferred to the whole carcass after 1 hour.

Table 1 Percentage of transferred Np from ligated jejunum after 1 hour

Np concentration	5.10^{-12} to 6.10^{-8} M	6.10^{-6} to 3.10^{-4}M	9.10^{-4} M
Number of animals analysed	10	15	6
% transferred Np ± 1 σ	5.2 ± 2.1	2.4 ± 1.3	0.91 ±0.31

3.2 Perfused jejunum

As the jejunum absorbs nearly 1 ml of water/10cm/hour from saline, perfusion of segments with a flow rate of 10 ml/hour was the only experimental procedure that maintains a constant Np concentration in the jejunallumen. The lumen volume was determined by measuring the liquid flowing out of the segments once perfusion was achieved. This volume was proportional to segment length and was considered to be 1 ml/10 cm (experimental value: 1.07 ml/10cm, $\sigma = 0.59$).

In these conditions, the amount of Np transferred was proportional to the time of perfusion and the length of segment. Table II shows, for different Np concentrations, the activity transferred per 10 cm per hour to the whole carcass.

Table II Percentage of transferred Np/10cm/hour from perfused jejunum segments

Np concentration	5.10^{-12} M	6.10^{-8} to 6.10^{-6} M	6.10^{-5} M
Number of animals analysed	16	9	5
% transferred Np $\pm 1\ \sigma$	2.66 $\pm$ 1.27	1.81 $\pm$ 0.71	1.33$\pm$0.44

Data shown in Tables I and II could account for a limited flux of Np through the jejunal epithelium that is estimated at about 4 nM/10cm/hour.

4 INFLUENCE OF VALENCY AND CHELATORS OF JEJUNAL Np TRANSFER

^{239}Np(IV) was prepared using hydroxylamine and HI in HCl successively as reducing agents. Valency state determinations of Np(IV) solutions in IM HCl and ^{239}Np(V) in HNO_3 (0.1 M) were made using the HTTA solvent extraction method[7].

As previously reported for low concentrations[8], Np(IV) (5.10^{-12} M) spontaneously transforms to Np(V) when the pH is raised to 7. This oxidation was abolished by adding citrate (1 mM) or 100 μM Ca DTPA.

Table III shows transfer values (percent/10cm/hour) for the different soluble forms of Np studied (5.10^{-12} M), perfused in saline (pH 7) into jejunal segments, and the tissue distributions of the transferred activity.

Table III Influence of valency and chelators on Np transfer through jejunum and tissue distribution of the transferred activity. At least 6 animals were analysed per group (perfused at 18 ml per hour)

	% Np transferred /10cm/hour	Tissue distribution as % of transferred Np			
		Liver	Femurs	1g blood	Kidneys
Nitrate (V)	2.75	5.62	3.06	0.41	5.69
	σ 1.23	σ 2.75	σ 0.71	σ 0.17	σ 1.69
Np(V)	0.77	5.88	3.07	0.44	5.07
+ 1mM citrate	σ 0.22	σ 1.61	σ 0.67	σ 0.18	σ 1.87
Np(V)	2.18	5.87	2.22	0.66	6.78
+ 100μM DTPA Ca	σ 0.52	σ 1.88	σ 0.38	σ 0.26	σ 2.69
Np(IV)	0.31	9.77	2.61	2.10	5.16
+ 1mM citrate	σ 0.12	σ 2.87	σ 0.71	σ 0.47	σ 1.34
Np(IV)	1.04	2.15	0.72	0.52	11.17
+ 100μM DTPA Ca	σ 0.40	σ 1.31	σ 0.42	σ 0.23	σ 5.25

In similar solutions, 2 to 3 times less Np(IV) is transferred through the jejunum than Np(V). The tissue distributions of transferred Np(V) appear constant while those of Np(IV) are significantly different and can be modified as a function of the chelators used.

5 CONCLUSION

The different experimental procedures used allow us to establish parameters for Np transfer through the jejunum. These are functions of concentration, valency and the presence of chelators.

These transfer measurements and the histological study show that the increased transfer of Np(V) observed *in vivo* for high ingested masses cannot be related to a toxic effect on the intestinal epithelium.

This mass effect could correspond to an increased amount of soluble Np(V) in the intestinal lumen.

For low ingested masses, Np(V) adsorption on gastrointestinal contents[4], as well as the expected reduction to Np(IV)[5], could account for the lower but constant f_1 observed.

6 REFERENCES

1. M.F. Sullivan, B.M. Miller and J.L. Ryan, Radiat Res., 94, 199, (1983).
2. H. Métivier, R. Masse and J. Lafuma, Radioprotection, 18, 13, (1983).
3. J.D. Harrison, D.S. Popplewell and A.J. David, Int. J. Radiat. Biol., 46, 269, (1984).
4. R.C. Thompson, Radiat. Res., 90, 1, (1982).
5. M.F. Sullivan, P.S. Ruemmler and J.L. Ryan, Radiat. Res., 100, 519, (1984).
6. R.P. Larsen, M.H. Bhattacharyya, R.D. Oldham and E.S. Moretti, Radiological and Environmental Research Division Annual Report, Argonne National Laboratory, Argonne, Illinois, ANL-82-65, Pt.II, pp 147-149, (1982).
7. J. Stary, in The Solvent Extraction of Metal Chelates, Edited by H. Irving (Pergamon Press, 1964) p 70.
8. J. Duplessis, M. Genet and R. Guillaumont, Radiochimica Acta, 21, 21, (1974).

UPTAKE OF RADIONUCLIDES FROM THE GASTROINTESTINAL TRACT IN RATS FED DIFFERENT FOODS

B. KARGAČIN AND K. KOSTIAL
Institute for Medical Research and Occupational Health,
M. Pijade 158, Zagreb, Yugoslavia

ABSTRACT

Experimental results showing the effect of different foods on the absorption and retention of strontium, lead, cadmium and mercury in suckling and adult rats are presented. Animals were on respective diets during a different time period before and after oral or intraperitoneal administration of ^{85}Sr, ^{203}Pb, ^{115m}Cd and ^{203}Hg. Rats fed on "human foods" had, in general, higher absorption values for all metals than animals on standard rat diet. Rat diet and its ingredients were found to be very effective in reducing metal absorption. This effect was observed both in suckling and adult rats but the degree of reduction was dependent on the age of the animal and was not the same for different metals. The results obtained could not be explained by differences in single food components. The bioavailability of metals from various diets is obviously a result of complex interactions occurring at the absorption sites.

1 INTRODUCTION

The digestive tract is the main route of entry of radionuclides into the body in conditions of environmental exposure. In such circumstances the body burden of radionuclides will primarily depend upon diet and age i.e. on the factors which greatly influence intestinal absorption of metals[1]. The purpose of this work is to present some studies concerning the effect of dietary factors on radionuclide uptake from the gastrointestinal tract. For this purpose we shall present data on the pharmacokinetics of lead, strontium, cadmium and mercury in rats fed on different diets which have been obtained in the Laboratory for Mineral Metabolism of the Institute for Medical Research and Occupational Health in Zagreb.

2 MATERIALS AND METHODS

All experiments were performed on albino outbred rats from the Institute's breeding farm, aged 1-2 or 6-8 weeks. Sucklings in litters reduced to six one day after birth were kept in individual cages with their mothers. In each litter some sucklings were artificially fed cow's milk (0.4 ml) administered with a dropper[2], others were given rat food or its ingredients (0.2 g - fish meal 37%, sunflower meal 28%, alfalfa 20%, cane molasses 12%, premix - a vitamin-mineral mixture 4%) administered with a glass spoon[3]. The animals received ^{85}Sr, ^{203}Pb, ^{115m}Cd or ^{203}Hg in milk, rat food or rat food ingredients over the 8 h artificial feeding period. Sucklings were returned to their mothers i.e. to milk diet immediately after administration of radionuclide.

Older animals (females) were housed in plastic cages in groups of 6-12 animals per cage. The length of dietary treatment varied according to the experiment. Animals were on different diets 3-7 days before the administration of radionuclide and 6 days after it, when they were killed. Only animals that were on human foods and received ^{85}Sr were on respective diets for 24 days and ^{85}Sr was administered on 18th day of the experiment. The radionuclides were given to older animals by gastric intubation.

The radioisotopes of ^{85}Sr, ^{115m}Cd and ^{203}Hg were supplied by the Radiochemical Centre Amersham, England, and that of ^{203}Pb by the Gustav Werner Institute, Uppsala, Sweden. The specific activity of ^{115m}Cd and ^{203}Hg was about 18,5 MBq/mg Cd or Hg and ^{85}Sr and ^{203}Pb were almost carrier free. The retention in the whole body, carcass (whole body after removal of the gastrointestinal tract) and gut was determined in a two-crystal scintiallation counter 6 days after radionuclide administration. All results were corrected for radioactive decay and geometry of the samples and expressed as percentages of the dose.

3 RESULTS AND DISCUSSION

The results of our first investigations into the effect of milk diet on lead absorption showed a surprisingly high retention of lead in sucklings fed on milk diet. Further investigations tried to reveal if the same effect could be observed also for other metals (strontium, cadmium and mercury) and in older animals. In sucklings on milk diet the retention was very high (Table I) - 35 per cent of the dose for cadmium, 48 per cent for

TABLE I The effect of milk on the retention of ^{85}Sr, ^{203}Pb, ^{115m}Cd and ^{203}Hg, administered orally, in 1 and 6-8-week-old rats

	1-week-old				6-8-week-old			
	milk (M)	rat diet (R)	M/R	Ref.[c]	milk (M)	rat.diet (R)	M/R	Ref.[c]
^{85}Sr[a]	74.2 (18)[b]	67.9 (17)	1.1	3	40.6 (6)	10.3 (14)	4	5
^{203}Pb	57.4 (17)			4	22.9 (24)	0.40(20)	57.3	6
^{115m}Cd	35.2 (18)	21.6 (18)	1.6	3	6.9 (10)	0.49(10)	14	7
^{203}Hg	47.7 (18)	7.7 (17)	6.2	3	2.7 (6)	0.64(6)	4	8

[a]Values presented as percentage of the dose in carcass, others in the whole body.

[b]Number of animals in parentheses.

[c]See references.

mercury and 74 for strontium. The administration of rat diet to the same age group significantly reduced radionuclide retention - the values were lower - 1.1 times for ^{85}Sr, 1.6 times for ^{115m}Cd and 6.2 times for ^{203}Hg. Older animals (6-8 weeks) on milk diet had also higher retention values of all metals than animals on control diet but never as high as in sucklings. These values were from 3 per cent for ^{203}Hg to 40 per cent for ^{85}Sr. The administration of standard rat diet caused in these animals a very high reduction of lead (57 times lower retention) and a lower reduction of strontium and mercury retention (4 times lower retention). Investigations were further concerned not only with the effect of milk but also of other foods commonly used in human nutrition. All animals on meat, bread and other "human" foods had a much higher whole body retention than control animals fed on rat diet (Table II). The retentions in animals fed on human diets were for ^{85}Sr 4-9 times, for ^{203}Pb 6-12 times and for ^{115m}Cd 4 times higher than in control rats.

TABLE II The effect of "human" diets (H) on the retention of ^{85}Sr, ^{203}Pb and ^{115m}Cd in 6-week-old rats

	^{85}Sr[a]		^{203}Pb		^{115m}Cd	
Ref.[c]	5	H/R	9	H/R	10	H/R
rat diet(R)	9.4 (5)[b]		0.44 (10)		1.8 (18)	
meat	85.2 (6)	9.1			8.3 (10)	4.6
bread	78.5 (6)	8.4	5.1 (10)	11.6	7.3 (10)	4.1
beans	56.3 (6)	6.0				
potatoes	43.8 (6)	4.7				
pig´s liver			3.0 (10)	6.8		

[a]Values presented as percentage of the dose in carcass, others in the whole body.
[b]Number of animals in parentheses.
[c]See references.

The retentions in the whole body and carcass of ^{85}Sr, ^{203}Pb, ^{115m}Cd and ^{203}Hg after intraperitoneal administration are presented in Table III. The values of radionuclide retention are practically the same in milk-fed and control 6-week-old animals, indicating that dietary treatment influences radionuclide metabolism primarily at the level of intestinal absorption. These studies gave a further proof of a very different bioavailability of metals in animals fed various foods. Since in all experiments rat diet caused the lowest cation absorption, we tried to identify particular components of rat diet that are responsible for the observed effect. We investigated the effect of several rat diet ingredients on radiostrontium absorption. We found five (fish meal, sunflower meal, alfalfa meal, cane molasses and premix) out of ten (yellow corn, ground wheat, brewers grains, soy-bean meal, powdered milk) ingredients to reduce strontium absorption even more than complete rat food[13]. When these five ingredients were supplemented to meat they reduced very effectively the retention of strontium, cadmium and mercury (Table IV). The ingredients of rat diet were more effective in reducing strontium and cadmium retention and as effective as complete rat food in reducing mercury retention. The administration of the same ingredients as a mixture to the sucklings caused, in relation to those on milk diet, a significant reduction in metal retention which was highest for mercury (Table V).

TABLE III The effect of milk on the retention of ^{85}Sr, ^{203}Pb, ^{115m}Cd and ^{203}Hg, administered intraperitoneally, in 6-8-week-old rats

	milk (M)	rat diet (R)	M/R	Ref.[c]
^{85}Sr[a]	69.0 (6)[b]	62.9 (6)	1.1	5
^{203}Pb	70.8 (6)	54.4 (8)	1.3	6
^{115m}Cd	82.3 (9)	82.4 (9)	1.0	7
^{203}Hg	64.3 (6)	56.5 (6)	1.1	8

[a]Values presented as percentage of the dose in carcass, others in the vhole body.
[b]Number of animals in parentheses.
[c]See references.

TABLE IV The effect of rat diet or its ingredients supplemented to meat on ^{85}Sr, ^{115m}Cd and ^{203}Hg whole body retention in 6-week-old rats (per cent dose)[c]

Diet	^{85}Sr		^{115m}Cd		^{203}Hg	
		1/2,3		1/2,3		1/2,3
1 meat (100%)	86 (15)[a]		5.2 (14)		0.72 (15)	
2 meat (75%) + rat diet (25%)	66 (15)	1.3	2.4 (15)	2.2	0.24 (15)	3
2 meat (75%) + rat diet (25%) ingredients[b]	20 (14)	4.3	1.6 (15)	3.3	0.24 (15)	3

[a]Number of animals in parentheses.
[b]Rat food ingredients (fish meal 9%, sunflower meal 7%, alfalfa 5%, cane molasses 3%, premix - a vitamin-mineral mixture 1%)
[c]See reference 11.

The high retention values of radionuclides in animals fed on human diets and significantly lower values in those on rat diet or ingredients cannot be easily related to differences in the daily intake of various food components, which have been shown to influence radionuclide absorption. In

TABLE V The effect of rat food ingredients on ^{85}Sr, ^{115m}Cd and ^{203}Hg whole body retention in suckling rats (per cent dose)[c]

Diet	^{85}Sr	1/2	^{115m}Cd	1/2	^{203}Hg	1/2
1 milk	78 (6)[a]		44 (9)		62.0 (6)	
2 rat food ingredients[b]	63 (12)	1.2	26 (18)	1.7	5.4 (10)	11.5

[a]Number of animals in parentheses.

[b]Rat food ingredients (fish meal 36%, sunflower meal 28%, alfalfa 20%, cane molasses 12% and premix 4%)

[c]See reference 12.

experiments with strontium rats on meat and bread diets with very different fat or carbohydrate intakes had almost the same strontium retention values[5]. Calcium intake, which is well known[14] to influence strontium absorption, was in rats on milk diet the same as in those on rat diet and strontium retention was significantly different. Also, the animals on similar dietary fibre intakes like animals on bread, beans and rat diet showed appreciable differences in strontium absorption. The retention of lead[6] and cadmium[7], likewise, could not be explained by differences in single dietary constituents. Solid diet reduces metal absorption possibly because its abrasive action either decreases metal deposition or accentuates the extrusion of the metal from the villi[15]. It might be that rat food or its ingredients produce ligands with a higher affinity for some metals than ligands present in the gut wall or produced during the digestion of human foods.

The bioavailability of metals from various diets is obviously a result of complex interactions occurring at the absorption sites. Diet might interfere in several steps of the absorption process. For cadmium and mercury dietary factors influenced primarily the transport from the luminal content into the gut and to a lesser extent the transport from the gut into the body.

Our results indicate that diet may significantly alter radionuclide absorption in neonates as well as in adults and therefore it presents an important factor in metal metabolism and toxicity. More data are, however, needed for the better understanding of the mechanism of metal uptake from the gastrointestinal tract.

4 REFERENCES

1. G.F. Nordberg, B.A. Fowler, L. Friberg, A. Jernelöv, N. Nelson, M. Piscator, H.H. Sanstead, J. Vostal and V.B. Vouk, Environ. Health Perspect., 25, 3, (1978).

2. K. Kostial, I. Šimonović and M. Pišonić, Nature, 215, 1181, (1967)

3. K. Kostial, I. Šimonović, I. Rabar and M. Landeka, Environ. Res., 25, 281, 1981.

4. K. Kostial, I. Šimonović and M. Pišonić, Nature, 233, 564, (1971).

5. K. Kostial, I. Šimonović, I. Rabar and M. Landeka, Period. Biol., 82, 229, (1980).

6. D. Kello and K. Kostial, Environ. Res., 6, 355, (1973).

7. D. Kello and K. Kostial, Toxicol. Appl. Pharmacol., 40, 277, (1977).

8. K. Kostial, I. Rabar, M. Ciganović and I. Šimonović, Bull. Environm. Contam. Toxicol., 23, 566, (1979).

9. K. Kostial and D. Kello, Bull. Environm. Contam. Toxicol., 21, 312, (1979).

10. I. Rabar and K. Kostial, Arch. Toxicol., 47, 63, (1981).

11. B. Kargačin and K. Kostial, Acta Biol., 10, 23, (1984).

12. K. Kostial, B. Kargačin and M. Landeka, Toxicol. Lett., 23, 163, (1984).

13. B. Kargačin and K. Kostial, Arh. hig. rada toksikol., 33, 185, (1982).

14. C.L. Comar and R.H. Wasserman, in Mineral Metabolism, Vol. II, edited by C.L. Comar and F. Bronner (Academic Press, New York, 1964), pp. 562-566.

15. L.B. Sasser and G.E. Jarboe, Toxicol. Appl. Pharmacol., 41, 423, (1977).

THE METABOLISM AND GASTROINTESTINAL ABSORPTION OF NEPTUNIUM AND PROTACTINIUM IN ADULT BABOONS

L.G. RALSTON,[1] N. COHEN,[1] M.H. BHATTACHARYYA,[2] R.P. LARSEN,[2]
L. AYRES,[1] R.D. OLDHAM,[2] AND E.S. MORETTI[2]

[1]Institute of Environmental Medicine, NYU Medical Center, Tuxedo, NY, USA
[2]Argonne National Laboratory, Argonne, IL, USA

ABSTRACT

The metabolism of neptunium and protactinium was studied in adult female baboons following intravenous injection and intragastric intubation. Neptunium-239, Np-237, and Pa-233 were prepared as either citrate-buffer, nitrate, or bicarbonate solutions with oxidation states of (V) and (VI). Samples of blood, urine, feces and autopsy tissues were measured by gamma spectrometry. Retention of neptunium and protactinium was determined _in vivo_ using whole and partial body gamma-scintillation spectrometry with [NaI-CsI(Tl)] detectors.

Immediately following intravenous injection (10^{-1} to 10^{-10} mg Np per kg body wt), neptunium cleared rapidly from blood, deposited primarily in the skeleton (54±5%) and liver (3±0.2%), and was excreted predominantly via urine (40±3%). For the first year post injection, neptunium was retained with a biological half-time of ~100 days in liver and 1.5±0.2 yr in bone. In comparison, injected protactinium (10^{-9} mg/kg) was retained in blood in higher concentrations and was initially eliminated in urine to a lesser extent (6±3%). _In vivo_ measurements indicated that protactinium was retained in bone (65±0.3%) with a half-time of 3.5±0.6 yr. The distribution and retention of protactinium in other tissues has not been determined at this time. Differences in the physicochemical states of the neptunium or protactinium solutions injected did not alter the metabolic behavior of these nuclides.

Fed and fasted baboons were administered solutions of Np(VI) bicarbonate (10^{-8} to 10^{-1} mg/kg) and Pa(V) citrate-buffer (10^{-9} mg/kg) by gavage. The gastrointestinal absorption value for neptunium in two fasted baboons, sacrificed at 1 day post administration, was determined to be 0.92±0.04%. Of the total amount of neptunium absorbed, 52±3% was retained in bone, 6±2% was in liver, and 42±0.1% was excreted in urine. The metabolism of neptunium followed oral and iv administrations was found to be similar. This observation was also true for baboons which had received oral and iv doses of protactinium. A method was developed to estimate GI absorption values for both nuclides in baboons, which were not sacrificed, by comparison of activities present in bioassay samples after injection and gavage. Absorption values calculated by this method for neptunium and protactinium in fasted baboons were 1.8±0.8% and 0.65±0.01%, respectively. Values for fed animals were 1 to 2 orders of magnitude less than those for fasted animals. Further experiments are currently underway to evaluate this assay technique.

1 INTRODUCTION

Considerable interest has been generated recently in re-evaluating the metabolic behavior of the actinide element neptunium.[1] This interest has been focused primarily on Np-237 ($T_{1/2}$=2.1x10^6 yr) which has been identified as one of the critical isotopes present in high-level nuclear waste at long-term repository sites.[2] In addition to neptunium, there is a growing concern for the increase in the production of protactinium, which is present as Pa-233 ($T_{1/2}$=27.4 days) in all nuclear waste containing its parent, Np-237, and as Pa-231 ($T_{1/2}$=3.2x10^4 yr) associated with the uranium-thorium breeder fuel cycle.[3] Furthermore, Pa-233 is a significant source of analytical interference in the <u>in vivo</u> measurement of Np-237 in man. In view of the paucity of studies concerning both neptunium and protactinium, further studies on their metabolism in mammals is essential for accurate risk assessments.

This present study on the metabolism of neptunium and protactinium in adult female baboons was designed to (1) determine the metabolic behavior of these nuclides after intravenous injection of solutions of differing physicochemical forms; (2) determine their gastrointestinal absorption factors (f_1 values) and metabolism in fed and fasted baboons; and (3) to develop a method for estimating GI absorption factors from bioassay samples which would not require the sacrifice of animals, and which could be applied for determining actinide burdens in cases of human exposure. The use of the baboon, a nonhuman primate, as a suitable subject for metabolism studies of actinides in man has been well documented.[4] Several studies, involving Am, Cm and U, have been successfully conducted at this laboratory.[5-7]

2 METHODS AND MATERIALS

2.1 Animals

Eleven adult female baboons (<u>Papio</u> <u>cynocephalus</u>, <u>anubis</u>, and hybrids), weighing between 11 and 18 kilograms, were used in this study. Animal board and husbandry were provided at New York University's Laboratory for Experimental Medicine and Surgery in Primates (LEMSIP) in Tuxedo, NY. Baboons were housed individually and maintained on a commercial diet. Water was provided ad libitum. Injection (im) of 0.1 mg/kg ketamine HCl was used to tranquilize animals prior to handling outside of confinement.

2.2 Preparation of Administration Solutions

Stock solutions of Np-237/Pa-233 and Am-243/Np-239 in 1 <u>N</u> nitric acid were obtained from Oak Ridge National Laboratories (Oak Ridge, TN). Solutions of Pa-233 were prepared by solvent extraction from Np-237 using diisobutyl carbinol (DIBC) as described by Sill.[8] Neptunium-239 was separated from Am-243 by cation exchange chromatography. The technique used to prepare solutions in the (V) and (VI) valance states is discussed in detail elsewhere.[9] Solutions prepared as Np(VI) or Pa(VI) contained 1 ppm chlorine as a holding oxidant. Both nuclides were administered (Table 1) in activity concentrations and chemical forms consistent with those used in actinide metabolism studies conducted at our laboratory and at others.[1]

2.3 Gamma Spectrometry and Counting Geometries

Four photon measurement systems were used to determine neptunium and protactinium gamma activities *in vivo* and in bioassay samples, including a low-level whole-body counter employing dual NaI-CsI(Tl) detectors, a Ge(Li), NaI(Tl) well, and a 3"x2" NaI(Tl) detector. Activity levels of Np-237, Np-239 and Pa-233 were calculated by integration of count rates in energy regions of interest between 60 to 120 keV, 220 to 280 keV, and 280 to 380 keV, respectively. Corrections were made to Np-237 determinations to account for the contribution of count rates from Pa-233. Additional alpha spectrometric measurements of Np-237 in some bioassay samples were conducted at ANL. Whole body counting of tranquilized and restrained baboons was accomplished using "meter-arc" and head count geometries. These systems and measurement techniques have been described in detail previously.[10]

TABLE I Administration protocols.

Isotope	# of Admin.	Chemical Form	pH	Oxid. State	Mode of Admin.	Dose (mg/kg)
Np-237	1	0.08M Na Citrate	3.2	V	Inj., IV	10^{-1}
Np-237	1	0.08M Na Citrate	3.2	VI	Inj., IV	10^{-1}
Np-239	1	0.08M Na Citrate	3.2	V	Inj., IV	10^{-10}
Np-239	1	0.08M Na Citrate	3.2	VI	Inj., IV	10^{-10}
Np-239	1	Nitrate	1.0	VI	Inj., IV	10^{-9}
Np-239	5	0.01M $NaHCO_3$	6.5-8.5	VI	Inj., IV	10^{-10}
Pa-233	2	0.08M Na Citrate	3.2	V	Inj., IV	10^{-9}
Pa-233	1	0.08M Na Citrate	3.2	VI	Inj., IV	10^{-9}
Np-237	3	0.01M $NaHCO_3$	6.5	VI	PO, Fasted	10^{-2}-10^{-1}
Np-239	10	0.01M $NaHCO_3$	6.5	VI	PO, Fasted	10^{-8}
Np-239	3	0.01M $NaHCO_3$	6.5	VI	PO, Fed	10^{-7}-10^{-6}
Np-239	1	Nitrate	3.0	V	PO, Fasted	10^{-8}
Pa-233	3	0.08-2.0M Na Citrate	3.2-6.5	V	PO, Fasted	10^{-9}
Pa-233	1	0.08M Na Citrate	3.2	V	PO, Fed	10^{-8}

2.4 Administration Protocols

Table 1 presents the experimental protocol used. Baboons were tranquilized and injected intravenously in the saphenous or femoral vein with 1 to 2 ml of administration solution. Solutions of neptunium or protactinium in volumes of 10 to 20 ml were introduced directly into the stomachs of fed or fasted (24 hr) baboons using a catheter inserted nasally. Due to the short radiological half lives of Np-239 and Pa-233, it was possible to perform multiple administrations of these isotopes in the same animal after sufficient time had passed for their biological clearance and decay. Data presented in this paper are expressed as a percent of the administered dose for each experiment conducted. Neptunium-239 and Pa-233 values have been corrected for radiological decay. Retention for all radionuclides is expressed as biological half-times, derived from least-squares, curve-fitting techniques. Total blood volume for each animal was estimated to be 7% of its total body weight. For clarity, some data has been omitted from the figures presented.

3 RESULTS

3.1 Metabolism of Neptunium and Protactinium After iv Injection

3.1.1 Whole Blood Retention

The retention of neptunium and protactinium in total blood volume as a function of time post injection is illustrated in Figures 1a and 1b, respectively. Separate measurements of cellular and serum components of whole blood samples indicated that greater than 95% of the activity in blood was associated with the serum component. Neptunium was observed to clear rapidly from blood and was retained with a half-time of 9±2 days for the interval 2 to 7 days post injection. No differences were noted in retention following the injection of neptunium solutions of varying chemical forms or oxidation states. However, within the first 24 hours, Np-237 was retained at a two fold higher concentration than Np-239, perhaps due to the differences in the initial neptunium masses injected (10^{-3} mg/kg Np-237 vs 10^{-10} mg/kg Np-239). In comparison to neptunium, Pa-233 was present in blood in higher concentrations and retained with a half-time of 6±1 days for the time interval from 2 days to 3 weeks post injection. No difference was observed in the retention of Pa(V) and Pa(VI).

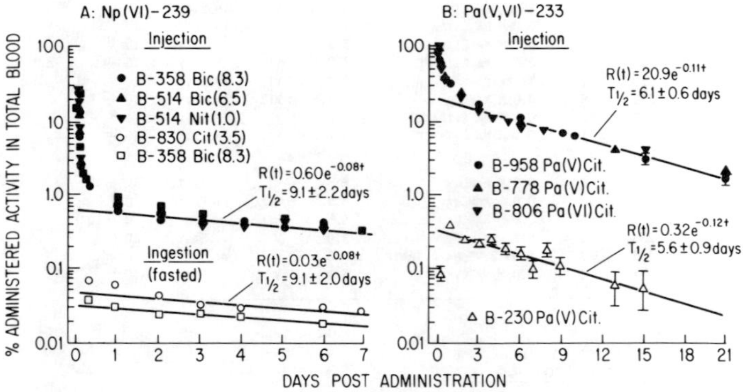

FIGURES 1a and 1b Total blood retention of Np and Pa.

3.1.2 Urinary and Fecal Excretion Patterns

Figures 2a and 2b show the elimination of neptunium and protactinium via urine as a function of time post injection. Urinary excretion was the primary route of elimination for both nuclides, especially during the first 24 hours post injection, regardless of the isotope, oxidation state, chemical form administered or volume of urine voided. For neptunium, a mean value of 33±10% (n=10) was recovered in urine within 24 hours. A cumulative amount of 45±7% was recovered by 2 months. The values of the percent protactinium recovered in urine were 6±3% for the first day and 15±4% cumulative excretion by day 21. Fecal excretion of both neptunium and protactinium was minimal. Cumulative recovery of neptunium after 27 days was 3±1%. The clearance of neptunium in feces could be described by a single exponential equation with 0.08% of the injected activity clearing with a half-time of ~100 days. Cumulative recoveries of protactinium were 0.03% and 3% by 1 day and 1 month post injection. The long-term retention half-time of protactinium in feces was ~300 days (0.05%). Elimination of neptunium and protactinium in feces was independent of the chemical form administered.

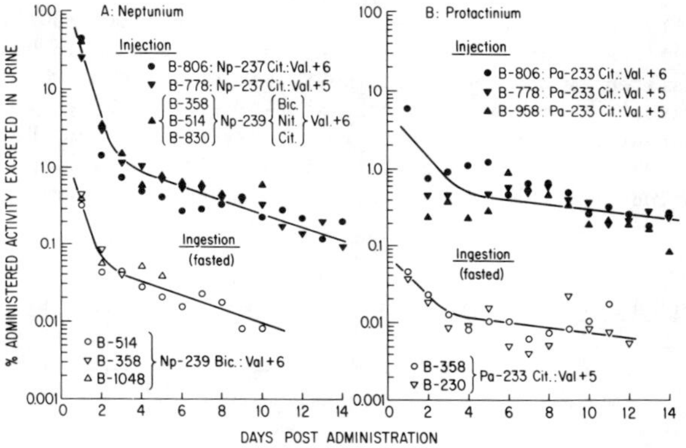

FIGURES 2a and 2b Urinary excretion of Np and Pa.

3.1.3 Whole Body Retention

Figures 3a and 3b show the *in vivo* retention of neptunium and protactinium as a function of time post injection. For the first year, the long-term retention of neptunium *in vivo* was 1.5±0.5 yr associated with 48±2% of the injected dose. Measurement of the skull indicated that neptunium deposited rapidly on bone surfaces and was retained with a single half-time of 1.5±0.2 yr. The long-term retention component of protactinium showed that 65±2% of the Pa-233 injected was retained with a half-time of 3.5±0.6 yr. Deposition and retention of Pa-233 in bone occurred in three phases: the

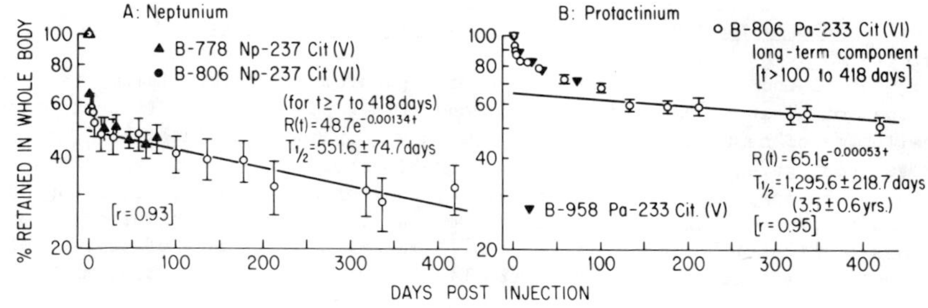

FIGURES 3a and 3b Whole-body retention of Np and Pa.

long-term component had a half-time of 3.0±0.5 yr. A mass balance of the activity excreted in urine and feces with the amount retained *in vivo* was in good agreement for both nuclides.

3.1.4 Tissue Distribution and Retention

Table 2 presents the data on the distribution and retention of neptunium in autopsy tissues acquired at 1 and 90 days post injection. Neptunium deposited initially in the skeleton with less involvement of the liver and kidneys. At 90 days post injection, neptunium retention in bone was still high and consistent with a retention half-time of ~1 year, whereas its retention in liver was low, indicating a half-time of ~100 days. The distribution and retention of protactinium in autopsy tissues has not yet been determined. However, as stated previously, *in vivo* whole body and head measurements indicate that the skeleton may be one of the primary sites of deposition.

TABLE II Distribution and retention of Np in tissues.

	Intravenous injection		Gavage		
Baboon #	B-1048	B-778	B-704	B-1050	B-704
Isotope	Np-239	Np-237	Np-239	Np-239	Np-237
Dose (μCi/kg)	0.09	0.08	12.6	3.6	1.7×10^{-3}
Dose (mg/kg)	4.1×10^{-10}	1.1×10^{-1}	5.4×10^{-8}	1.5×10^{-8}	2.5×10^{-3}
Chem. Form	0.01M Bic.	0.08M Cit.	0.01M Bic.	0.01M Bic.	0.01M Bic.
Oxid. State	VI	V	VI	VI	VI
Sac. (days)	1	90	1	1	33
Fed/Fast	-	-	Fasted	Fasted	"Fasted"
Tissue	Percent of injected activity (% ± SD)		Percent of gavage activity (% ± SD)		
Liver	2.6 ±0.3	0.75±0.03	0.065±0.003	0.037±0.002	0.026±0.001
Kidney	0.5 ±0.03	0.03±0.001	-	0.011±0.004	0.001±0.0004
Other	1.8 ±0.5	0.19±0.03	-	0.007±0.002	-
Skeleton	54±5	42±2	0.45 ±0.025	0.510±0.028	0.150±0.004
Tot. Retained	57±5	43±2	0.515±0.025	0.547±0.029	0.176±0.004
Urine	40±3	53±3	0.376±0.024	0.397±0.025	0.150±0.020
Feces	0.01±0.001	4±1	-	-	-
Tot. Absorbed			0.891±0.035	0.944±0.038	0.33 ±0.020

3.2 Metabolism of Neptunium and Protactinium After Gavage

3.2.1 Whole Blood Retention

Figures 1a and 1b show the retention of neptunium and protactinium in the blood of fasted baboons at selected times post gavage. The retention half-time of neptunium in blood was 9±2 days for the time interval 2 to 7 days post gavage. The percent concentrations of neptunium in the 8 and 24 hour blood samples of fed baboons were 1 to 2 orders of magnitude less than those for fasted animals. No differences were observed in the blood retentions of high (10^{-1} mg/kg) or low (10^{-10} mg/kg) mass amounts of Np(VI) bicarbonate administered. The retention half-time of protactinium in the blood of fasted baboons was 6±1 days for the time interval 2 to 15 days post gavage. At the dose administered, protactinium could not be measured in the blood of a fed baboon, indicating that the GI absorption of Pa-233 in this animal was at least a factor of 10 lower than the absorption occurring in fasted animals.

3.2.2 Urinary and Fecal Excretion Patterns

Figures 2a and 2b show the urinary excretion curves for neptunium and protactinium after gavage. Cumulative excretion of neptunium in the urine of fasted baboons by 1 and 7 days was 0.38±0.09% (n=8) and 0.42±0.07%, respectively. Values for fed baboons were 0.003±0.003% and 0.04±0.02% for the same times. Cumulative excretion of Pa-233 was 0.04±0.01% by day 1 and 0.10±0.01% by day 7. No Pa-233 activity could be determined in the urine of a fed baboon. In general, greater than 95% of the gavaged dose of either neptunium or protactinium was recovered in feces 1 week after administration. Cross-contamination of activity in urine with activity in feces was avoided in most experiments and rarely, if ever, occurred with the first day urine samples.

3.2.3 Systemic Distribution, Retention and Absorption

Data on the distribution, retention and absorption of neptunium after gavage are presented in Table 2. The GI absorption value, i.e., the summation of the amount retained in the liver and skeleton plus the cumultive amount excreted in urine, for neptunium in two fasted baboons was 0.92±0.04%. A GI absorption value of 0.33±0.02% was obtained for a third animal at sacrifice on day 33 and suggested that this baboon may have been only partially "fasted" at the time of administration. The distribution of neptunium in each tissue relative to the total amount absorbed was 52±3% in the skeleton, 6±2% in liver, 1.2±0.4% in kidney, and 42±0.1% eliminated in urine for day 1, and 46±3% in bone, 8±0.6% in liver, 0.3±0.1% in kidney, and 46±7% eliminated in urine by day 33. Determinations of the distribution and retention of neptunium in fed baboons, and of protactinium in fed and fasted animals, are currently in progress.

3.2.4 Estimation of GI Absorption Factors

As a result of the noted equivalence of the metabolism of neptunium or protactinium after iv and oral administrations, a method[11] was used to estimate GI absorption values for both nuclides without necessitating animal sacrifice. This method involved a comparison of activity values measured in 8 hour and 1 day bloods, 1 day urines, and biopsy samples of liver and caudal vertebrae obtained after injection and gavage. Table 3 shows that GI absorption values for neptunium and protactinium calculated by this method.

TABLE III Estimated GI absorption values.

Isotope	Oxid. State	Feeding Regimen	Estimated GI absorption values for Np and Pa[a] (% ± SD)				Mean Absorption Values (% ± SD)
			Bloods	Urine	Liver Biopsy	C. Vertebrae	
Np-239	V,VI	Fasted	2.7 ±0.9 (11)[b]	1.1 ±0.2 (7)	1.7±0.4(1)	-	1.8 ±0.8
Np-237	VI	Fasted	2.5 ±1.8 (4)	1.2 ±0.4 (1)	-	1.5±0.4(1)	1.8 ±0.7
Np-239	VI	Fed	0.07±0.03(2)	0.01±0.002(3)	-	-	0.04±0.05
Pa-233	V	Fasted	0.65±0.30(3)	0.64±0.11 (2)	-	-	0.65±0.01

(a) See text; (b) Number of determinations.

4 DISCUSSION AND CONCLUSIONS

4.1 Metabolism of Neptunium and Protactinium After iv Injection and Gavage

Data reported in this study show the similarity in the metabolism of neptunium, administered either intravenously or orally, as solutions of differing chemical forms and doses. These data also suggest that the fractions of the administered dose of neptunium transferred from blood to the skeleton, liver, and urine will be 0.50, 0.05, 0.01, and 0.45, respectively. The retention half-times for neptunium in bone and liver were calculated to be 1-2 years and ~100 days, respectively, for periods of up to 1 year post administration. The gastrointestinal absorption factors for neptunium were 1×10^{-2} (1%) for fasted baboons administered Np(VI) bicarbonate with 1 ppm chlorine, and 1 to 2 orders of magnitude less for fed baboons. Table 4 presents a comparison of the transfer coefficients and f_1 values for soluble neptunium compounds suggested by this study with those recommended by the ICRP[12,13] and Thompson.[1] Our data is comparable to those listed in this table with the major exception that our half-times for bone and liver are much shorter. Additional studies on the long-term retention of neptunium in animals is indicated.

Data presented in this study also shows the similarity in the metabolism of injected and orally administered protactinium. Although preliminary, it suggests that the transfer coefficient to bone is ~0.65, with a retention half-time of 3.5 yr. The f_1 value for protactinium in fasted baboons was estimated to be ~1%. Current ICRP values for protactinium are 0.40 (100 yr) bone, 0.15 (10 and 60 days) liver and 0.43 in urine, with an f_1 value of 0.1%. Further studies on the metabolism of protactinium are currently in progress.

TABLE IV Suggested metabolic parameters for "soluble" neptunium in Man.

TISSUE	ICRP 1960	ICRP 1980	THOMPSON*	THIS STUDY
SKELETON	0.45 (200 Y)	0.45 (100 Y)	0.60 (100 Y)	0.50 (~1-2 Y)
LIVER	0.05 (150 Y)	0.45 (40 Y)	0.15 (40 Y)	0.05 (~100 D)
KIDNEY	0.03 (175 Y)	-	-	-
EXCRETED	0.47	0.10	0.35	0.45
F_1 VALUE**	1×10^{-4}	1×10^{-2}	1×10^{-3}	1×10^{-2}

* Ref. (1), p. 25; ** GI absorption value.

4.2 Estimation of GI Absorption Values

Mean GI absorption values estimated for neptunium and protactinium using the iv/oral method (Table 4) are comparable to those values reported for these nuclides in this study and in others.[1,14] If valid, this method may be applicable for determining actinide burdens in cases of human exposure.

5 REFERENCES

1. R.C. Thompson, Radiat. Res., 90, 1-32 (1982).
2. B.L. Cohen, Health Phys., 42, 133-143 (1982).
3. P.M. Bryant, in Biological Implications of Radionuclides Released from Nuclear Industries, Vol. I, IAEA, Vienna, 1-25, 1979.
4. N. Cohen and M.E. Wrenn, in Medical Primatology, Part III (Karger, Basel, 1972), pp. 226-236.
5. R.A. Guilmette, N. Cohen and M.E. Wrenn, Radiat. Res., 81, 100-119 (1980).
6. T. Lo Sasso, N. Cohen and M.E. Wrenn, Radiat. Res., 85, 173-183 (1981).
7. J. Lipsztein, N. Cohen, M.E. Wrenn, N.P. Singh and S. Rachidi Marão, Health Phys., 43, 124-125 (1980).
8. C. Sill, Analyt. Chem., 38, 1458-1463 (1966).
9. M.H. Bhattacharyya, R.P. Larsen, R.D. Oldham, E.S. Moretti and M.I. Spaletto, NUREG Document (in press, 1985).
10. N. Cohen and L. Ralston, Radioactivity Studies, New York University Medical Center, Institute of Environmental Medicine, Progress Report to the U.S. Department of Energy, COO-3382-21, June 1983.
11. J.A. DeGrazia, P. Ivanovich, H. Fellows and C. Rich, J. Lab. Clin. Med., 66, 822-829 (1965).
12. ICRP, Report of Committee II on Permissible Dose for Internal Radiation (Pergamon Press, Oxford, 1960).
13. ICRP, Limits for Intake of Radionuclides by Workers, Publication 30 (Pergamon Press, Oxford, 1979).
14. J.D. Harrison and J.W. Stather, Radiat. Res., 88, 47-55 (1981).

This investigation was supported by DOE Contract No. DE-AC02-76EV03382, awarded by the U.S. Department of Energy, and by Center Grants No. ES00260, awarded by the National Institute of Environmental Health Sciences, and No. CA13343, awarded by the National Cancer Institute, DHHS.

INFLUENCE OF CHEMICAL FORM, FEEDING REGIMEN, AND ANIMAL SPECIES ON THE GASTROINTESTINAL ABSORPTION OF PLUTONIUM[a]

M. H. BHATTACHARYYA,* R. P. LARSEN,* N. COHEN,+ L. G. RALSTON,+ R. D. OLDHAM,* E. S. MORETTI,* and L. AYRES+

*Argonne National Laboratory, Argonne, IL, USA
+Institute of Environmental Medicine, NYU Medical Center, Tuxedo, NY, USA

ABSTRACT

We evaluated the effect of chemical form and feeding regimen on the gastrointestinal (GI) absorption of plutonium in adult mice at plutonium concentrations relevant to the establishment of drinking water standards. Mean fractional GI absorption values in fasted adult mice were: Pu(VI) bicarbonate, 15×10^{-4}; Pu(IV) bicarbonate, 20×10^{-4}; Pu(IV) nitrate (pH2), 17×10^{-4}; Pu(IV) citrate, 24×10^{-4}; and Pu(IV) polymer, 3×10^{-4}. Values in fed adult mice were: Pu(VI) bicarbonate, 1.4×10^{-4}; Pu(IV) polymer, 0.3×10^{-4}. Pu(VI) is the oxidation state in chlorinated drinking waters and Pu(IV) is the oxidation state in many untreated natural waters.

To assess the validity of extrapolating data from mice to humans, we also determined the GI absorption of Pu(VI) bicarbonate in adult baboons with a dual-isotope method that does not require animal sacrifice. Fractional GI absorption values obtained by this method were $23 \pm 10 \times 10^{-4}$ for fasted baboons (n=5) and $1.4 \pm 0.9 \times 10^{-4}$ for fed baboons (n=3). We have so far validated this method in one baboon and are currently completing validation in two additional animals.

At low plutonium concentrations, plutonium oxidation state [Pu(VI) vs. Pu(IV)] and administration medium (bicarbonate vs. nitrate vs. citrate) had little effect on the GI absorption of plutonium in mice. Formation of Pu(IV) polymers and animal feeding decreased the GI absorption of plutonium 5- to 10-fold. The GI absorption of Pu(VI) bicarbonate in both fed and fasted adult baboons appeared to be the same as in fed and fasted adult mice, respectively.

---

[a]Work supported by the U.S. Nuclear Regulatory Commission under Fin No. A2218 and the U.S. Department of Energy under Contract No. W-31-109-ENG-38.

1. INTRODUCTION

We are studying the gastrointestinal (GI) absorption of plutonium under conditions relevant to establishing drinking water standards. The primary oxidation state we have studied is Pu(VI), the state present in chlorinated drinking water.[1] The medium is dilute bicarbonate, the major anion in Lake Michigan and many other drinking waters. Plutonium concentrations in administered solutions are low, 1×10^{-10} M, the molar concentration of ^{239}Pu at the maximum permissible concentration for plutonium in drinking water (5 pCi/ml; 0.18 Bq/ml).[2]

The International Commission on Radiological Protection (ICRP) is evaluating data to recommend a GI absorption value for use in establishing standards for oral exposure to plutonium in an environmental setting.[3] Previous ICRP recommendations have pertained to occupational exposures only.[4] Two areas that need to be considered in evaluating GI absorption data for establishing standards are (1) the effect of chemical form on the GI absorption of plutonium and (2) the extrapolation of data obtained in laboratory animals to humans. These two areas are addressed in the study reported here.

Our results compare the GI absorption of Pu(VI) in bicarbonate medium with that of Pu(IV) in various administration media used by other investigators.[5-10] We also report results of a study in progress on the GI absorption of Pu(VI) bicarbonate in both fed and fasted adult baboons. These results will provide insight into extrapolation of rodent data to humans.

2. MATERIALS AND METHODS

2.1 Plutonium Solutions

Four types of solutions of plutonium were prepared as described in detail elsewhere.[11] These solutions were approximately 5×10^{-10} M in plutonium and had the following compositions:

(1) Pu(VI) in 0.01 M $NaHCO_3$, 1 ppm chlorine, pH 8.3

(2) Pu(IV) in 0.01 M $NaHCO_3$, 0.01 M NaI, pH 8.3

(3) Pu(IV) in 0.01 M HNO_3, pH 2

(4) Pu(IV) in 0.17 M citrate buffer, pH 4.5

The plutonium isotope was $^{237/239}Pu$, ^{236}Pu, or ^{238}Pu. The amounts of plutonium in the administered solutions, as well as the percentages in the IV and VI states, were determined by the lanthanum fluoride method.[12] The percentage of ultrafilterable plutonium (% UF) was determined by the procedure of Lindenbaum and Westfall.[13] This test gives a measure of the degree of polymerization of Pu(IV) in solution. A high % UF value indicates a low percentage of polymeric plutonium.

2.2 Plutonium Determinations

Blood, urine, and soft tissues were digested in concentrated HNO_3 prior to analysis for ^{236}Pu or ^{238}Pu. Bone samples were ashed in a muffle furnace at 600°C, dissolved in concentrated HNO_3, and diluted to a known volume with 6 M HNO_3. The amount of ^{236}Pu or ^{238}Pu was determined by alpha spectrometric isotope dilution with ^{242}Pu as the isotopic diluent. The amounts of ^{237}Pu administered and the amounts retained by the mice were determined with a 2-inch-thick, 4-inch-diameter sodium iodide crystal detector by counting the neptunium K x-rays emitted in the decay of ^{237}Pu.

2.3 Experiment Protocols

Effect of chemical form of plutonium. Adult female B6CF1/ANL mice, 80 to 90 days old, were divided into seven groups (A-G) and each mouse was orally administered a particular isotope and solution of plutonium. Mice in groups A-E were fasted for 24 h, from 5 p.m. on Day 0 to 5 p.m. on Day 1; at 9 a.m. on Day 1, during the inactive phase, each mouse was given 0.2 ml of plutonium solution by gavage. Groups F and G received 3 ml of plutonium solution in drinking water tubes overnight (along with food) from 5 p.m. on Day 0 to 9 a.m. on Day 1. Amounts of plutonium administered were approximately 500 Bq ^{237}Pu or 80 Bq ^{236}Pu. Mice were sacrificed on day 6, and the amounts of ^{237}Pu or ^{236}Pu in the liver and the skinned, eviscerated carcass of each mouse were determined. The amount in the whole skinned, eviscerated carcass was calculated from the amount in the headless portion, as described by Bhattacharyya et al.[14]

Effect of animal species. This experiment was a collaborative one between M. Bhattacharyya and R. Larsen of ANL and N. Cohen and L. Ralston of New York University Medical Center. Several isotopes of plutonium, uranium, and neptunium were administered to each baboon. Only the plutonium portion of the study is reported here.

Four adult female baboons (Papio cynocephalus, Papio anubis, and hybrids, 12-15 kg) were housed individually in steel metabolism cages at the Laboratory for Experimental Medicine and Surgery in Primates (LEMSIP) in Tuxedo, New York. The four baboons (B1-B4) were tranquilized with ketamine hydrochloride (0.1 mg/kg, intramuscular) and administered 10 ml of a $^{239}Pu(VI)$-0.01 M $NaHCO_3$ solution by nasogastric intubation at 10 a.m. on Day 0. For baboons B1, B3, and B4, this ^{239}Pu solution was administered 1.5 h after the 8:30 a.m. meal (fed animals). Baboon 2 was deprived of food for 14 h (no morning feeding) at the time of the ^{239}Pu administration to simulate a "baboon without breakfast".

Starting at 10 a.m. on Day 13, all four baboons were deprived of food for 26 h. At 10 a.m. on Day 14 (24 h fast), each baboon was again tranquilized and administered 10 ml of a $^{236}Pu(VI)$-0.01 M $NaHCO_3$ solution by nasogastric intubation. Simultaneously each baboon received 1 ml of a $^{238}Pu(VI)$-0.01 M $NaHCO_3$ solution by intravenous injection into the right femoral vein. The amounts of plutonium administered to each baboon were: ^{239}Pu (orally), 720 to 9000 Bq; ^{236}Pu (orally), 1400 to 7000 Bq; ^{238}Pu (intravenously), 100 to 160 Bq. Samples of blood were collected at several times after each plutonium administration. All urine and feces were collected from the time of the ^{239}Pu administration on Day 0 until Day 45. On Day 18, a urine sample was obtained by catheterization from all but baboon B1. On Day 45, baboons B1, B2, and B4 were sacrificed, while biopsies of liver and caudal vertebrae were removed from B3 without sacrificing the animal.

The amounts of ^{239}Pu, ^{238}Pu, and ^{236}Pu in samples of blood, urine, liver, and caudal vertebrae from all four baboons were determined by alpha spectrometric isotope dilution. GI absorption values were calculated with the dual-isotope method described by DeGrazia et al.[15] for determining the GI absorption of calcium in humans. For each simultaneous administration of ^{236}Pu orally and ^{238}Pu intravenously, four to six values of GI absorption were calculated from (1) the oral/intravenous(iv) activity ratio in individual samples of blood, urine, liver, and caudal vertebrae and (2) the amounts of each isotope administered:

$$\text{fractional GI absorption} = \frac{S236}{S238} \times \frac{A238}{A236}$$

where S236 and S238 are, respectively, the amounts of ^{236}Pu and ^{238}Pu in a sample S of blood, urine, liver, or caudal vertebrae and A238 and A236 are, respectively, the amounts of ^{238}Pu administered intravenously and ^{236}Pu administered orally. For the ^{239}Pu administrations on Day 0, two to three values of GI absorption were calculated from the oral/iv activity ratios ($^{239}Pu/^{238}Pu$) in samples of liver and caudal vertebrae. We validated the dual-isotope method for Baboon B1 by comparing the GI absorption value calculated for the fasted baboon with that directly determined from analysis of the orally administered isotope excreted in urine and that retained by the whole baboon at sacrifice. Further validation of the method by direct analysis of the remaining two sacrificed baboons is in progress.

3. RESULTS

3.1 Effect of Chemical Form of Plutonium

The fractional GI absorption of plutonium in fasted mice was approximately 2×10^{-3} (0.2%) independent of the plutonium oxidation state, Pu(VI) vs. Pu(IV) (Groups A vs. B, Table I) and of the administration medium for Pu(IV), bicarbonate vs. nitric acid vs. citrate (Groups B vs. C vs. D). In contrast, the absorption of polymerized Pu(IV) was five- to sevenfold less than unpolymerized plutonium in both fasted (Groups B vs. E) and fed (Groups F vs. G) mice. The fractional GI absorption of Pu(VI) in fasted mice, 1.5×10^{-3} (Group A), was 10-fold greater than in fed mice, 1.4×10^{-4} (Group F).

3.2 Effect of Animal Species: Mouse vs. Baboon

Individual values for the fractional GI absorption of Pu(VI) in fasted, adult baboons are shown in Table II. The mean value for all animals was 2.3×10^{-3} (0.23%), almost the same as that in fasted mice (Table I, Group A). The mean value of Pu(VI) in fed adult baboons was 1.4×10^{-4} (0.014%), 16-fold lower than in fasted animals. Again the value in fed baboons is essentially the same as that in fed mice (Table I, Group F).

As shown in Table II (fasted animals, last line), the fractional GI absorption of Pu(VI) in one overnight-fasted "baboon without breakfast" was 2.8×10^{-3} (0.28%), 20-fold greater than in fed baboons and similar to that in 24-h fasted baboons.

For Baboon B1, the fractional GI absorption value directly determined for the fasted animal by analysis of ^{236}Pu excreted in urine and that retained in the baboon at sacrifice was 2.0×10^{-4}. This value is nearly identical to the mean value, 2.2×10^{-4}, calculated with the dual-isotope method (Table II).

TABLE I Influence of chemical form on the gastrointestinal absorption of plutonium in fed and fasted mice.

Experiment	Group Designation	Feeding Regimen	Oral Plutonium Solution: % UF	Oxidation State: VI	Oxidation State: IV	Medium	n	Fraction Retained ($x10^4$)[a]: Total	Fraction Retained ($x10^4$)[a]: Liver
Pu(VI)	A	Fasted	85	12	88	0.01 M $NaHCO_3$	20	15 ± 3	3 ± 1
Pu(IV), high % UF	B	Fasted	70	100	0	0.01 M $NaHCO_3$	21	20 ± 2	3 ± 1
Pu(IV), nitric acid	C	Fasted	53	100	0	0.01 M HNO_3	12	17 ± 3	4 ± 1
Pu(IV), citrate buffer	D	Fasted	84	100	0	0.17 M citrate	8	24 ± 5	5 ± 1
Pu(IV), low % UF	E	Fasted	10	97	3	0.01 M $NaHCO_3$	3	3 ± 1	0.9 ± 0.3
Pu(VI)	F	Fed	99	0	100	0.01 M $NaHCO_3$	12	1.4 ± 0.2	0.4 ± 0.1
Pu(IV), low % UF	G	Fed	25	99	1	0.01 M $NaHCO_3$	8	0.31 ± 0.02	0.10 ± 0.01

[a]Values are means ± SE for the number of mice shown, n. Mice were sacrificed 6 days after oral plutonium administration.

4. DISCUSSION

4.1 Effect of Chemical Form of Plutonium

Studies of the behavior of plutonium in fresh water systems have shown that fallout plutonium in solution (that not bound to sediments) is in the form of Pu(V) or of Pu(IV) complexed to humic acids.[16,17] It has also been shown that the chlorination of drinking water results in the oxidation of soluble plutonium to Pu(VI).[1] Our studies have focused on the GI absorption of Pu(VI) because most drinking waters consumed by large populations are chlorinated. We have shown that the fractional GI absorption of Pu(VI) in fasted mice is 15×10^{-4} and in fed mice is 10-fold lower, 1.4×10^{-4} (Table I). These values need to be considered in deriving a standard for exposure to plutonium in drinking water.

To be relevant to the establishment of drinking water standards, studies of plutonium GI absorption must, as a first approach, involve administration of chemical forms of plutonium present in drinking water. However, we need to know if results of studies of the GI absorption of plutonium in chemical forms other than those in drinking water, e.g., Pu(IV) in nitric acid or in citrate buffer, can be used to set drinking water standards. Our results demonstrate that the GI absorption of Pu(IV) is the same as that of Pu(VI) if the administered Pu(IV) is not polymerized (Table I). Under these same conditions, the GI absorption of Pu(IV) is also independent of administration medium: 0.01 M bicarbonate, 0.01 M nitric acid, or 0.17 M citrate buffer (Table I). These conclusions could not have been reached from our data, however, had we not

TABLE II Gastrointestinal absorption of plutonium in fed and fasted adult baboons.

Animal Number	Isotope	Oral Plutonium Solution: Oxidation State IV	VI	% UF	Fractional GI Absorption ($x10^4$)
FASTED ANIMALS[a]					
B1	^{236}Pu	1	99	100	2.2 ± 0.2 (4)[b]
B2	^{236}Pu	20	80	44	11 ± 2 (6)[b]
B3	^{236}Pu	8	92	83	62 ± 10 (5)[b]
B4	^{236}Pu	32	68	74	14 ± 3 (5)[b]
B2	^{239}Pu	7	93	95	28 ± 10 (3)[b]
				Mean ± SE (n)	23 ± 10 (5)[c]
FED ANIMALS					
B1	^{239}Pu	19	81	100	3.3 ± 1.5 (2)[b]
B3	^{239}Pu	66	34	83	0.49 ± 0.03 (2)[b]
B4	^{239}Pu	27	73	97	0.47 ± 0.05 (3)[b]
				Mean ± SE (n)	1.4 ± 0.9 (3)[c]

[a]Baboons were fasted for 24 h prior to the ^{236}Pu administrations. Baboon B2 was fasted for 14 h prior to the ^{239}Pu administration to simulate a "baboon without breakfast."

[b]Value presented is the mean ± SE for the number of calculated GI absorption values, n, shown in parentheses. Calculated values were obtained as described in the text.

[c]Summary value for all fasted or all fed animals is the mean ± SE for the number of animals, n, shown in parentheses.

characterized our solutions of Pu(IV) with respect to polymer formation: the GI absorption of Pu(IV) in fasted mice with one solution was sevenfold greater than with a second solution, the latter being the one containing polymers (10% ultrafilterable, Table I). Analogous results were obtained with fed mice, again because we were not able to reproducibly prepare solutions of Pu(IV) in 0.01 M $NaHCO_3$ medium that were not polymerized. This problem does not arise with solutions of Pu(VI), because this form of plutonium is not subject to polymerization, regardless of the method of preparation or solution conditions.

The mean fractional GI absorption values in fed adult rats reported in the literature for Pu(IV) in 0.01 M nitric acid show an approximate 10-fold range, from a low of 3×10^{-5} obtained in experiments of Weeks et al.[6] to a high of 48×10^{-5} reported by Sullivan et al.[8] This 10-fold range is what might be expected if the lowest GI absorption value were obtained with solutions of Pu(IV) that contained polymers and the highest value with solutions that did not. Support for this explanation comes from the observation that our value for Pu(IV) absorption in fed adult mice when the Pu(IV) was polymerized, 3×10^{-5} (25% UF solution, Table I), is the same as the low value for fed adult rats reported by Weeks et al.[6] The similarity of these values indicates that the Pu(IV) in the 0.01 M HNO_3 solutions of Weeks et al. may have been polymerized to an extent similar to that for the plutonium in our 25% UF solution. In addition, the value obtained by Sullivan et al.[8] for the GI absorption of Pu(VI) is 28×10^{-5}, similar to the high GI absorption value for Pu(IV), 48×10^{-5}.

The possibility also exists that Pu(IV) solutions prepared in citrate buffer, where a complexing anion is present to prevent polymerization, can contain polymers. GI absorption values reported by Weeks et al.[6] for Pu(IV) in citrate buffer also show a 10-fold range, from 3×10^{-4} to 4×10^{-3}.

We conclude from the above observations that the GI absorption values reported in the literature determined for Pu(IV) rather than Pu(VI) must be from studies in which the Pu(IV) solutions were characterized and shown to be low in polymers if they are to be used as primary data for the establishment of drinking water standards. Because most investigators do not analyze their Pu(IV) solutions for polymers, the most reliable GI absorption values for the establishment of drinking water standards come from studies of Pu(VI) absorption, the form of plutonium in chlorinated drinking waters. Information is still needed on the GI absorption of other forms of plutonium in natural waters, e.g., Pu(IV) complexed to humic acids. We are currently conducting studies in this area.

4.2 Effect of Animal Species: Mouse vs. Baboon

We became interested in comparing the GI absorption of plutonium in mice vs. baboons when we observed that the GI absorption values of two other metals, cadmium and lead, were 10-fold lower in our ANL mice than in humans and that GI absorption values for these metals in baboons (for lead) and monkeys (for cadmium) were the same as in humans.[11] Early results reported here from an experiment in progress demonstrate that, unlike those of cadmium and lead, the GI absorption values of Pu(VI) in fed and fasted baboons appear to be the same as those in our fed and fasted ANL mice, respectively. These results, once finalized, will allow us to extrapolate the GI absorption data that we and others have obtained in mice and rats to humans with greater confidence.

Good agreement between the GI absorption value of ^{236}Pu in fasted baboon B1 determined by the dual isotope method, 2.2×10^{-4}, and the value determined by analysis of tissues and excreta, 2.0×10^{-4}, validates in one baboon the dual isotope method for determining the GI absorption of plutonium. Baboons B2 and B4 have also been sacrificed, and samples of tissues and urine are currently being analyzed to provide a GI absorption value by direct assay to be compared with the calculated values reported here. If results are positive, we will have validated a useful non-sacrifice method for determining plutonium GI absorption values based on assay of blood and urine samples only, allowing multiple measurements of GI absorption to be made sequentially in the same animal.

5. CONCLUSIONS

Our results on the GI absorption of Pu(VI) in mice and baboons indicate that GI absorption values that need to be considered in setting drinking water standards are 2×10^{-3} (0.2%) for fasted adults and 2×10^{-4} (0.02%) for fed adults. Additional studies are needed to determine the relevance of GI absorption values in fasted animals to the establishment of drinking water standards. The GI absorption of plutonium complexed to humic acids in drinking water should also be considered. Studies in these areas are currently under way in our laboratory.

6. REFERENCES

1. R. P. Larsen and R. D. Oldham, Science, 201, 1008-1009 (1978).
2. Standards for Radiation Protection, Energy Research and Development Administration Manual, Washington, DC, April 1975, Chapter 0524.
3. R. C. Thompson, Biology Department, Pacific Northwest Laboratory, Richland, WA, personal communication (1984).
4. International Commission on Radiological Protection Publication 30, Limits for Intakes of Radionuclides by Workers, Pergamon, Oxford (1979).
5. J. Katz, H. A. Kornberg, and H. M. Parker, Am. J. Roentgenol. Radium Therapy Nuclear Med., 73, 303-308 (1955).
6. M. H. Weeks, J. Katz, W. D. Oakley, J. E. Ballou, L. A. George, L. K. Bustad, R. C. Thompson and H. A. Kornberg, Radiat. Res. 4, 339-347 (1956).
7. M. F. Sullivan, J. L. Ryan, L. S. Gorham and K. M. McFadden, Radiat. Res., 80, 116-121 (1979).
8. M. F. Sullivan, Health Phys., 38, 173-185 (1980).
9. J. W. Stather, J. D. Harrison, P. Rodwell and A. J. David, Phys. Med. Biol., 24, 396-407 (1979).
10. J. W. Stather, J. D. Harrison, H. Smith, P. Rodwell and A. J. David, Health Phys., 39, 334-338 (1980).
11. M. H. Bhattacharyya, R. P. Larsen, R. D. Oldham, E. S. Moretti and M. I. Spaletto, NUREG/CR-4208, in press (1985).
12. G. H. Coleman, The Radiochemistry of Plutonium, p. 17, Publication NAS-NS-3058, (National Academy of Sciences, National Research Council, Washington, DC, 1965).
13. A. Lindenbaum and W. Westfall, Int. J. Appl. Radiat. Isot., 16, 545-553 (1965).
14. M. H. Bhattacharyya, R. P. Larsen, H. C. Furr, D. P. Peterson, R. D. Oldham, E. S. Moretti and M. I. Spaletto, Health Phys., 48, 207-213 (1985).
15. J. A. DeGrazia, P. Ivanovich, H. Fellows and C. Rich, J. Lab. Clin. Med., 66, 822-829 (1965).
16. M. A. Wahlgren and K. A. Orlandini, pp. 757-774, Migration in the Terrestrial Environment of Long-Lived Radionuclides from the Nuclear Fuel Cycle, Proc. IAEA Symp. IAEA-SM-257-89, Knoxville, TN (1981).
17. D. M. Nelson, W. R. Penrose, J. O. Karttunen and P. Mehlhaff, Environ. Sci. Technol., 19, 127 (1985).

THE CHEMICAL FORMS OF PLUTONIUM IN THE GASTROINTESTINAL TRACT

D. M. TAYLOR, J.R. DUFFIELD and S. A. PROCTOR

Kernforschungszentrum Karlsruhe, Institute for Genetics and Toxicology, Postfach 3640, D-7500 Karlsruhe 1 and University of Heidelberg, D-6900 Heidelberg, F.R.G.

ABSTRACT

In order to study the chemical form in which plutonium, and other actinides may occur in the various parts of the human gastro-intestinal tract, a simple model system has been developed. This system consists of a "mouth", in which reactions with human saliva may be studied, a "gastric" system containing a simulated gastric juice and an alkaline "duodenal" compartment. Centrifugation and gelpermeation chromatography on Sephadex G-50 were used to separate "insoluble/hydrolysed", high molecular weight (protein-bound), and low molecular weight plutonium. Except when the plutonium was present as a complex with DTPA, the proportions of all three components were found to change markedly during the transition from "mouth" to "duodenal" conditions. In the presence of a tea infusion or potato juice over 70% of the plutonium was in insoluble form, but in the presence of orange juice about 70% of the plutonium remained as soluble low molecular weight complexes in the mouth and duodenum. It is concluded that the absorption of plutonium from the gastrointestinal tract is more likely to be influenced by the ligands present at the site of absorption than by the chemical form in which the plutonium enters the mouth.

INTRODUCTION

The extent to which plutonium is absorbed from the human gastrointestinal tract, and the influence of dietary factors and the chemical form of the ingested radionuclide on its subsequent absorption are important questions for the assessment of risk from occupational or environmental exposure.

The mechanism of absorption of plutonium through the intestinal mucosa is not known, neither is there any information about the chemical form and speciation of plutonium, or other actinides, in the various parts of the gastrointestinal tract. The investigation of these parameters in animals is difficult and direct human studies with plutonium are not possible. Thus it was decided to investigate some aspects of the speciation of plutonium in a simple in vitro model of the upper human gastrointestinal tract.

The model used consists of three parts, a "mouth" system in which reactions with fresh human saliva may be studied; a simulated "stomach" system in which the mouth contents react with a synthetic gastric juice and a "duodenal" system in which the gastric contents are neutralised. This report describes preliminary studies with plutonium but the model may also be applied to other actinides such as neptunium.

EXPERIMENTAL

The basic experimental protocol is as follows: Fresh human saliva, 3 to 5 mL, is mixed with a few nCi of plutonium-239 plus any other substance it is desired to study and the pH is checked with a pH meter. After 15 to 30 min incubation a sample of the mixture is centrifuged at 6000g to determine the sedimentable fraction and the supernatant is analysed on a small column of Sephadex G-50, eluted with 0.14M NaCl, this separation yields the proportions of "protein-bound", "low-molecular weight" and "hydrolysed" (retained on the Sephadex) material. The remainder of the "mouth contents" is then mixed with an equal volume of a synthetic gastric juice (0.2g pepsin, 400mg sodium bicarbonate and 16 mMoles HCl in 100mL water) and allowed to stand for 30 min before being analysed in exactly the same way as the mouth contents. The pH of the gastric mixture is kept between pH2 and 3. The residual gastric contents are then brought to pH 8 with solid sodium bicarbonate and allowed to stand for 30 minutes before being analysed in the same way as the other systems. All samples are assayed for radioactivity by liquid scintillation counting.

RESULTS AND DISCUSSION

The results of studies carried out with plutonium nitrate and citrate are shown in Table 1. The citrate concentration used, 25mM, was chosen since it is about one tenth that expected in natural citrus fruit juice, such as orange or lemon juice. The results suggest that quite extensive changes occur in the gastric juice, including dissociation of the low-molecular-weight complexes present in the mouth and the formation of a greater proportion of insoluble species. In the alkaline conditions of the duodenum low molecular weight complexes are reformed but the gel chromatographic profiles suggest that they may not be the same as those originally present in the mouth contents. More detailed studies, using Sephadex G-15 or other gels, are needed to identify the various complexes formed.

To examine the effects of some individual foodstuffs the experiments were repeated with the addition of either a strong infusion of tea or a sample of a commercial orange juice to the saliva-plutonium mixture and analysing the mixture as described above. The proportions of the radioactivity recovered in the various fractions are shown in Figure 1 and the chromatographic profiles found in the mouth, stomach and duodenal systems with the tea mixture are shown in Figure 2.

TABLE 1 The fractionation of plutonium in saliva, simulated gastric juice and duodenal contents.

Fraction	Percent of total plutonium					
	SALIVA pH=7		GASTRIC JUICE pH=2		DUODENAL JUICE pH=8	
	Nitrate	Citrate	Nitrate	Citrate	Nitrate	Citrate
Insoluble (sedimentable) Material	60	29	57	87	11	18
High molecular weight (protein)	0	13	3	5	3	0
Low molecular weight	24	54	36	7	84	81
Hydrolysed species (retained on Sephadex)	16	4	3	0	0	0
Total Recovery	100	100	96	100	98	99

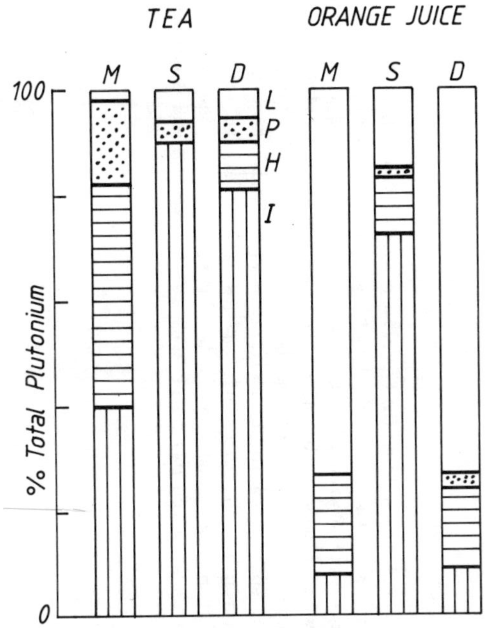

FIGURE 1 The effect of tea or orange juice on the fractionation of plutonium in the simulated mouth (M), stomach (S) and duodenum (D). I=insoluble H=hydrolysed, P=protein-bound, L=low-molecular weight.

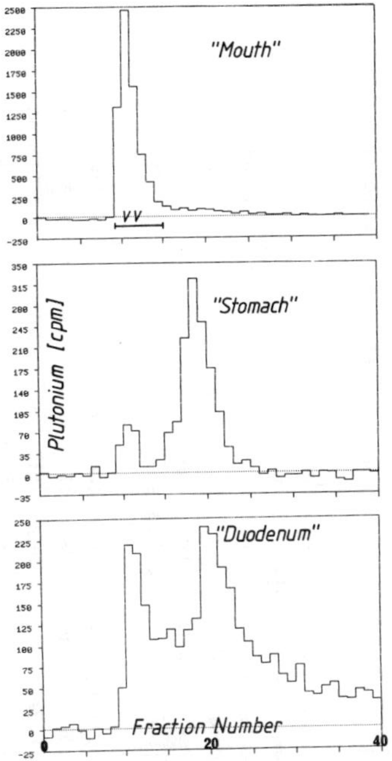

FIGURE 2 Chromatographic profiles of the soluble fractions from the mouth, stomach and duodenum the model system containing a tea infusion. Chromatography on Sephadex G-50 eluted with 0.14 M NaCl. vv=void volume.

The distribution shown in Figure 1 suggest that in the presence of an infusion of tea over 80% of the plutonium is present as insoluble or hydrolysed species in all three compartments. The chromatographic profiles of the soluble fractions from the tea study, which are shown in Figure 2, are markedly different in mouth, stomach and duodenum which lends support to the view that extensive dissociation occurs in the stomach with new, and not necessarily identical complexes being formed or reformed in the duodenum.

In the study with orange juice, there is a dramatic increase in the amount of insoluble/hydrolysable plutonium in the stomach but two thirds of this material was resolubilised, probably by the bicarbonate, in the duodenal compartment. However, the chromatographic profiles are similar for all three compartments showing a single fairly broad peak which elutes before plutonium citrate and which may be a plutonium carbonate complex.

In another study a soluble extract of raw potato was used in place of orange juice, about 50% of the plutonium was found in the soluble/hydrolysed fractions of all three compartments but the chromatographic profiles showed marked differences between mouth, stomach and duodenum.

A further experiment in which the plutonium was present as a DTPA complex (30uMoles DTPA) showed that all the radioactivity remained in the soluble fractions of the mouth, stomach and duodenal compartments and all three chromatographic profiles were similar to that expected from a plutonium-DTPA complex.

Further studies using a more sophisticated model and a wider range of single or mixed food components are planned.

CONCLUSIONS

The general conclusions from this preliminary study using a very simple model system, are that, except when plutonium is present as a chelate which is stable over a wide pH range or, most probably, if it is present as a highly insoluble substance such as an oxide, the chemical speciation undergoes marked changes during the passage from the mouth to the sites of absorption in the small intestine. Thus the chemical form in which plutonium is ingested appears less likely to influence the absorption than the food and other residues, or the endogenous complexing ligands, which may be present in the stomach and duodenum.

THE SPECIATION OF IODINE IN THE ENVIRONMENT

ROBERT A. BULMAN

National Radiological Protection Board, Chilton, Didcot, Oxon OX11 ORQ

ABSTRACT

The speciation of iodine in the environment is discussed under the following topics: (i) sea surface to atmosphere, (ii) chemistry in bulk seawater, (iii) iodine in rocks, (iv) iodine in soils, (v) iodine in plants and (vi) iodine in solidified wastes.

1. Sources of Radioactive Iodine

No other fission product exhibits the same degree of diversity of chemical forms as that exhibited by iodine. Its potential for cycling through the environment is high and in the succeeding pages I will look at the speciation of iodine as it is cycled through the environment. Fortunately the diversity of its chemistry has generated a large volume of literature and lead to its use as a monitor of the redox state by chemical oceanographers.

It is estimated that natural processes such as spontaneous fission of U and spallation reactions of Xe in the upper atmosphere contribute a steady state concentration of 10^{-14} g of ^{129}I per g of ^{127}I[1]. (The supply of ^{129}I formed in the primordial solar system is now exhausted, cf $t_{\frac{1}{2}}$ 16 Ma to age of Earth ca 4.5 aeons). It has been estimated that by the year 2000 the worldwide level of ^{129}I will have increased fivefold as a result of anthropogenic activities[2]. The levels of production of ^{129}I from thermal reactors in the European Communities has been compiled[3].

The geochemistry of iodine has been extensively studied because of the nature of goitre and because fission processes release radioactive isotopes of iodine - altogether 27 isotopes are known of which 9 are produced in the fission cycle but there is only one stable isotope, ^{127}I. The significance of the radiological hazard of radioiodine in the environment has long been appreciated[4]. ^{131}I ($t_{\frac{1}{2}}$ 8.04d) is of major concern in any release of fresh fission products with the air-pasture-cow-milk foodchain recognised as the critical exposure pathway. Because ^{131}I has a short half-life it is easy to regulate its release to the environment but this is not so with ^{129}I, $t_{\frac{1}{2}}$ 16 Ma, which can be expected to contaminate the environment eventually, regardless of whatever initial steps are taken to contain it. Because of this potential for contamination of the environment some understanding of the speciation of iodine in the environment is necessary.

2. CHEMISTRY OF IODINE IN THE ENVIRONMENT

2.1 Sea surface to atmosphere

Iodine is known in oxidation states -1, +1, +3, +5 and +7; the most important environmental ones are 0, -1 and +5. The influence of pH and redox potential upon chemical form is given in Fig. 1[5]. Generally, iodine will be present in aqueous systems as iodide and iodate[6] although smaller amounts may exist as organo-iodine compounds and especially in the surface films of seawater[7]. (Other organoiodine forms are diiodotyrosine, found in some plants[8], and thyroxine, the principal form of iodine in animals).

Extensive studies have shown that over half of the atmospheric iodine is in a gaseous form[9]. Atmospheric iodine is a major component of the global iodine cycle and an atmospheric enrichment process for iodine must occur as the ratio of iodine to chlorine is much greater (typically in the range 100-1000) than the ratio determined in seawater. Obviously a mechanism more complicated than evaporation of sea spray is active. It has been suggested that this enrichment arises from photochemical oxidation of iodide[10]. Although laboratory studies support such a mechanism it has not yet been demonstrated that there is elemental iodine in sea water. Garland and Curtis have demonstrated that ozonised air passed over seawater generates airborne iodine, they propose that a substantial amount of atmospheric iodine may arise in a similar way[11]. In contrast Korzh[11a] suggests uncharacterized iodoorganics may play an essential role in iodine enrichment in aerosols. By employing ^{22}Na and ^{131}I in sea water, he has demonstrated that iodine transfer from sea water exceeded that of sodium by a factor of 1.1×10^2.

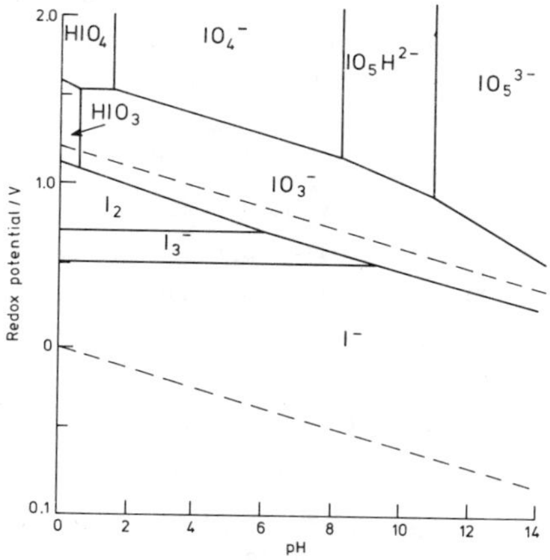

FIGURE 1 Pourbaix diagram for iodine in water at 25^{0}C. Domain of stability of water (---). Reproduced with permission.

Several reports suggest that the biological production of iodomethane represents a major route of transfer from the sea to the atmosphere[12-14]. Iodomethane is the only organic derivative of iodine yet found in the atmosphere although it is possible that other low molecular organic forms may also be present[15]. One study has shown that 50% of the iodine in foam and spray was present in the particulate organic form, in contrast this form accounted for only a few percent in bulk seawater[16]. Interest in the formation of marine aerosols, produced by bubble eruption through surface-active material on the sea surface, has shown that this could lead to the transfer of iodine to the atmosphere if the sea surface microfilm was enriched in iodine[17,18]. Truesdale has considered that there may be interchange of iodine between particulate and gaseous forms such as the free radicals I and IO which may react with ozone, nitric oxide, nitrogen dioxide and carbon monoxide[19]. These reactions could give hypoiodous acid (HOI), a species recently detected in the atmosphere[20] and hydroiodic acid. Analysis of rainwater collected in rural locations of the UK has shown 45% as iodate and 55% as iodide[21]. In other studies it has been shown that in ultrafiltered rainwater (nominal pore diameter <0.45 μm) from a coastal site, iodate represented 48% of the iodine. In contrast, in samples collected 70 km inland this value fell to 5%[22].

Speciation of iodine influences the dry deposition of the element: in its elemental form the deposition velocity is approximately 1 cm/sec[20,23] but for iodomethane the deposition velocity drops to ≈0.01 cm/sec or less[24,25]. For hypoiodous acid the deposition velocity lies between these two extremes[20].

In the fallout from nuclear weapon test programmes most of the radioiodine was found associated with particulate matter[26,27]. In contrast only a small fraction of the ordinary iodine of the atmosphere was found associated with particulate materials.

2.2 Chemistry in Bulk Seawater

The biophilic nature of iodine and the redox potential of iodide-iodate interconversion has prompted exploitation of these phenomena in studies of marine anoxic environments[28,29]. In marine sediments, iodine is initially associated with planktonic or other organic material and is subject to release and diagenetic redistribution during organic matter decomposition[29-31]. In the open oceans, iodate is the predominant species and generally its concentration correlates with either phosphate, nitrate plus nitrite concentrations, or the apparent oxygen utilization in deep water. Iodide profiles characteristically bear an inverse relationship to iodate. Iodide concentration is at a maximum concentration at the sea surface and becomes undetectable below the euphotic zone. It has been suggested that surface water iodide is formed by enzymatic reduction of iodate by nitrate reductase[31]. In anoxic areas iodine exists solely as iodide and this is fully expected from theoretical considerations[32] and finally confirmed for the Cariaco Trench, the Black Sea[28] and the intermittently anoxic Saanich Inlet[33].

2.3 Iodine in Rocks

The chemistry of iodine in rocks is of interest for the obvious reason of predicting the release of iodine from geologic disposal of solidified radioactive wastes. The relatively large ionic radius of iodide (0.22 nm) does not permit it to replace fluoride and hydroxide and consequently its occurrence in minerals is as a minor constituent, possibly as a fluid inclusion[34]. Its biophilic nature points to an association with organic matter when present in sedimentary rocks. So far iodine is known to occur in only a few mineral forms such as marshite (CuI), miersite ([Ag,Cu]I), iodargite (AgI), lautarite (Ca[IO_3] $2H_2O$).

In igneous rocks the distribution of iodine is likely to be determined by factors such as its large ionic radius, its possible occurrence in melts in non-ionic forms[35] and its possible chalcophilic character[36]. The iodine content of intrusive rocks ranges from 140 10^{-9} in granite to over 2000 10^{-9} in hydrothermally altered porphyrite[37]. Of particular note is the high level of iodine in carbonatites. In contrast to the other halogens, iodine displays a uniformity of distribution in intrusive rocks and this would indicate that its distribution is not markedly influenced by chemical composition[38].

In the extrusive rocks there is little variation in the iodine content of the various rock types[37]. Of interest is the high level of iodine reported for some obsidians, natural glasses somewhat similar to granite but with a higher silica content. Natural glasses such as obsidians and tektites have been studied as part of a programme in the prediction of the long term stability of vitrified radioactive wastes[39]. The loss of iodine from volcanic rocks by leaching is less than 20% of the total[38], a figure which is sufficiently high to make containment in rocks difficult. Of course, this high mobility is to be expected for an anion.

2.4 Iodine in Soils

There is little information on the behaviour of iodine during weathering of rocks; generally it might be expected that much of it will be released on weathering of the rocks as rapidly solubilized compounds[36,40]. Evidence to support this hypothesis comes from the dry valleys of Antarctica where sodium iodate has been found along with other products of the weathering of rocks[41]. Presumably iodine was released from rocks and underwent oxidation to the iodate more or less *in situ* in the absence of washout by rainfall. A ready leachability of iodine from tertiary marine sediments has been assumed because of the high iodine contents of rivers draining such areas[42].

The iodine content of soils is much greater than the rocks from which the soil is derived, the typical enrichment factor being 20-30[43]. Several authors have suggested an enrichment via atmospheric sources[41,44]. On the basis of an annual rainfall of 80 cm and an iodine content of 1.5/μg, Whitehead has suggested that wet deposition would contribute 12 g of iodine/ha/year [44]. Läg and Steinnes have shown a correlation of iodine and bromine content of soils of the marine littoral[45].

Soils containing significant amounts of fine material or organic matter are exceptionally rich in iodine. The degree of immobilization in organic soils can be assessed by the general unavailability of peat-bound iodine[43]. The residence time of iodine in the "soil solid" depends upon several factors but it could be at least as long as the life-time of soil humates, about 750 years[46]. Acidic conditions in soils favour the loss of iodine by leaching, whereas carbonate soils tend to act as a barrier to iodine migration. The clay fractions of soils which are noted for their ability to fix cations, fix iodine, a feature of particular note for illite[37].

In soil solutions inorganic iodine will be present principally as iodide but iodate could predominate in arid alkaline soils[47]. Vinogradov reported that around 50% of the iodine in arid soils is present as water soluble iodine[43] but under more normal conditions only a small proportion of iodine in soil is present in water soluble forms[48,49]. In soil solution iodine appears to be in equilibrium with that in the solid phase[48]. The fate of iodomethane in soil is not known but a recent study has indicated that it may act as a potential metal mobilizing agent in anoxic environments[50].

Investigations into the fate of iodine in soils have been extensive[51-54]. Several negatively charged radioiodine species have been detected in aqueous extracts[52]. Investigations into the association of iodine with humates asnd fulvates in the A_1 horizon of sod-podzolic heavy loam soil and leached medium loamy chernozem have shown several processes to be involved[54]. Radioiodine-labelled water-insoluble humates of 21,000 and 7900 daltons were detected in the solid phase of the chernozem. In the sod-podzolic soil the radioiodine was associated with water-insoluble humates of 1800 and 2800 daltons. An uptake of iodine into water-soluble organic-substances of < 5000 daltons was also noted. A biochemically mediated association of iodine with some soil components has been established[53].

2.5 Iodine in Plants

An essential status for iodine for higher plants awaits demonstration. In the strongly reducing conditions of rice paddies excess iodide can cause 'Akagare disease'[54]. Iodide is more readily taken up by plant roots than iodate[49] but most of the iodide taken up by roots is retained in the roots. In some plants iodine ($<0.1\%$) may become incorporated into amino acids such as diiodotyrosine, which was found in the germinating seeds of barley and mung beans[8]; in lettuce seedlings it would appear that iodine remains in an unbound inorganic form[56].

Gaseous iodine is taken up by plants through the stomata and through the leaf surface, about 60% and 40%, respectively for grass[57]. No more than 5% of the iodine entering by stomata is transferred elsewhere in the plant[58]. However, in one study where plants were exposed to artificially high atmospheric levels of iodine and hydrogen iodide, high levels were also found in roots and other tissues[59]. Iodine, if weakly associated with particulate material deposited on plant leaves, will slowly penetrate leaf tissue but, in general, it will stay associated with particulates until dislodged by rainfall or wind.

2.6 Iodine in Solidified Wastes

Disposal options for ^{129}I can be considered primarily as the problem of isolating the radionuclide in a geologically stable form in a material as inert as possible. But underlying these considerations is an understanding of the interaction of species such as iodide and iodate with other inorganic materials. As mentioned above, 20% of the iodide in volcanic rocks is lost by leaching[38] and it is difficult, therefore, to conceive a material which could immobilize ^{129}I once contamination with water has occurred. The difficulties encountered in vitrifying iodides has led to an investigation of the possibility of isolating the radionuclide in cements[60] and halophosphate glasses[61], materials which as yet do not prove completely satisfactory. Scheele et al found silver iodide, followed by iodine-loaded silver mordenite, and the iodates of barium, calcium, mercury and strontium had the best leach resistance from Portland III cement[60]. Coutare and Seitz have investigated the interaction of the anions of iodine with kaolinite, haematite and pelagic red clay in buffered solutions and seawater and suggested that pelagic red clay could be considered as an appropriate host for ^{129}I[62].

3. Conclusions

Although there is a fairly extensive literature on the environmental chemistry of iodine there are still some areas where we need to improve our knowledge about the factors which are involved in its cycling. Perhaps not unexpectedly one of the major gaps in our knowledge is its cycling in the oceans. The speciation events which control its release from the sea surface microfilm are unknown as well as the kinetics of the release. Somewhat better known is the chemistry of iodine in ocean floor muds[37], but again the kinetics of its movement remains unstudied. These are areas where perhaps we should be directing some of our research interests. If iodine is to be associated with pelagic red clays, for instance, we need to know something about the life-time of biogeopolymers on the ocean floor.

4. REFERENCES

1. R R Edwards, Science, 137, 851 (1962).
2. G Haury and W Schikarski, in Global Chemical Cycles and their Alterations by Man, edited by W. Stumm (Proc. Dahlem Conference, Berlin, 1977), pp.165-188.
3. H A C McKay, I F White and P Miquel, Radioactive Waste Management Nuclear Fuel Cycle, 5, 81 (1984).
4. H J Dunster, H Howells and W L Templeton, in Proceedings Second United National International Conference on the Peaceful Uses of Atomic Energy, Geneva 1-13, Sept. 1958.
5. H J M Bowen, Environmental Chemistry of the Elements. (Academic Press, London, 1979), p.26.
6. K Sugawara and K Terada, J. Earth Sci., Nagoya Univ. 5, 81 (1957).
7. R A Duce, Amer. Chem. Soc., Div. Water and Waste Chem., 7, 79 (1967).
8. L Fowden, Physiol. Plant., 12, 657 (1959).
9. R A Duce, J W Winchester and T W van Nahl, J. Geophys. Res., 70, 1775 (1965).
10. Y Miyake and S Tsunogai, J. Geophys. Res., 68, 3989 (1963).
11. J A Garland and H Curtis, Report AERE-PR EMS 7, AERE Harwell, 1979.

11a. V D Korzh, Atmos. Environ. 18, 2707 (1984).

12. J E Lovelock, R J Maggs and R J Wade, Nature, 241, 194 (1976).
13. P S Liss and P G Slater, Nature, 247, 181 (1974).
14. W L Chameides and D D Davis, J. Geophys. Res., 85, 7383 (1980).
15. T E Graedel, Chemical Compounds in the Atmosphere. (Academic Press, New York, 1978), p.440.

16. G A Dean, N.Z. J.Sci., 6, 208 (1963).

17. F Y B Seto and R A Duce, J. Geophys. Res., 77, 5339 (1972).

18. J L Moyers and R A Duce, J. Geophys. Res., 77, 5229 (1970).

19. V W Truesdale, The Transfer of Iodine from Marine Surfaces to the Atmosphere - a Review Report. Institute of Hydrology, Wallingford, Oxon, 1980.

20. P G Voillequé and J H Keller, Health Phys., 40, 91 (1981).

21. S D Jones, Studies on the Speciation of Iodine in Rain and Freshwaters, Ph.D. Thesis, University of Wales, Aberystwyth, 1981.

22. J B Luten, J R W Woittiez, H A Das and C L De Ligny, J. Radioanal. Chem., 43, 175 (1978).

23. F A Tikhomirov and I M Ryzhova, Sov. At. Energy, 51, 457 (1981).

24. D H F Atkins, R C Chadwick and A C Chamberlain, Health Phys., 13, 91 (1967).

25. K Heineman, K J Vogt and L Angeletti, in Physical Behaviour of Radioactive Contaminants in the Atmosphere (IAEA, Vienna, 1974) pp.249-259.

26. A E J Eggleton, D H F Atkins, L B Cousins, Health Phys., 9, 1111 (1963).

27. R W Perkins, Health Phys., 9, 1113 (1963).

28. G T F Wong and P G Brewer, Geochim. Cosmochim. Acta, 41, 151 (1977).

29. W J Ullman and R C Aller, Geochim. Cosmochim. Acta, 47, 1423 (1983).

30. B N Price and S E Colvert, Geochim. Cosmochim. Acta, 41, 1769 (1977).

31. S Tsuonagi and T Sase, Deep-Sea Res., 16, 489 (1969).

32. G T F Wong, Mar. Chem., 9, 13 (1980).

33. S. Emerson, R E Cranston and P S Liss, Deep-Sea Res., 26, 859 (1979).

34. C W Correns, in Physics and Chemistry of the Earth, Vol.1, edited by L H Ahren, K Rankama and S K Runcorn (Pergamon Press, London, 1956).

35. L N Kogarko and L A Gulyayeva, Geochem. Int., 729 (1965).

36. K Rankama and T G Sahama, Geochemistry (Chicago University Press, Chicago, 1950).

37. K H Wedepohl, Handbook of Geochemistry, Vol.II/4, (Springe-Verlag, Berlin, 1978).

38. M Yoshida, K Takahaski, N Yonehara, T Ozawa and I Iwaski, Bull. Chem. Soc. Japan, 44, 1844 (1971).

39. G J McCarthy, Scientific Basis of Nuclear Waste Management (Plenum Press, New York, 1979).

40. V M Goldschmidt, Geochemistry (Oxford University Press, Oxford 1954).

41. G W Gibson, N.Z. J.Geol. Geophys. 5, 361 (1962).

42. G S Konovalou, Dokl. Acad. Sci. SSSR, 129, 912 (1976).

43. A P Vinogradov, The Geochemistry of Rare and Dispersed Chemicals in the Soil, 2nd ed (Consultants Bureau, New York, 1959).

44. D C Whitehead, J. Appl. Ecol., 16, 269 (1979).

45. J Läg and E Steinnes, Geoderma,16, 317 (1976).

46. M Schnitzer and S U Khan, Humic Substances in the Environment, (Marcel Dekker, New York, 1972).

47. G A Fleming, in Applied Soil Trace Elements, edited by B E Davies (John Wiley & Sons, Chichester, 1980), pp.199-234.

48. A Saas, C. R. Seances Acad. Agric. Fr. 67, 161 (1981).

49. D C Whitehead, J.Sci.Food Agric., 24, 547 (1973).

50. J S Thayer, G J Olson, F E Brinckman, Environ. Sci. Technol. 18, 726 (1984).

51. V A Prokhorov, A B Shchukin, O M Kvetnaya and P Sh.Malkovich, Sov. Soil Sci., 10, 663 (1978).

52. F A Tikhomirov, I T Moiseev and T U Ruisina, Agrokhimiya, 11, 115 (1980); Chem.Abstr. 94, 29338.

53. H Behrens, in Environmental Migration of Long Lived Radionuclides, Proc. Symp. Knoxville, 27-31 July 1981, (IAEA, Vienna, 1982) p.27.

54. S U Kasparov, Deposited Document 1982 VINITI, 286-83, 59-81; Chem.Abstr. 100, 137991b.

55. M. Tann, T. Yamomoric, M. Inoe and K. Yuita, Bull Toyamme Agric. Exp. Station 8, 55 (1977).

56. J W Cowan and M Esfahani, Nature, 213, 289 (1967).

57. E H Markee, Report ERL-ARL-29. US National Oceanic and Atmospheric Administration Technical Memorandum.

58. F P Hungate, J F Cline, R L Ukher and A A Selkders, in Biology of Radioiodine, edited by L K Bustad (Pergamon Press, New York, 1964), pp.79-86.

59. W C Hanson, in 'Radioecology', Proc. National Symp, Colorado, edited by V Schultz and A W Kelment (Reinhold, New York, 1961), pp.581-601.

60. R D Scheele, L L Burger and K D Wiemers, in Geochemical Behaviour of Disposed Radioactive Waste, edited by G S Barney, J D Navratil and W W Schultz (Amer. Chem. Soc., Washington DC, 1984), pp.373-387.

61. J P Malagani, A Wasniewski, M Doreau, G Robert and R Mercier, Mat. Res. Bull, 13, 1009 (1978).

62. R A Coutare and G Seitz, Nucl. Chem. Waste Manag. 4, 301 (1983).

SPECIATION OF RADIOIODINE IN AQUATIC AND TERRESTRIAL SYSTEMS UNDER THE INFLUENCE OF BIOGEOCHEMICAL PROCESSES

H. BEHRENS

Gesellschaft für Strahlen- und Umweltforschung mbH München, Institut für Radiohydrometrie, D-8042 Neuherberg, F.R. Germany

ABSTRACT

In aerated surface and soil water iodide undergoes chemical reactions which bind the iodine on dissolved organic compounds. The reactions seem to be instigated by microbial activity. In detail they appear as an iodination of organics effected by extracellular enzymes. The reactions tend to reach equilibria determining the distribution of iodine on different species. In soil/water systems iodine is bound on undissolved organics in the same manner. Humic substances seem to play a considerable role in binding the converted iodine. Iodate is not formed under these oxidizing reactions. It is found to be reduced and thus involved in the above processes. Under reducing conditions no conversion of iodide occurs; rather a retransformation of organically bound iodine to iodide was found in anaerobic systems.

1 INTRODUCTION

The behaviour and the fate of radionuclides in the environment is basically connected with their chemical state on the one hand and with the chemical conditions of the systems on the other. Thus a thorough understanding of chemical processes in environmental systems and resulting speciation of radionuclides is indispensable for the discussion of their behaviour, e.g. in radioecological respect.

This report deals with investigations on the chemical conversion of radioiodine which was added in the form of I^- or IO_3^- to samples of surface fresh water or to soil/water systems. In aerated systems usually transformation of I^- in other chemical forms (except IO_3^-) and strong fixation of the radioiodine by soils was observed while IO_3^- seemed to be included in the processes after an evident reduction to I^-. In contrast to that under strictly anaerobic conditions, I^- proved to be stable.

In the course of the investigations, inorganic reactions including isotope exchange and photochemical or radiochemical conversion were excluded as a cause for the observed reactions. Rather, the evidence showed that microbial activity should be responsible for this behaviour of radioiodine in environmental systems. The following observations among others allow for this conclusion:

- The ability of samples of surface fresh water or soil water to convert I^- is lost if the samples are preheated to boiling before the addition of radioiodide. Fixation of radioiodine, added in the form of I^-, by soils, does not occur if the material is autoclaved prior to the experiment. These observations can be seen in the context of inactivation of microbial activity by heat.
- In some samples of deep groundwater normally beeing inactive to converte I^-, after initial periods of inactivity, up to several weeks or sometimes months, a beginning of iodide conversion was observed. Similarly, iodide conversion in autoclaved surface water could be instigated by incubation with small amounts of untreated surface water, soil or even with normal house dust. These findings indicate that the observed conversion of I^- can be linked to the growth and interaction of microorganisms in the respective water samples.

In the following, a selection of experimental results is given to characterize the conditions of the reactions as well as the speciation of iodine in the systems in question.

2 EXPERIMENTS

2.1 Methods

The distinction between I^- and other radioactive iodine compounds which formed in surface and soil water, was made by coprecipitation of the I^- with AgCl, maintaining an excess of Cl^- over Ag^+. Under this condition considerable amounts of other iodine compounds than I^- remained in the filtrates[1]. Separation and characterisation of other iodine species in solutions was achieved by thin layer chromatography[2], continuous electrophoresis (method by Hannig[3]) and size exclusion chromatography. The sorption behaviour of radioiodine in soil/water systems was investigated in batch shaking tests and in soil column tests.

2.2 Observation of iodide conversion in surface and soil water

When radioiodide is added to surface or soil water a fraction of it becomes non-precipitable with AgCl. After some time equilibria are attained (Figure 1, a) in which generally 15 - 25 % of the iodine remains as I^- or at least in a form which behaves like I^- in the precipitation reaction and in

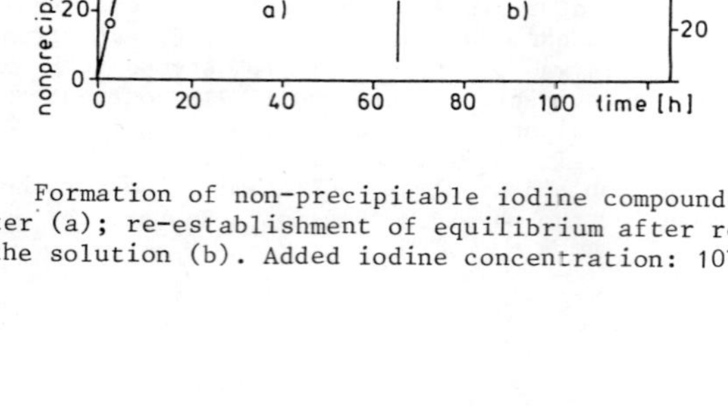

FIGURE 1 Formation of non-precipitable iodine compounds from I^- in river water (a); re-establishment of equilibrium after removal of I^- from the solution (b). Added iodine concentration: 10^{-8} M.

all other tests applied. Transformation of I^- proceeds at the highest observed rates if its concentration is on the order of 10^{-8} M, which is also the concentration of naturally occuring iodine in fresh water (0.5 - 2 ppb). When the concentration of I^- added is higher by one or several orders of magnitude, the reaction rate is lower, however the same equilibrium state is attained (Figure 2). It seems worth noting that, in contrast to the differing reaction rates at different iodine concentration levels the turnover rates of the formation of non-precipitable iodine are all about the same value, indicating a certain capacity of the system for the transformation of I^-. The existence of an equilibrium is shown by removing the I^- fraction from the solution (Figure 1, b); after some time the same state of equilibrium between I^- and converted iodine as before is re-established, indicating the release of I^- out of the converted forms of iodine.

2.3 Effects of water sterilization by filtration and of heating the sterile filtrates

Surface and soil water that were filtered through a 0.2 µm membrane filter showed the same ability of iodide conversion as the untreated samples, sometimes with a moderate reduction of the conversion rates. Storing the filtrates in darkness for several months resulted in a slight reduction of the iodide converting ability merely, which decreased drastically when the samples were exposed to intensive light. When the filtrates were heated for some minutes at elevated temperatures before adding the radioiodide, the activity of the filtrates to converte I^- decreased with increasing temperatures (Figure 3).

These experiments indicate that the substances causing the iodide conversion can occur in the aqueous solutions independently from microorganisms. This fact can be explained with the iodide conversion by extracellular enzymes which can be destroyed or denaturated by light or by heat.

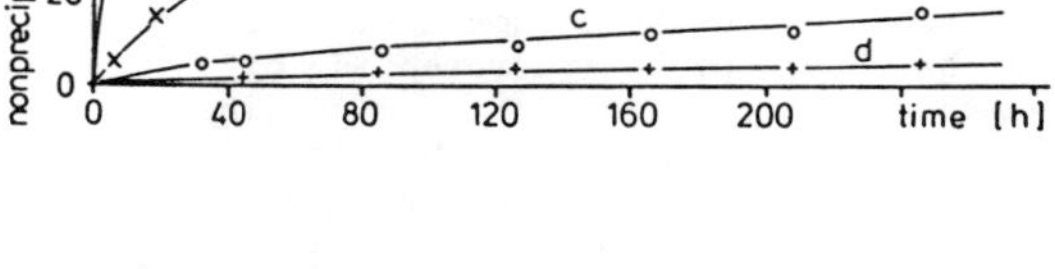

FIGURE 2 Formation of non-precipitable iodine compounds from I^- in river water. Iodine concentrations: a) carrier-free; b) 10^{-7} M; c) 10^{-6} M; d) 10^{-5} M.

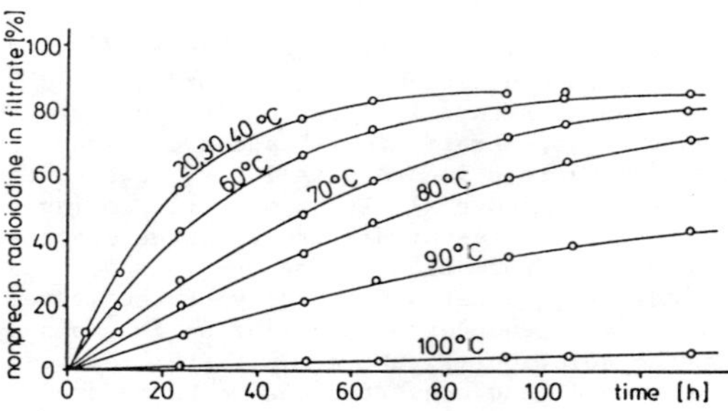

FIGURE 3 Formation of non-precipitable iodine compounds in sterile filtrated river water that has been heated for 2 minutes to the temperatures given in the figure. Added iodine concentration: 10^{-8} M.

2.4 Redox behaviour of the converted iodine

Addition of sulphite to samples of surface or soil water caused a partial retransformation of converted iodine to I^-. As soon as the sulphite disappears from the solution by selfoxidation, the same equilibrium as before will be re-established. Exclusion of air oxygen, e.g. by flushing the water samples with nitrogen gas caused stability of the added I^- which however was converted in the described way after reaeration of the samples. Further information on the redox behaviour was obtained from soil/water batch and column tests (cf. 2.6, 2.7).

2.5 Chromatography and electrophoresis

Thin layer chromatography under simultaneous electrophoretic separation of surface and soil water yielded, beside some unconverted I^-, several anionic radioiodine containing spots which could not be related to known species if iodine[3]. Continuous electrophoretic separations showed the converted iodine to occur in a broad band of anions with different electrophoretic mobilities, lower than that of I^-.

Size exclusion chromatograms of surface and soil water containing converted radioiodine were run under simultaneous detection of radioactivity and UV absorbance (254 nm) in the column effluents. Good correlations were obtained between absorbance peaks (representing organic substances) and radioactivity peaks, indicating that the radioiodine was associated with organic substances (Figure 4). Peaks a) and b) can be attributed to humic substances with molecular weights in the range of 1000 to 10 000.

2.6 Soil/water batch tests

Batch shaking tests were run with a variety of soils and aqueous solutions, using mostly a solid to liquid ratio of 0.25 g/ml. Radioiodine that was added as I^- was generally sorbed by soil materials at relatively fast rates. Simultaneously in the batch solutions a fraction of the radioiodine was converted in the manner described above. After the phase of rapid sorption,

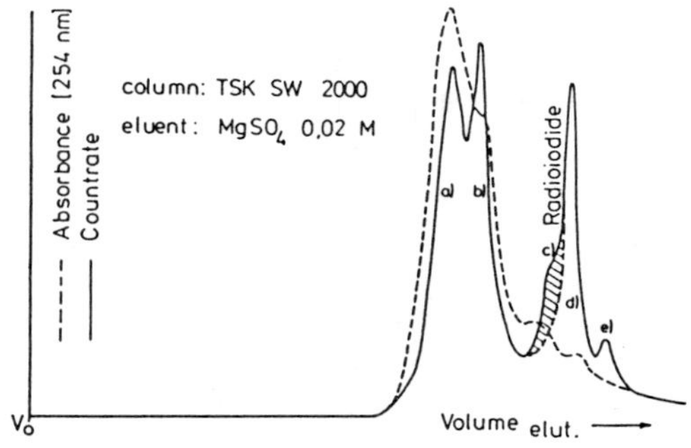

FIGURE 4 Size exclusion chromatography of iodine compounds that have been formed from radioactive I^- in soil water; the hatched area corresponds to I^- in equilibrium with the other iodine compounds.

another phase with a lower rate of sorption follows (Figure 5, curve a) in which a decrease of the converted radioiodine in the solution together with a further decrease of iodide concentration can be observed. Thus finally a good deal of the previously dissolved converted radioiodine is also sorbed by the soils.

Relatively high K_D-values of radioiodine, ranging from several hundred to more than a thousand, were found in batch equilibria with organic soils. In contrast to the experiments with untreated soils, no conversion of I^- in the batch solutions and no corresponding sorption of radioiodine occured in batches with autoclaved soils.

Under anearobic conditions, generated by oxygen exclusion at the beginning or during the experiments, in batch tests a drastic release of iodine out of the converted and sorbed formes under bach-formation of I^- was observed (Figure 5, curves b_1 and b_2). A relation of iodine release to nitrate reduction in anaerobic batches was found (Figure 5, curves c_1 and c_2).

2.7 Soil column tests

In soil column tests, under water saturated as well as unsaturated conditions, generally strong fixation of radioiodine which was applied as I^- occured (example in Table I). Obviously, chemical conversions of I^- as described in the preceeding chapters, will be responsible for that. In contrast, in columns with autoclaved soils, no sorption of radioiodine was observed, leaving the columns in unconverted form as I^- and thus indicating the influence of speciation on sorption and migration processes.

IO_3^- and iodine bound by organics as obtained in transformation of I^- in surface or soil water, showed lower rates of sorption by soils than iodine in the form of I^- did, suggesting involvement of further reactions like reduction of IO_3^- or deiodination of organics in the sorption process.

3 DISCUSSION

The experiments demonstrate that in aquatic and terrestrial environments speciation of iodine should be related to microbial activity. Looked at it more closely, the direct action of microorganisms seems to play a minor role. As can be derived from the processes in the aqueous phase, the reactions appear as an oxidative transformation of iodine into organic bonds, mediated by extracellular enzyme systems. Enzymic iodination processes are well known to occur in biological systems and they are also applied in vitro for production of organic iodine compounds. Such iodinating

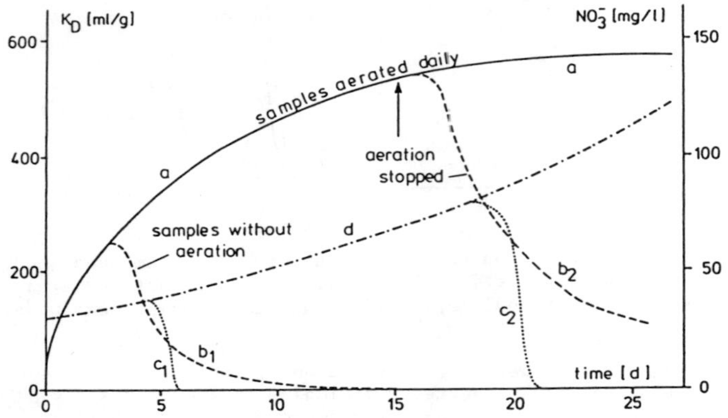

FIGURE 5 Iodine distribution coefficients (K_D) in soil/water batch. a) aerated samples; b_1) non-aerated samples; b_2) aeration stopped after 15 days; d, c_1, c_2) NO_3^- concentrations in samples corresp. to a, b_1, b_2.

TABLE I Distribution of radioiodine in a soil column after 7 days of water percolation under water unsaturated condition; injection of radioiodide into the column inlet on the 3rd day of a 10 days run.

Column length: 200 mm — Flow rate: 10 ml/cm^2·d
Soil type: brown earth

Depth of layer mm	Fraction of total inj. radioiodine %
0 - 5	77.75
5 - 10	18.60
10 - 15	1.78
15 - 20	0.69
20 - 25	0.32
25 - 30	0.19
30 - 35	0.12
35 - 40	0.09
40 - 45	0.06
45 - 50	0.05
50 -100	0.18
100 -200	0.04

enzyme systems consist of peroxidases and components generating H_2O_2, e.g. as a by-product in the oxidation of glucose with help of glucose oxidase. All these prerequisits exist in the environmental systems in question.

The iodination processes appear to be reversible to a large extent. This is suggested by the release of I^- from the organic fractions under anaerobic and other reducing conditions. So far, the actual distribution of iodine among different chemical forms in environmental systems is given by an equilibrium of processes binding iodine in organic materials as well as releasing it continuously. Thus with time, newly introduced radioiodine can mix with the stable resident iodine in the systems.

As a result of the above processes speciation of iodine in surface fresh water and water/soil systems showes I^- as the only inorganic component beside a variety of organic iodine compounds. In the water phase generally about 15 - 20 % of the iodine is found as I^-. In organic soils more than 99.9 % of total iodine is associated with undissolved, obviously organic matter, based on iodine distribution coefficients of several hundreds ml/g, while the fraction of dissolved iodine is made up of I^- and organic iodine compounds according to the above ratio (Figure 6). The organic compounds that bind iodine in the conversion processes are not yet fully identified. However, humic sustances seem to play a considerable role in binding the converted iodine.

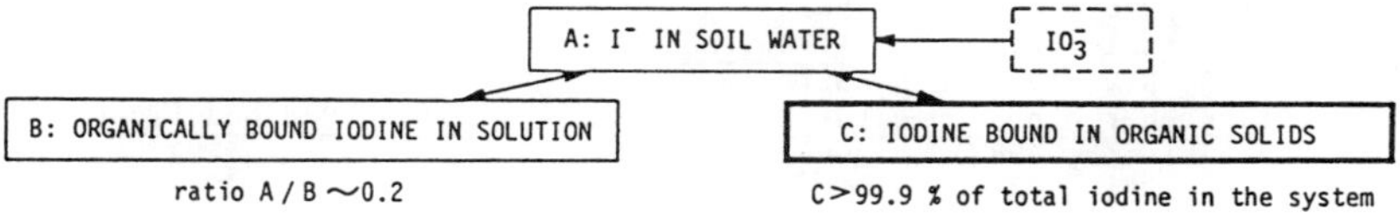

FIGURE 6 Schematic of iodine distribution in soil/water systems.

These considerations are in agreement with most of the results of many investigations on the behaviour of iodine in environments reported thus far. Bonding of iodine to organic material has become evident during the last decades in studies on sorption of iodine by soil through the work of Raja[4], Wildung[5], Whitehead[6-8], Saas[9,10], Szabóra[11], Prokhorov[12], Tikhomorov[13] and others. Reduction of IO_3^- in soils was shown in several studies[7,11,12]. Tikhomorov[13] analysed the distribution of organically bound radioiodine in soils with respect to different types of organics showing humic and fulvic acids as the dominant acceptors. Radioecological observations of a decreasing bioavailability of radioiodine newly introduced into soils made by Saas[9] and Schüttelkopf[14] coincide with the proposed mechanism of a gradual mixing of the radioiodine with the stable iodine content of the soils.

Finally, the above processes and resulting speciation of radioiodine may also be of importance concerning its migration in soils. Complete mixing of added radioiodine and existing stable iodine is a requirement for the effectiveness of a dynamic compartment model[15], suggesting fallout iodine residence times in surface soils of several thousands of years. Residence times on the same order can also be derived from batch distribution coefficients for radioiodine in soils measured in our experiments ranging from several hundreds to more than a thousand ml/g.

4 ACKNOWLEDGMENT

This work was performed with partial support by the Deutsche Forschungsgemeinschaft

5 REFERENCES

1. H. Behrens, Environmental Migration of Long-lived Radionuclides (IAEA Vienna, 1982), p. 27
2. H. Behrens, Z. Anal. Chem., 236, 396 (1968)
3. K. Hannig, Z. Anal. Chem., 181, 244 (1961)
4. M.E. Raja, K.L. Babcock, Soil Sci., 91, 1 (1961)
5. R.E. Wildung, R.C. Routson, R.J. Serne, T.R. Garland, BNLW-1950, PT2, UC-48, 37 (1974)
6. D.C. Whitehead, J. Sci. Fd. Agric., 24, 547 (1973)
7. D.C. Whitehead, J. Soil Sci., 25, 461 (1974)
8. D.C. Whitehead, J. Soil Sci., 29, 88 (1978)
9. A. Saas, Radioprotection, 2, 205 (1976)
10. A. Saas, Séance du 21. Jan. 1981 (Academie d'Agriculture de France, (1981), p. 161
11. T. Szabová, S. Palágyi, Pol'nohospodarstvo, 22, 673 (1973)
12. V.M. Prokhorov, A.B. Shchukin, O.M. Kvetnaja, P.Sh. Malkovich, Pochvovedenye, 11, 33 (1978)
13. F.A. Tikhomorov, S.V. Kasparov, B.S. Prister, V.G. Sal'nikov, Pochvovedenye, 2, 54 (1980)
14. H. Schüttelkopf, 6. Jährl. Kolloquium Proj. Nukleare Sicherheit 28.-29. Nov. 1978 (KFK Karlsruhe)
15. D.C. Kocher, Environmental Migration of Long-lived Radionuclides, (IAEA Vienna, 1982), p. 669

DISCUSSION

BONOTTO:

Does autoclaving affect the organic substances present in water and their capability of binding iodine?

BEHRENS:

Of course autoclaving means a heavy stress to organic materials present in water and soil and will "denature" them to some extent so that a change but no abolition of iodine uptake will result. The full disappearance of iodine binding by organics will thus more reflect full inactivation of microbial and correlated activity.

IODINE-125 AND IODINE-131 IN THE THAMES VALLEY AND OTHER AREAS

J. R. HOWE, M. K. LLOYD AND C. BOWLT*

Ministry of Agriculture, Fisheries and Food, Central Veterinary Laboratory, Weybridge, Surrey, KT15 3NB (United Kingdom); *Department of Radiobiology, Medical College of St. Bartholomew's Hospital, Charterhouse Square, London, EC1M 6BQ (United Kingdom).

ABSTRACT

Part of the Iodine-125 and Iodine-131 waste from hospitals and research centres is discarded down drains and passes through sewage and water reclamation works into the river system. Relatively high concentrations of radioiodine occur in outfalls that discharge into the river Thames, lower levels are found in the mainstream river and less still in the reservoirs and tap water supplies abstracted from the river. The pathway from waste to drinking water could account for the low levels of Iodine-125 found in the thyroid glands of some farm animals and human beings in the Thames valley.

1 INTRODUCTION

Whilst monitoring the ^{129}I ($T_{\frac{1}{2}}$ = 1.6 x 10^7 yr) content of thyroid glands from farm animals at varying distances from nuclear-fuel reprocessing plants it was noticed that low levels of radioiodine were present in some animals far distant from nuclear installations and radionuclide production sources. In particular it was found that thyroids from control sheep housed at the Central Veterinary Laboratories (C.V.L.), Weybridge, Surrey, contained similar activities to those grazing around Windscale, Cumbria, 300 miles to the north of Weybridge. The shorter half-life of the Weybridge samples eventually identified the radioiodine present as ^{125}I ($T_{\frac{1}{2}}$ = 60 d) which, like ^{129}I, emits photons close to 30 keV.

The possibility of self-contamination of specimens was carefully investigated and ultimately eliminated. The origin and environmental pathways of such ^{125}I, and also ^{131}I, found in the Thames valley and elsewhere is the subject of this paper.

2 METHOD

Activities of samples were measured in a calibrated NaI (Tl) well crystal with thin walls which have high signal-to-background ratios for the ~30 keV gamma-photons from ^{125}I but are less efficient for ^{131}I. With sufficient lead screening the limit of detection for counting periods of 10^3 minutes was about 5 mBq (0.14 pCi) for ^{125}I and 80 mBq (2.16 pCi) for ^{131}I.[1]

Small thyroids preserved in formalin were counted whole, but larger glands were macerated and dried before measurement. With plant and water samples the radioiodine was radiochemically extracted and precipitated as silver iodide before counting.

3 TRACING THE SOURCE AND PATHWAY

3.1 Farm Animals

The extent of this pollution was indicated by finding measurable amounts of radioiodine in not only sheep, cows, pigs, goats, horses and rabbits kept at the C.V.L. as well as in some animals from surrounding farms but also in about a third of the glands measured from sheep and cows in the East Midlands and the South-East, i.e. Leicestershire, Nottinghamshire, Derbyshire, Oxfordshire, Berkshire, Kent and Surrey.

3.2 Swans

A major clue in tracing the source of the radioiodine was the almost chance finding that some mute swan thyroids contained more than 100 times as much activity as farm animals. These were from dead birds being measured for lead poisoning. Rather surprisingly in view of its short half-life, some of these swans also contained ^{131}I ($T_{\frac{1}{2}}$ = 8 d). Significantly, out of about 400 swans examined nearly all of the positive specimens derived from the rivers Thames[2] and Trent[3] or their tributaries, with very few from elsewhere.

On the 240 km stretch of the Thames from Lechlade to Putney about 80% of thyroids contained detectable ^{125}I and about 5% contained ^{131}I. For the Trent the values were 70% and 13% for ^{125}I and ^{131}I respectively.

3.3 Water Weeds

The immediate source of the relatively high levels of radioiodine in swans was sought in their food. Swans are largely sub-aqueous feeders, particularly on water plants. Various species were checked and found to contain only very low levels of ^{125}I and ^{131}I except the mixed filamentous algae, an interwoven mass of variable amounts of Cladophora, Spirogyra and other species, known as "blanket weed". Traces of the algae are often found on post-mortem analysis of the contents of swan's gizzards and this is probably the main source of radioiodine in swans.

An important finding was that the higher levels of radioactivity in blanket weedoccurred immediately downstream of some of the sewage treatment outlets emptying into the rivers and their tributaries. Further sampling showed that radioiodine was passing through all the sewage works studied.

3.4 Water

It now seems clear that the ^{125}I and ^{131}I present in the swans and the blanket weed derives from waste from research establishments and hospitals which is disposed of down the drains. Effluent water from several sewage outfalls was analysed and shown to contain ^{125}I and ^{131}I. Not surprisingly the levels varied from place to place and also with time at any one site. For instance, a 7-day bulk sample from the Sandford (Oxford) outfall contained 60 mBq/l of ^{125}I and 35 mBq/l of ^{131}I as iodides, whereas a single spot sample contained 14 mBq/l and 60 mBq/l of ^{125}I and ^{131}I respectively.

The surface water of the River Thames provides a large proportion of the potable water supply to London and its south western approaches. Figure 1 shows the water abstraction points in relation to sewage outfalls. To establish the sort of concentration on radioiodine present in the water used for domestic supply a series of 25 litre samples from the Thames, the water intake lines, filter beds and a tap on the domestic water supply were analysed. Most of the ^{131}I and a few of the ^{125}I concentrations were below the limits of detection but values up to 5 mBq of ^{125}I as iodide per litre were found. Drinking water to Weybridge derives from the Thames via the North Surrey Water Company works and the mean of 5 separate determinations of tap water from here was 1.1 mBq/l of ^{125}I as iodide. It is concluded that this low, but significantly positive, value accounts for the activity present in farm animals in the areas which are supplied with water from the domestic supply. More recently levels of activity up to 5 times this amount have been found in the iodate fraction from tap water.

Water extracted from the Thames is stored for different times in reservoirs (2½ - 100 days) and mixed, in some cases, with varying amounts of bore-hole water before distribution. This is assumed to account for part, if not all, of the regional variation of ^{125}I in animal thyroids.

Fig. 1 Some sewage treatment works and domestic supply water abstraction points on the River Thames and its tributaries

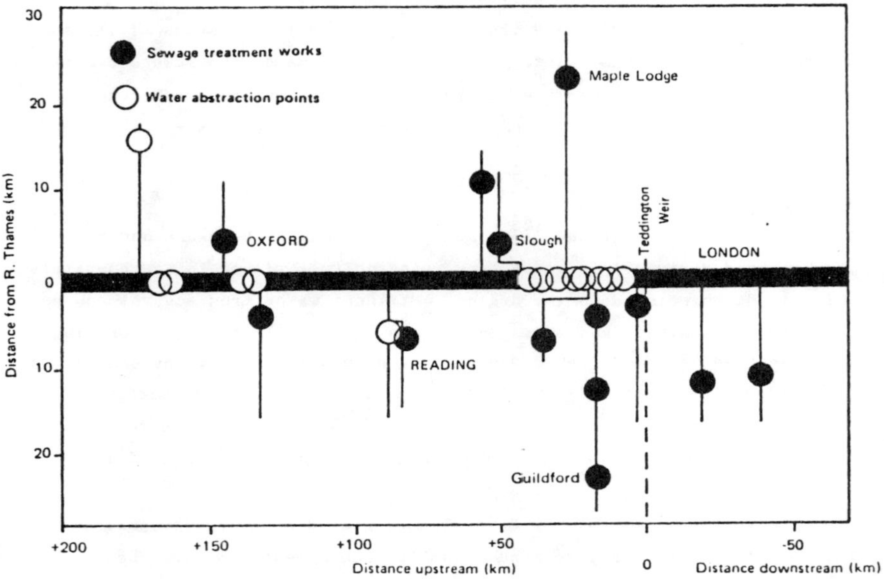

4 HUMAN THYROID LEVELS

If the farm animals derived their radioiodine from domestic drinking water the corollary is that it ought to be present in humans using the same supply. The expected levels are too low to be detected externally in living persons and autopsy samples are necessary.

Thyroids from people who had lived in an area of S.E. London and an area of N.W. London have been analysed. About a third had detectable amounts of ^{125}I varying up to 6 mBq/g (fresh wt). It is known that the contribution of Thames water in the drinking supply to the S.E. London area varies between 50 - 100%.

Thyroids from people who had lived in an area of N.W. Surrey, however, all gave positive results with concentrations up to 9 mBq/g (f.w) and mean of 3.5 mBq/g (f.w). This area is supplied from the Walton abstraction point with a delay time of only 2½ days. The values found in farm animals from the C.V.L. in this area ranged from 4 - 80 mBq/g (f.w) but taken over a different time period to those of the humans.

These levels contribute very small radiation doses to man and animals since they are well below the acceptable limit for members of the public.

5 SPECIATION IN THE ENVIRONMENTAL PATHWAY

The results presented here lead to the conclusion that ^{125}I is being taken up by human thyroids via the sewerage→ river→ drinking water supply route. It is assumed that most of the radioiodine originates in the waste from laboratories as unreacted iodide and iodine incorporated into proteins. As this mixture passes through the reducing conditions of the sewage system the liquid effluent, containing relatively high and variable levels of iodide, is separated from the sludge with little partition of iodine onto the solid phase (circa 6 μgI/g dry wt)[4]. The bacteria used in the treatment may break down some protein iodine into simpler soluble polypeptides and iodoamino acids and this process is likely to continue in the river system. The sewage effluent from towns and cities results in elevated levels of both stable and radioactive iodine in the river[4] although concentrations of either are unlikely to exceed a few micrograms of iodine or a few millibequerels of ^{125}I per litre of water.

Iodate might be expected to appear under the mild oxidising conditions that exist in the cleaner parts of the river and is highly likely to be the dominant inorganic species present after the chlorination process used in preparing potable water. However, concentrations to be measured are at the limits of normal detection methods which makes it difficult to approportion between iodide and iodate. Also there are analytical difficulties with ion-exchange resins because other species, particularly nitrate which may be present at 10,000 times greater concentrations, are more strongly exchanged by the resin.

At the other end of the chain the bacterial breakdown under the intensely reducing conditions of the animal stomach and gut, result in the production of methyl iodide and inorganic iodide prior to absorption of the latter and its incorporation into the thyroid gland. Some preliminary measurements have been made on the different inorganic forms of radioiodine in sewage effluent, river water and drinking water. Activity has been found in both the iodide and iodate form but in variable proportions. No work has been done on the organic iodine in water.

6 REFERENCES

1. J. R. Howe, M. K. Lloyd, A. E. Hunt and F. G. Clegg, Health Phys., 46, 244 (1984).
2. J. R. Howe and M. K. Lloyd, Sci.Tot.Envir. (In Press).
3. J. R. Howe and A. E. Hunt, Sci.Tot.Envir., 35, 387 (1984).
4. D. C. Whitehead, J.App.Ecol., 16, 269 (1979).

ABSORPTION OF HYPOIODOUS ACID BY PLANT LEAVES

J.GUENOT, C.CAPUT AND Y.BELOT
Commissariat à l'Energie Atomique.Institut de Protection et de Sureté Nucléaire
BP 6 - 92260 FONTENAY-AUX-ROSES ;

F. BOURDEAU
Electricité de France, Direction de l'Equipement/S.E./
30, Avenue de Wagram, F 75008 PARIS, FRANCE

ABSTRACT :

Deposition of hypoiodous acid to leaves of sunflower (Helianthus annuus L.) was measured in a laboratory exposure chamber, under well-defined conditions of humidity, temperature and illumination. Transpiration measurements were done using a dew-point hygrometer and were used to deduce stomatal opening. For comparison, deposition of molecular iodine and methyl iodide were also investigated. The results showed that, at relative humidities of 80-95 per cent, the stomatal resistance controlled the rate of absorption of hypoiodous acid and that the cuticular absorption was negligible. The rate of deposition is about ten times smaller than that of molecular iodine and much greater than that of methyl iodide which is very poorly taken up by leaves. Because hypoiodous acid does not deposit on external tissues, as elemental iodine does, it may be inaccessible to removal by rain and may have a longer biological half-life.

1 INTRODUCTION

We use the provisional name of hypoiodous acid (HOI) to design a form of iodine, distinct from elemental iodine and organic iodine, which has been observed in a humid atmospher [1-2-3] and which has the properties of hypoiodous acid [4-5]. Numerous authors, using a selective sampling system [5-6-7], have demonstrated the existence of this form in atmospheric discharges of 131 I under normal conditions [8-9]. Other authors have referred to the possibility of an emission of HOI under accident conditions [10]. There are few data on its transfer between the atmosphere and the vegetation, other than some preliminary data obtained by Voillequé and Keller [11]. Unfortunately, these data were obtained on only a single sample of vegetation under conditions which were not well known, using HOI which may have contained traces of elemental iodine. The purpose of the experiments reported here was to study the transfer of HOI between the atmosphere and plant leaves under well-defined conditions of stomatal resistance of the leaves and relative atmospheric humidity, using carefully purified HOI.

2 EQUIPMENT AND METHODS

2-1 Experimental chamber

The studies were performed on sunflowers (*Helianthus annua L.*) aged between 2 and 3 weeks. These plants were exposed to gaseous HOI in an open system, adapted from Lascève and Couchat [12].The aerial part of the plant is isolated in a 4-litre spherical glass chamber maintained at a constant temperature. The light is provided by a 400W Osram HQI lamp supplying several hundred W/m^2 at the plant level. The chamber is swept by an air flow of 200 litres/hour, at constant temperature and humidity and containing the form of iodine studied. The air brought into contact with the plant is homogenised by means of a microfan. The transpiration of the plant is determined by measuring the difference between the concentrations of water vapour at the inlet and outlet of the chamber by means of a dew-point hygrometer.

2-2 Generation and determination of the iodine compounds:

The labelled HOI is generated using a method adapted from Keller [5] and Kabat [13]. One millilitre of a solution of carrier-free ^{123}I in the iodide form is introduced in a 250 ml flask with 50 µl of 0.1 M iodide solution, 50 µl of 0.02 M iodate solution and 250 µl of 2.5 N acetic acid. The volume is then adjusted to 100 ml with a buffer solution of pH7 and the mixture is heated to 80°C. A slow stream of nitrogen (1 to 15 litres/hour) is bubbled through the resulting solution, cooled to condense excess of water vapor, filtered through ten 50 mesh copper screens - previously etched with hydroiodic acid- and finally mixed with the stream of air feeding the exposure chamber. The copper screens serve to remove the elemental iodine from the nitrogen stream.

The air leaving the exposure chamber passes through a system for selective collection of the various iodine forms. The sampler comprises, from upstream to downstream, one filter to collect particulate iodine, six specially treated copper screens to collect elemental iodine, two layers of a specific HOI sorbent to collect HOI and finally two layers of activated charcoal to collect organic species. Each layer was 4.2 cm diameter and 1.5 thick. The specific HOI absorbent is a commercially available material developed and tested by Kabat [7-14].

2-3 Experimental procedure

After the plant has been exposed for 3 hours, the leaves are cut, their area determined and then counted for their ^{123}I content. The rate of HOI uptake by leaves is defined as the quantity of iodine taken up per unit time and per unit leaf area, divided by the concentration of iodine in the air, it has the dimensions of a velocity.
At the end of the experiment, the leaf conductance for water, which is related to the opening of the stomatal pores on the leaf surface, is also calculated. The conductance for water is equal to the quantity of water transpired per unit time and unit leaf area, divided by the difference between the water vapor concentrations inside and outside the leaf, assuming the air inside the leaf is at 100% r.h.
The experiment described above was repeated several times under different conditions of stomatal opening of the leaves and under realistic conditions of temperature and humidity. Different stomatal opening were obtained by modifying the intensity of illumination.
Additional experiments were carried out to determine the aerodynamic conductance of the boundary layer of the leaves for water vapor. This was done by measuring the evaporation rate of a film of water deposited on the underside of real or artificial leaves. The aerodynamic conductance is equal to the quantity of water evaporated per unit time and unit leaf area divided by the difference between the water vapor concentrations on the surface of the film and in the surrounding air of the experimental chamber. The water vapor concentration on the surface of the film is assumed to be equal to the saturated vapor concentration of air at leaf temperature.

3 RESULTS

The major part of ^{123}I is found on the HOI absorbent. Traces of iodine on copper screens are entirely attributable to a slight HOI sorption onto copper [13]. It can then be assumed that the labelled mixture introduced in the exposure chamber is free of elemental iodine, and contains 98-99% of HOI and 1-2% of organic forms which do not trouble because of their very low uptake by plants leaves [15-16]. The mass concentration of HOI per unit volume of air was approximately the same for all the tests. This concentration, calculated from the quantity of carrier added to the source and from the output of the generator, was about 30 µg/m^3 of air for the majority of the tests and 0,6 µg/m^3 for 2 complementary tests.
Table 1 collates the conditions of the various tests and the results obtained. It will be noted that the humidity of the air was maintained at a fairly high level (71-92%) and the temperature was about 18°C for the first 8 tests and about 25°C for the others. The uptake rate for HOI and the leaf water vapour conductance were calculated as indicated above.
Two attempts to measure the aerodynamic conductance of the leaves for water vapor gave values of 3.9 and 3.4 cm/s. It will be noted that these values represent at least 4 times the leaf conductance for water given in Table 1. The leaf conductance for water measured here is therefore limited not by aerodynamic transport through the boundary layers of the leaves but rather by transport through the leaves surface itself. It is thus possible to assume that the leaf conductance for water given in Table 1 is substantially equal to the conductance of the stomata for the transfer of water vapor through the epidermis of the leaf.

TABLE 1: Summary of the results of measurement of the rate of HOI uptake by sunflower leaves
* HOI concentration 0.6 µg/m3. For the others, 30 µg/m3.

	Date	température (°C)	relative humidity (%)	HOI uptake rate (cm/s)	Conductance H2O (cm/s)
Light	23/5	17.5	72	0.173	0.440
	13/6	17.0	84	0.258	0.793
	28/6	17.5	88	0.167	0.400
	03/7	17.5	76	0.161	0.518
	05/7	18.0	74	0.179	0.484
	11/7	18.7	92	0.116	0.520
	12/7	25.8	71	0.158	0.472
	19/7	25.5	71	0.293	0.982
	24/7	25.5	72	0.227	0.775
	7/11 *	19.5	85	0.190	0.540
	8/11 *	19.0	88	0.458	1.468
obscurité	19/6	16.5	82	0.103	0.265
	21/6	18.0	80	0.056	0.266
	31/7	21.5	83	0.058	0.164

The rate of HOI uptake has been plotted on the graph of Figure 1 as a function of the leaf conductance for water vapor.

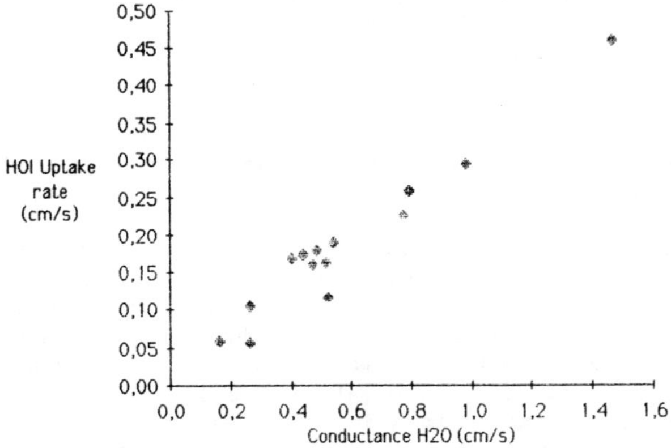

FIGURE 1: The uptake of HOI by sunflower leaves as a function of leaf conductance for water vapour.

These two parameters are linearly correlated with a coefficient of correlation of 0.95. The intercept of the regression line with the ordinate axis, which is equal to 0.013, is small and not significant. The slope of the regression line is equal to 0.298. This correlation indicates that the HOI uptake rate is controlled by the resistance of stomata to HOI diffusion, and that the cuticular adsorption is negligible. The slope value represents the ratio of HOI and H20 diffusion coefficients. Thus HOI diffusion coefficient can be calculated from the values of the slope and of the H20 diffusion coefficient. It is found to be equal to be 0.075 cm2/s.

4 DISCUSSION

The above results should be compared to similar data obtained for the two others species of gaseous iodine, i.e. elemental iodine I2 and methyl iodide ICH3. Guenot [17], then Garland [18] have shown that I2 uptake by bean leaves occured by diffusion through stomata and by sorption onto external tissues of leaf. As relative humidity becomes greater than 60-70%, stomatal conductance no longer has a dominant influence on the uptake rate and the second mechanism can be considered as the chief route of I2 uptake. Authors who have studied ICH3 deposition onto plant leaves reported some very low uptake rate, 3 orders of magnitude smaller than for I2, this result being attributed to the very low reactivity of ICH3 with leaf tissues [15-16].

It can be seen that the way of uptake is very different for each of the above gaseous iodine species. To establish a quantitative comparison, we have measured in the same conditions the uptake rate of elemental iodine and of methyl iodide by sunflower leaves. Table 2 collates the results we have obtained and also the main factor affecting the uptake rate of each species.

TABLE 2: Comparison of the uptake rate by sunflower leaves of the iodine species I2, HOI, ICH3.

Gaseous iodine species	Uptake rate by leaves	Main factor affecting the uptake rate
Elemental iodine	up to 5 cm/s	Relative humidity
Hypiodous acid	up to 0,5 cm/s	Stomatal conductance
Methyl iodide	$0.5.10^{-4}$-2.10^{-4} cm/s	Affinity for leaf tissues

These results, obtained for isolated plants, can be used to evaluate the transfer of iodine species to a plant canopy. The transfer is usually expressed by the deposition velocity Vg, which is equal to the quantity of iodine deposited per unit area of canopy and per unit time, divided by the concentration by volume at a reference height[19]. It should be also expressed by the transfer resistance $r = 1/Vg$. This resistance is equal to the sum of three resistance in series. The first is that of the turbulent layer between the reference height and the top of the canopy. The second is that of the boundary layers which develop in contact with the leaves [20]. The third corresponds to the ability of leaves to trap the iodine species. It depends on the uptake rate measured above and on the leaf area index, defined as the area of leaf per unit area of soil. A calculation of these resistances and of the resulting Vg values was made for a typical herbaceous canopy, using mean values for foliar uptake rate (I_2: 2cm/s, HOI: 0.2cm/s, ICH_3: 1µm/s). This leads to the following Vg values (I_2: 1.3 cm/s, HOI: 0.4 cm/s, ICH_3: 3µm/s). It can be observed that the ratio of I_2 and HOI deposition velocities is much smaller than the ratio of their foliar uptake rates. This is due to the effect of aerodynamic resistances which limit the transfer of I_2 to canopies in spite of its high affinity for foliar surface.

5 CONCLUSIONS

Working with very carefully purified hypoiodous acid, we have shown that this gaseous species was taken up by sunflower leaves only by diffusion through stomata and not by sorption on the external tissues of the leaves as molecular iodine is. In our experiments, deposition of I_2 was ten-times greater that of HOI. But this difference is damped when these results are applied to a plant canopy, taking into account the aerodynamic resistances and there is no large difference between HOI and I_2 behaviour at the canopy level.
A rather more important consequence is that iodine deposited under HOI form may be much less removable by rain than when deposited under I_2 forms, and so, have a longer biological half-life. This point should be verified by further experiments.

6 REFERENCES

1 A. C.Chamberlain, A. E. J. Eggleton, W. J. Megaw, and J. B. Morris, 1963, *J. of Nuclear Eng., Parts A/B*, 17, 519-550.
2 W. E. Browning, R. E. Adams, and J. Wilhem, 1964, *Nuclear Safety Program Semiannual Progress Report for Period Ending June 30, 1964, ORNL-3691*, p.68.
3 I. E. Nakhutin, N. M. Smirnova, G. A. Laushkina, and G. A. Loshakov, 1969, *At. Energy (USSR)*, 26, 390-391.
4 F. O. Cartan, H. R. Beard, F. A. Duce, and J. H. Keller, 1968, *Proc. of the 10th AEC Air Cleaning Conference*, CONF-680821.
5 J. H. Keller, F. A. Duce, D. T. Pence and W. J. Maeck 1970, *Proc. 11th AEC Air Cleaning Conf.* (Edited by M. W. First and J. M. Morgan, Jr.), AEC Rep. CONF-700806, Harvard Air Cleaning Laboratory, Boston.
6 W. A. Emel, D. C. Hetzer, C. A. Pelletier, E. D. Barefoot and J. F Cline, 1977, *Proc. 14th ERDA Air Cleaning Conf.* (Edited by M.W. First) ERDA Rep. CONF-760822, Harvard Air Cleaning Laboratory, Boston.
7 M. J. Kabat 1976, *14th ERDA Nuclear Air Cleaning Conf.*, Sun Valey.
8 C. A. Pelletier, J. E. Cline, E. D. Barefoot, R. T. Hemphill, P. G. Voillequé and W. A. Emel 1978, *EPRI Rep. NP-939*, EPRI Palo Alto.
9 P. G. Voillequé 1979, *EPRI Rep.* NP-1269.
10 M. J. Kabat, 1980, *16th DOE Nuclear Air Cleaning Conf.*, San Diego.
11 P. G. Voillequé and J. H. Keller, 1981, *Health Phys.* 40, 91-94.
12 G. Lascève et Ph. Couchat, 1980, *Ann. agron.*, 31 (3), 273-283.
13 M. J. Kabat, 1983, Personal communication

14 M. J. Kabat, 1983, *Technical Specification S 61*, R. C. Contamination Inc. , 149 Greyabbey Trail, Scarborough, Ontario, Canada M1E 1W2.

15 D.H.F. Atkins, R.C. Chadwick and A.C. Chamberlain, 1967, *Health Phys.* , 13, 91-92

16 Y. Nakamura and Y. Ohmono, 1980, *Health Phys.* 38, 307-314

17 J. Guenot , C. Caput, Y. Belot, F. Bourdeau and L. Angeletti, 1982, *Proceedings of the Joint Radiation Protection Meeting on Radiological Impact of Nuclear Power Plants on Man and the Environment*, Lausanne, September 1981.

18 J. A. Garland and L. C. Cox, 1984, *Atmospheric Environment*, 18 (1), 199-204.

19 A. C. Chamberlain, 1966, *Proc. Roy. Soc.*, A.290, 236-260.

20 A. C. Chamberlain and R. C. Chadwick, 1966, *Tellus*, 18, 226-237.

VOLATILIZATION OF IODINE FROM SOILS AND PLANTS

R. E. WILDUNG, D. A. CATALDO, AND T. R. GARLAND

Pacific Northwest Laboratory
Richland, Washington, USA 99352

ABSTRACT

Elevated levels of ^{129}I, a long-lived fission product, are present in the environment as a result of nuclear weapons testing and fuel reprocessing. To aid in understanding the anomalous behavior of this element, relative to natural I (^{127}I), in the vicinity of nuclear fuel reprocessing plants, preliminary laboratory-growth chamber studies were undertaken to examine the possible formation of volatile inorganic and organic I species in soil and plant systems.

Inorganic ^{129}I added to soil was volatilized from both the soil and plant during plant growth, at average ratios of 2×10^{-3}%/day soil and 9×10^{-3}%/day foliage, respectively. Volatilization rates from soil were an order of magnitude less in the absence of growing roots. Less than 2% of soil or plant volatiles was subsequently retained by plant canopies.

Volatile I, chemically characterized by selective sorption methods, consisted principally of alkyl iodides formed by both soil and plant processes. However, plants and soils containing actively growing roots produced a larger fraction of volatile inorganic I than soil alone.

1 INTRODUCTION

The long-lived isotope ^{129}I is a naturally-occurring radioisotope of I, but elevated levels are present in the environment as a result of nuclear weapons testing and fuels reprocessing.[1] The ratio of ^{129}I to ^{127}I, the primary stable isotope of I, is estimated currently to be on the order of 10^{-14}; and ^{129}I has been suggested as a tracer for natural I in the environment.[2] Iodine concentrations differ markedly in the terrestrial environment depending, in part, on parent geological materials and on weathering and depositional processes. However, surface soils contain generally elevated concentrations of I, compared to their parent rocks; and it has long been hypothesized[3] that much of the I in surface soils is derived from atmospheric sources that, in turn, are replenished from marine waters. Research on the terrestrial behavior of I has thus focused on deposition and surface retention processes.[4,5]

Recent studies have suggested that the behavior of ^{127}I and ^{129}I may differ in the vicinity of nuclear reprocessing facilities. For example, plant concentration ratios (pg/g plant ÷ pg/g soil) for ^{129}I and ^{127}I ranged from 0.5 to 4 and 0.08 to 1.2, respectively, for mature plant species in a semi-arid environment.[6] Direct atmospheric releases from nuclear reprocessing facilities at the study site terminated 10 years prior to measurement. After cessation of direct releases, ^{129}I from fallout (estimated from measured ^{96}Sr and ^{137}Cs levels and relative fission yields) contributed only $2.5 \times 10^{-3}\%$ of the total soil ^{129}I levels. Therefore, elevated ^{129}I concentrations in plants relative to ^{127}I resulted from differences in chemical form and availability to plants of the I isotopes in soils, or from secondary atmospheric sources of ^{129}I.

The volatilization of I added to soil has been demonstrated and postulated to result from the soil formation of I_2 or CH_3I.[7,8] The latter is present in the earth's atmosphere,[9] and has been postulated as a major avenue for transfer of I from the ocean to the atmosphere.[10] Plants have also been shown to transform and volatilize certain elements after root absorption from soil; examples include S, Se and Hg.[11-13] The current study was therefore undertaken to assess these secondary sources of ^{129}I to the atmosphere. The objectives were 1) to examine the possible volatilization of radioiodine from soils and plants during plant growth on soils amended with inorganic radioiodine; and 2) to chemically distinguish the volatile inorganic and organic I forms.

2 MATERIALS AND METHODS

2.1 Plant Uptake of Iodine from Soil

Soils representing a range in physicochemical properties[14] and endogenous ^{129}I concentrations were brought to 67% field capacity, planted with soybean (*Glycine max* cv. Williams), and maintained in HEPA-filtered growth chambers for a period of 40 days. Growth conditions included a 16-hr light cycle (500 $\mu Em^{-2}\ s^{-1}$), a day/night temperature cycle of 26/22°C, and 50% RH. Soil and plant samples were freeze-dried and analyzed for ^{129}I by means of neutron activation.[15]

2.2 Volatility of Iodine from Soils and Plants

The volatility of I was determined in enclosed soil/plant chambers fitted with gas flow trains through which volatiles were drawn by vacuum

(Figure 1). The soil/plant chamber (A) consisted of two half-shells and an intermediate plate that permitted the soil and plant compartments to be physically separated once the stem was sealed to the intermediate plate with plumber's putty. The volatiles produced separately by the soil and plant canopy compartments (A) were drawn into two additional chambers (B,C) to evaluate canopy retention. Each half-chamber had an air inlet and outlet port. The I volatiles were removed from inlet air with silica gel, and the outlet was connected to the gas train. Flow rates were maintained at 30 to 40 mL/min. Soil moisture was maintained by water addition through a third port; additions were based on water usage in a separate control system. The soil, 600 g Ritzville silt loam (20% moisture), was uniformly amended with 16 ng ^{125}I and planted with soybean. The isotope ^{125}I was used because of its ease of detection at masses approximating environmental levels of ^{129}I. After 21 days of plant growth, the soil and plant were fitted into the chamber, and ^{125}I emissions were monitored for 14 days. At the end of this period, the plant canopy was also analyzed for total ^{125}I.

2.3 Forms of Volatile Iodine

The separation and quantification of inorganic and organic I volatiles released from soils and plants were accomplished with the use of sequential sorbents in the gas train. A series of sorption traps were initially examined for their effectiveness in sorption of inorganic ^{125}I and $CH_3{}^{125}I$. Of the materials tested, dry alfalfa tissue and paraffin had little or no affinity for organic I. Activated charcoal and XAD-2 resin were effective in trapping organic I, but their usefulness was markedly reduced by the high moisture levels of chamber air. Columns containing silica gel or soil were most effective in sorbing CH_3-I under

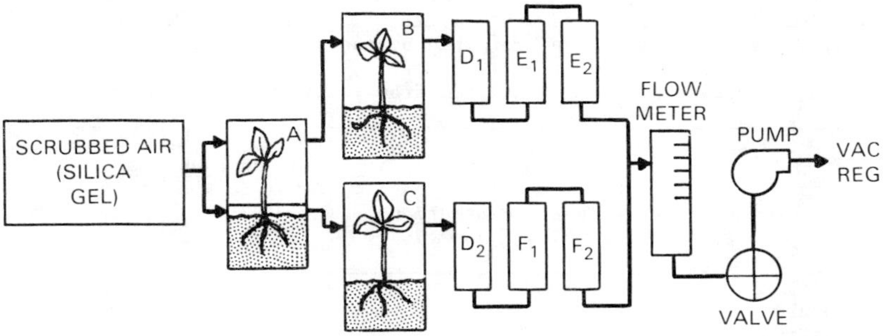

A - PLANT AND SOIL CONTAINING ^{125}I

B - PLANT CANOPY FOR EVALUATING RETENTION OF VOLATILES FROM PLANT

C - PLANT CANOPY FOR EVALUATING RETENTION OF VOLATILES FROM SOIL

$D_{1,2}$ - RED PHOSPHORUS SUSPENSION

$E_{1,2}$, $F_{1,2}$ - SILICA GEL

FIGURE 1 Apparatus for evaluation of iodine volatilization from soils and plants and subsequent plant retention.

high humidity conditions. However, both of them also sorbed inorganic I volatiles such as HI and I_2. To avoid these problems, volatiles produced by plant canopies and soils were first passed through red phosphorus suspension traps (Figure 1; $D_{1,2}$) that effectively removed inorganic volatiles and not alkyl I.[16] THe effluent from this trap was then passed through silica gel to retain organic volatiles (Figure 1, $E_{1,2}$, $F_{1,2}$).

3 RESULTS AND CONCLUSIONS

3.1 Plant Uptake of Iodine from Soil

The soils employed contained from 0.17 to 22.9 pg ^{129}I/g, a 135-fold range in concentration (Table I). Following the uptake period, plant tissues contained 62 to 85 pg ^{129}I/g, a relatively narrow range considering the diversity of soil properties and the range in soil ^{129}I concentrations. However, in five of the nine soils in which soil ^{129}I concentrations were less than 800 pg/g, total ^{129}I present in the plant exceeded that present in the soil. This strongly suggested a gas phase cycling of ^{129}I between treatments, with either the soil or vegetation representing the source of ^{129}I. The ^{129}I level in the growth chamber air was not measured during this experiment.

3.2 Volatility of Iodine from Soils and Plants

The results of investigations to separate the soil and plant sources and canopy retention of I are outlined in Table II. Of the ^{125}I added to soil, 0.028% was volatilized over the 14-day period. Slightly more than 1% of this total was retained by the plant canopy. The plant, which contained 27 pg ^{125}I after growth for 21 days on ^{125}I-amended soil, also participated in I volatilization following placement in the chamber. Although the source plant canopy contained substantially less ^{125}I than the soil, the fraction released was a factor of 20 higher. Of the 0.12% released, 2% was intercepted and retained by the second canopy.

TABLE I Uptake of endogenous ^{129}I from soils by soybeans.

Soil	^{129}I Levels in Soils and Plants			
	Soil Concentration (pg/g)	Plant Concentration (pg/g)	Total Soil (pg)	Total Plant (pg)
Burbank	11.5	80	4,380	687
Esquatzel	22.9	85	9,618	683
Lickskillet	2.16	66	734	902
Palouse	0.18	78	85.4	685
Ritzville	6.5	85	2,939	1,190
Rupert	10.6	62	5,565	621
Umapine	0.17	81	66.3	843
Walla Walla	1.26	77	617	860
Warden	0.74	70	281	683

TABLE II Volatilization of iodine from soils and plants: Retention by a plant canopy (apparatus described in Figure 1).

Components	Total I Concentration (pg)	Fraction of Total[a] (%)
Amended soil	16,000	
Recipient plant canopy	0.057	0.0004
Total volatile	4.45	0.028
Plant grown in amended soil	27	
Recipient plant canopy	0.0007	0.0026
Total volatile	0.033	0.12

(a) After 14 days of plant growth

Analyses of the gas train at 2-day intervals during growth indicated little difference in rate of release. Average volatilization rates amounted to 0.002%/day and 0.009%/day for the soil and the plant canopy, respectively. Thus, it would appear that soils and plants may act as significant sources of volatile radioiodine.

3.3 Forms of Volatile Iodine

The form and rate of loss of I volatile from the soil and plant during plant growth and from soil controls are outlined in Table III. The I that was volatilized from the plant canopy and the soil/root system contained approximately 25% inorganic I, likely HI and/or I_2, and 75% organic I. Iodine that was volatilized from soil incubated without plant roots was markedly different in composition. Only 2 to 3% of the volatile I was inorganic I. Increasing the soil I concentration by a factor of 100 did not affect the distribution between inorganic and

TABLE III Form and rate of loss of volatile iodine.

Source System	Inorganic Species (%)	Organic Species (%)	Rate of Volatilization (pg/kg/day)
Plant canopy alone (27 pg/g)	25	75	2.4
Soil (33 pg/g) with root	29	71	0.5
Soil alone (93 pg/g)	3	97	0.05
Soil alone (10.6 ng/g)	2	98	11.4

organic forms. The dominance of organic species in the volatile fraction from soil requires further investigation. However, the reaction mechanisms may not be unique to soils. Washed quartz sand systems behaved similarly, suggesting a nonbiological surface-mediated reaction. The abiotic formation of alkylated derivatives (CH_3I, C_2H_5I) from inorganic I has been demonstrated,[17] but the source of C (organic residuals in reagents or atmospheric C) and the reaction mechanisms were unclear.

Increasing the soil I concentrations 100-fold (Table III) resulted in a 230-fold increase in volatilization rate, suggesting a saturation of soil sorption capacity over this concentration range. At the lower concentration levels, the rate of I volatilization from soils with roots absent was substantially below that for soils containing roots. This, combined with the fact that the proportion of inorganic forms is higher for shoots and soils where roots are present, suggests different volatilization mechanisms for plant and soil systems. Plant metabolic processes have been previously implicated in the production of volatile inorganic species, and I volatilization may involve systems known to be present in plants for the volatilization of excess S as H_2S.[11] The plants appear to be more effective (per unit mass) in conversion of I to volatile forms than the soils (Table III), although this observation should be tempered by the fact that soil mass markedly exceeds plant biomass in most environments.

4 ACKNOWLEDGMENTS

This work is supported by the U.S. Department of Energy under contract DE-AC06-76RLO 1830.

5 REFERENCES

1. T.P. Kohman and R.R. Edwards, Report No. NYO-3624-1, Carnegie Institute of Technology, Pittsburgh, Pennsylvania, 1966.
2. R.R. Edwards, Science, 137, 851, (1962).
3. V.M. Goldschmidt, Geochemistry, edited by A. Muir, (Clarendon Press, Oxford, 1954).
4. K. Heinemann and K.J. Vogt, Health Physics, 39, 463, (1980).
5. J.A. Garland and L.C. Cox, Atmospheric Environment, 18, 199, (1984).
6. T.R. Garland, D.A. Cataldo, K.M. McFadden, R.G. Schreckhise and R.E. Wildung, Health Physics, 44, 658, (1983).
7. D.C. Whitehead, J. Appl. Ecology, 16, 269, (1979).
8. A. Saas and A. Grauby, Health Physics, 31, 21, (1976).
9. H.B. Singh, L.J. Salas and R.E. Stiles, Environ. Sci. Technol., 16, 872, (1982).
10. P.S. Liss and P.G. Slater, Nature, 247, 181, (1974).
11. H. Rennenberg, Ann. Rev. Plant Physiol., 35, 121, (1984).
12. P. Rosenfeld and O.A. Beath, Selenium in Geobotany, Biochemistry, Toxicity, and Nutrition, (Academic Press, London, 1964).
13. D.D. Gay, EPA-600/3-76-049, NTIS, Springfield, Virginia, 1976.
14. R.E. Wildung, T.R. Garland, K.M. McFadden and C.E. Cowan, in Scientific Seminar on the Behaviour of Technetium in the Environment, Cadarache, France, October 21-26, 1984, (Elsevier, Amsterdam, in press).

15. M.H. Studier, C. Postmus, Jr., J. Mech, R.R. Walters and E.N. Sloth, J. Inorg. Nucl. Chem., 24, 755, (1962).
16. E.P. Samsel and J.A. McHard, Ind. Eng. Chem. Anal. Ed., 14, 750, (1942).
17. I. Miyanaga, A. Kasai and K. Imai, in Environmental Behaviour of Radionuclides Released in the Nuclear Industry, IAEA-SM-172/11, (International Atomic Energy Agency, Vienna, 1973), pp. 157-165.

RADIOISOTOPES SPECIATION AND BIOLOGICAL AVAILABILITY IN FRESHWATER

O.L.J. VANDERBORGHT

Radioisotope Metabolism Laboratory, S.C.K.-C.E.N.
B-2400 Mol, Belgium &
Biology Department, University of Antwerpen, UIA,
B-2610 Wilrijk, Belgium

Abstract

Supposed speciation can increase by one order of magnitude the biological availability of actinides in freshwater, such an increase being exceptional. Raw modeling purposes can usefully be satisfied with the existing data. More precise information is necessary for exact estimations of radioactivity hazards in case of severe local radiocontaminations in specific freshwater bodies.

Continuing effort is needed to bridge the analytical knowledge about speciation with its biological end-effects : a multidisciplinary approach must be emphasized. Information on the metabolic behaviour of actinides (with neptunium and curium) is urgently needed to better understand and forecast the biological cycling of these radioisotopes in freshwater organisms, and thus also their potential hasard to man.

This symposium thus is very useful to provide information on the link between speciation and biological or ecotoxicological end-effects, including the metabolic pathways of different physicochemical forms of a specific radionuclide. Bioaccessability of organisms for nuclides has to be checked as carefully as bioavailability of nuclides.

1. IS SPECIATION IMPORTANT FOR BIOAVAILABILITY OF ENVIRONMENTALLY IMPORTANT RADIOISOTOPES ?

That uptake and assimilation of some drugs and poisons can be strongly influenced by their physicochemical form, is known since long in medicine. The uptake of As can thus differ by a factor of 10^3 ; Co, Cr and Fe will be more available when given as cobalamin, the glucose tolerance factor and heme iron respectively, etc.

The importance of such speciation came into the ecotoxicological field some 15 years ago mainly by the Minamata mercury intoxication study revealing the dramatic effects of methylated Hg as compared to its inorganic forms. A broad interest developed since than into the interaction of speciation and biological effects of environmental contaminants or trace elements. One of the meetings devoted to this subject was the 1981 NATO conference on "Trace element speciation in surface waters and its ecological implications" (edited by G.G. Leppard in 1983 at the Plenum Press). Speciation in ecology and geochemistry received ample attention during the 1979, 1981 and 1983 CEC International Conferences on "Heavy metals in the environment", the CEC-IAEA meeting on speciation in Ispra (1981), and in at least some 100 further papers and public reports on speciation.

The subject seems thus well documented. The number of syntheses of knowledge on speciation is overwhelming, and it can seem redundant to present another one to this audience that is well informed on and well aware of the speciation problems.

Anyhow, scrutinizing all this material revealed that most of the work went towards the analytical aspects of speciation. Motivation for the research is mostly based on the supposed predominant role of speciation in the ultimate integration of the metals, trace elements or pollutants in biological cycles and in the contamination of human population groups. The obtained results demonstrate that such stimulation not only started, but also substantially increased the knowledge of the complexity of interactions to which metals and other (before considered as simple ionic or elemental) pollutants, are subjected once they are released in water, soil or air.

In contrast to this, the experimental confirmation of such direct relationship between biological availability and well-defined speciation is rather exceptional. It could be that this is, once more, due to the difficulty of having a polydisciplinary team-work on the subject establishing the interlinking of physical and chemical speciation, the geochemical cycling and the biological end-effects. It is indeed important to stress such need in the field that is interesting all of us these days ; need that supposes also broad interests of the funding organisms and in which a team of a few scientists will too rapidly restrict itself in its own safe borderlines.

The purpose of this review will thus eventually be to emphasize the yet open need of real experimental evidence of speciation effects on biological availability.

2. TRACE ELEMENT AND RADIOISOTOPE SPECIATION

Research on trace element speciation could give some indications on possible outcomes for and specific differences with speciation studies of radioisotopes and especially of transuranics in freshwater environments. I indeed think that we all agree on the importance of environmental behaviour of "associated" elements if we study radioactive isotopes that are artificial (such as Am and Pu) or rather scarcely present in the natural environment. It became increasingly evident that particular Pu does not exist at 10^{-10} molar concentrations, but that this Pu can be associated with particles of other elements or organic solids or colloids. This fact illustrates both similarity and discrepancies between transuranic speciation and trace element speciation. In trace element studies and even in ecotoxicological research on the most toxic metals such as Cd the lowest concentrations at which chronic effects can occur are in excess of 1 µg/l. Batley[1] rightly states "that there is little point in attempting to measure speciation where the total metal concentrations are one or two orders of magnitude below this". In experiments with high tracer levels of actinides, we are working at still 3 levels of magnitude lower, and in environmental samples we go down by 4 more... . But in radioecological risk estimations, these low levels can be of utmost importance. Batley[1] also points to the inherent impossibilities of chemical-mathematical modeling of the heterogeneous interactions of metal species with the mixed organic and inorganic colloïdal and particular phases of most natural systems.

3. WHAT'S SPECIAL IN FRESHWATER SPECIATION ANALYSIS ?

We are all quite familiar with the difficulties of separating physical species such as "particulate", "colloidal", "dissolved" ; and a number of authors use the galant expression of "operationally defined" analysis of speciation to overcome these difficulties.

Filtration, ultrafiltration (suction or positive N_2-pressure), dialysis (preferentially equilibrium dialysis), electrodialysis, electrophoresis, high voltage electrophoresis, centrifugation, ultracentrifugation, gel filtration, coprecipitation, specific solvent extraction, ion exchange resins, chelating resins, polymeric adsorbents, electrodeposition, ... are applied as well in freshwater speciation analysis as in the other

TABLE I. Some physical parameters of solutes, colloids, and particles likely to exist in freshwater (after Steinnes[2] and Batley[1]).

Chemical form binding the metals	Approximate diameter (nm)	Molecular weight range
Small dissolved compounds		
Hydrated metal ions	< 1	< 200
Inorganic complex		
Small organic molecules	1-4	< 200
Larger dissolved compounds		
Amino acids	2-4	
Fatty acids	2-6	
Fulvic acids	2-6	$200 - 10^4$
Polyhydroxo-complexes	5-10	
Polysilicates	5-10	
Colloidal material		
Humic acids		
Inorganic colloids (metal hydroxides, clay minerals)	10-100	$10^4 - 10^6$
Suspended particulate material		
Inorganic & organic particles	> 100	$> 10^6$

environments (marine, soil, biological fluids...), and other papers in this symposium are devoted in detail to these methodologies.
Some difficulties appear anyhow more severely in freshwater medium, due to its relatively dilute characteristics, its lower buffering capacity, and the relatively higher content of humics supposed to be of a different composition.

This is already true for the most simple approach of speciation analysis, based on the size separation and in which the 0,45 µm pore size filters are giving a first but undecisive splitting in what is sometimes ment to be "soluble" and "particulate" material. A glance to table 1 where some colloidal and particulate materials are given with their approximate diameters, can induce some hesitations about these definitions of (non)-filtrable materials.

The possible loss of highly adsorbing species on or in the filters is often overlooked and can be a serious error when handling freshwater samples that mostly contain less saturating material (Ca^{++}...) than seawater and soil extracts. The most simple and straightforward measure of such filter adsorbtion is the use of two overlaying filters. The underlaying filter contains only adsorbed material, the upper filter adsorbed and particulate radioisotopes. Substraction of both values yields the net particulate material. To allow equal contacts of filtered solution with both filters, a large pore size (glass wool...) material can be put in between both filters, to allow an equal wetting of the two facing sides of the filters. Filtering has to be done either by positive pressure or by gentle suction to prevent significant evaporative deposition of radionuclides on the lower surface of the 2nd filter.

4. EQUILIBRIA DISRUPTED BY THE ANALYSIS

How fragile the equilibria between species can be in freshwater is illustrated by a few examples.

Hydrous iron oxide is known to strongly adsorb a number of radioisotopes of Ru, Am, Pu etc. But hydrous iron oxide itself flocculates at pH 6.6 ; except when humic acid is added. The humic acid maintains the hydrous iron oxide in colloidal form. More generally one can state that organic ligands will stabilise inorganic colloids (Batley[1]). On the other hand, trace metal adsorption on particles will be enhanced in natural waters by complex macromolecular ligands, probably through the changing of the surface orientations of metal binding sites on inorganic (e.g. iron oxide) particles (Davis[3]).

The use of "slow" separation methods induces a supplementary factor of gradual shifting of species during the analysis. This could be the reason for disparate findings on Am-speciation on ion-exchange columns, and of the reluctance to use the slow classical dialysis bags.

The more dilute the medium, the more labile co-existing species could be, and the higher the risk to accumulate artefacts when applying sequentially a variety of separation techniques on the same sample. This general statement is to favour the application of different analyses directly on different samples.

5. ORGANICS IN SOLUTION

The role of humic acids (= HA) in speciation probably is as difficult to analyze as the humic acids themselves. The soil-derived HA differ in origin and structure from the marine ones (Harvey[4]). Most HA complexes are likely to be strongly sorbed on gel permeation material, thus giving a serious drawback to this technique (Steinnes[2]). No agreement exists on how to analyze HA, as most of the techniques for DOC (dissolved organic carbon) that can be used together with speciation analysis yield low values for the HA, due to incomplete destruction. The need to use natural waters to check any results obtained with the "synthetic" ones (Maarten Smies[5]) is especially valid for HA speciations.

Most of the organic complexing capacity is provided by autochtonous plant produced material. Thus 30-50% of all fixed carbon is released as soluble extracellular products in a number of algae (Giesy[6]). In some lakes, much (up to 7 mg/l) of the organic colloidal material is present as adhesive ribbon-like units with a diameter of about 5 nm, and formed by a diversity of plant species. The production of these fibrils is stimulated by phosphor deficiency. The transformation of Mn-ions into manganese oxides is influenced by the fibrils, and Fe added in various forms showed iron oxides associated to the fibrils (Leppard G.G.[7]). These observations point to the possible role of such fibril-like material in the binding of e.g. actinides, but nothing is in fact known about their effect of speciations on bioavailability, even for trace elements.

The same holds anyhow for a good deal of the research on speciation as well of trace elements as of environmental radioisotopes. A high degree of sophistication is developed in the analysis of speciation ; but some of the conclusions of the authors about the relevance of speciation analysis for practical and pragmatic bioavailability predictions or estimations are relevant (capitals are our own) : "Increasingly, one is impressed by the PROBABLE important roles of solids in binding and redistributing trace elements" (Leppard[7]) ; "The natural change of physicochemical conditions MUST HAVE consequences for trace element speciations and associated biological processes" (Maarten Smies[5]) ; "Analytically-defined speciation of the dissolved state is only a PARTIAL ANSWER to the interactions with organisms... . The voltametric measurement ... will have RELEVANCE IN AN INDIRECT MANNER for biologists" (Nürnberg[8]) ; " .. only VERY SOPHISTICATED TECHNIQUES can enable us to reach some conclusive statements on the mechanism of trace element absorption in aquatic biota. Nevertheless, this is NOT WHAT THE ECOLOGIST REALLY NEEDS ..., he needs ... a simple analytical tool to check... the potential hazard of a chemical in an ecosystem" (Baudo[9]) ; "... sediments may reduce bioavailability or be a source of bioavailable heavy metals. Chemical analysis of the abiotic pollutant reservoirs DOES NOT LEAD TO AN ADEQUATE ENVIRONMENTAL RISK ASSESSMENT ..." (Marquenie[10]).

6. SPECIATION AND BIOAVAILABILITY OF ACTINIDES

Freshwater has received less attention as a potentially contaminated environment in actinide radioecology because most of such waste output from reprocessing goes to sea. The inland installation of reprocessing plants and shallow land burial of slightly α-contaminated radioactive wastes could be a more direct source, disposal in geological layers a rather indirect source of actinides in freshwater ecosystems.

Their relatively elevated content of humic acids, directed attention to these acids in freshwater actinide research, and a discussion on this point was made by Vangenechten[11]. The actual effects of humic acids on actinide behaviour has been questionned in estuarine sediments (Bulman[12]). In alkaline freshwater ponds, the organic carbon cycling was suggested to be the primary mechanism in Am- and Cm-cycling. Pu(V) or a mixture of Pu(V,VI) was indicated to be the prevalent oxidation states in solution (Trabalka[3]). The Pu-EDTA complexes are recognised as being of high stability and high migrability in subsurface runoff ; whereas polar organic compounds such as palmitic and stearic acids could be involved in ^{90}Sr and ^{137}Cs migration (Kirby[14]).

Charge-form speciation analysed by ion-exchange resins agreed well with the MINTEQ geochemical model calculations for those elements that do not form organic complexes in groundwater (^{54}Mn, ^{144}Ce, ^{131}I, ^{24}Na, ^{137}Cs, ^{99}Mo, ^{99}Tc, ^{151}Pm, ^{239}Np). The complexation by organics gave rise to discrepancies with the speciation model for ^{65}Zn, ^{60}Co, ^{59}Fe and possibly ^{51}Cr (Jenne[15]). As ^{59}Fe and ^{241}Am show very similar adsorption and particle formation patterns in a very extended range of freshwater types (Vangenechten J.H.D., personal communication), this discrepancy could be extrapolated to Am. The straightforward paper of Nelson[16] on Pu speciation, concludes i.a. that the reduced Pu (III,IV) prevails over the oxidized ones (V,VI) in waters with high dissolved organic carbon (more than a few $mg.l^{-1}$). Low particle densities favour also the reduced forms. Adsorption to natural particles is mainly by these reduced forms. The K_D of Pu is lower in waters with higher colloidal organic carbon content.

As it was the case for most research on trace element speciation, these recent papers on actinide speciation spent most of their effort on the physical and chemical analysis of the actinides, and less or no <u>simultaneous</u> effort went to bioavailability research. Supplementary <u>difficulties</u> induced by the extreme variability of freshwater bodies in contrast to the marine environment, can be partly responsible for this one-sided approach. The available data on biological fixation of Pu and Am partly reflect this variability ; appendix 1 compiled by J. Bierkens illustrates this with about 100 data on concentration factors in fresh-water organisms for Pu and Am. Here also is a lack of simultaneous effort in the papers to relate these data to speciation.

Adsorption of actinides on the exoskeleton makes up one of the main routes of entry to the macrofauna ; the effects of speciation on such adsorption was studied by J. Bierkens[17]. He concludes that humic acids and still more a trivalent metal (Al) decrease Am-adsorption. The desorption rate of Am is also increased by Al. The formation of a bidentate organic complex is postulated, bridging between the Am in solution and the exoskeleton. Thiels[18] could induce a tenfold higher uptake of Am in a freshwater gastropod in acidic oligotrophic waters compared to mesotrophic and slightly basic ones. In conditions nearer to the natural ones met in ten quite different European freshwaters, she obtained only a two-fold difference in Am-fixation. It was the highest in waters with the lowest Na and K-content.

The gastric uptake can also be increased by an order of magnitude when Am is intubated together with an acid solution of humates in a freshwater crayfish (Bierkens, this symposium).

Microbial activity is quite often claimed to be a potential factor for actinide mobilisation from (fresh)-water sediments, but in sandy sea- and freshwater sediments there was no such solubilising action seen neither in the K_D nor in the bioavailability of Am (Vanderborght[19]).

Experimental freshwater ecotoxicology on speciation of neptunium and curium is scarce or lacking, we feel it necessary to include these radio-isotopes in any research involving the very-long-term risk assessments.

Emphasis must be laid on physiological side-effects when bioavailability is tested in rather severe conditions such as extreme pH and concentration ranges, organic colloids etc. The final effects can be due by speciation of the nuclides,but also to a change of the "BIOACCESSABILITY" of the living organism. Thus, the rapid and massive secretion of mucus by a number of animals in such conditions can indeed induce a different set of adsorption and absorption characteristics of the organisms without necessarily affecting the isotope speciation.

7. REFERENCES

1. G.E. Batley, Current status of trace element speciation studies in natural waters. In : Leppard[20], p. 17-36, (1983).
2. E. Steinnes, Physical separation techniques in trace element speciation studies. In : Leppard[20], p. 37-47, (1983).
3. J.A. Davis and J.O. Leckie, Effect of adsorbed complexing ligands on trace metal uptake by hydrous oxides. Environ. Sci. Technol., 12, 1309, (1978).
4. G.R. Harvey, D.A. Boran, L.A. Chesal and J.M. Tokar, Structures of marine fulvic and humic acids. Mar. Chem., 12, 119-132, (1983).
5. M. Smies, Biological aspects of trace element speciation in the aquatic environment. In : Leppard[20], p. 177-193, (1983).
6. J.P. Giesy, Biological control of trace metal equilibria in surface waters. In : Leppard[20], pp. 195-210, (1983a).
7. G.G. Leppard and B. Kent Burnison, Bioavailability, trace element associations with colloïds and an emerging interest in colloidal organic fibrils. In : Leppard[20], p. 105-122, (1983b).
8. H.W. Nürnberg, Voltametric studies on trace metal speciation in natural waters, part II ; In : Leppard[20], p. 211-230.
9. R. Baudo, Is analytically-defined chemical speciation the answer we need to understand trace element transfer along a trophic chain ? In : Leppard[20], p. 275-290, (1983).
10. J.M. Marquenie, W.C. De Kock and P.M. Dinneen, Bioavailability of heavy metals in sediments. CEC int. conf. Heavy metals in the environment, Heidelberg, p. 944-947, (1983).
11. J.H.D. Vangenechten, S. Van Puymbroeck, J. Bierkens, G. Van Keer and O.L.J. Vanderborght, Speciation of ^{241}Am in freshwater systems : effects of inorganic and organic substances. In : The environmental transfer to man of radionuclides released from nuclear installations, Proc. CEC-seminar Brussels 17-21 October 1983, V/7400/84, p. 173-192 (Luxembourg).
12. R.A. Bulman and A.L. Reed, Association of Pu and Am with bio-geopolymers from Ravenglass silts. In : Int. CEC Conf. Heavy metals in the environment, Heidelberg, p. 1235-1237, (1983).
13. J.R. Trabalka, M.A. Bogle and T.G. Scott, Actinide behaviour in a freshwater pound. CONF-8311110-6.
14. L.J. Kirby and W.H. Rickard, Chemical species of migrating radionuclides at commercial shallow land burial sites. PNL-4432-9, (1984).
15. E.A. Jenne, C.E. Cowan and D.E. Robertson, A comparison of analytical charge-form and equilibrium thermodynamic speciation of certain radionuclides. PNL-SA-12093, p. 10, (1984).
16. D.M. Nelson, R.P. Larsen and W.R. Penrose, Chemical speciation of Pu in natural waters. CONF-8311110-5, p. 40, (1983).
17. J. Bierkens, J. Vangenechten, S. Van Puymbroeck and O. Vanderborght, Influence of humic acids and Al on the fixation of ^{241}Am on the exoskeleton of Astacus leptodactylus. In : The environmental transfer to man of radionuclides. Proc. CEC-seminar Brussels 17-21 October, 1983, V/7400/84, Luxembourg, p. 307-320, (1984).
18. G. Thiels, Biological availability of ^{241}Am for a freshwater

gastropod, related to the water physicochemistry. Ph.D. thesis, Univ. of Antwerpen, U.I.A ; Biology Dept., B-2610 Wilrijk-Antwerpen (Belg.).

19. O.L.J. Vanderborght, S. Van Puymbroeck, J. Gerits, J. Vangenechten, J. Bierkens, Microbial sterility and salinity : effects on ^{241}Am distribution and transfer coefficients in sandy marine and freshwater sediments. Behaviour of long-lived radionuclides in the marine environment, CEC, eds. Cigna & Myttenaere, p. 317-325, (1984).

20. G.G. Leppard, Editor of Trace element speciation in surface waters, NATO conference 1981, Plenum Prees, N.Y. & London, ISBN 0.306.41269.1, (1983).

8. APPENDIX 1

CONCENTRATION FACTORS (= CF) FOR Am & Pu IN FRESH WATER ORGANISMS.

J.G.E. BIERKENS - University of Antwerp & SCK Mol

Bacteria : (M.H. Smith et al., 1976)

Aeromonas hydrophyla — CFAm after 72h = 3.2×10^2

Algae : (Emery et al., 1974 ; Emery & Klopfer, 1976)

Tetraspora	CF 239,240Pu = 8000	CFAm = 8900
Cladophora	CF 239,240Pu = 2600	CFAm = 2400

(M.H. Smith et al., 1976)

Scenedesmus obliquus	CFAm (72 hours) = 1.4×10^5
Selenastrium capricornutum	2.9×10^5
Chlorella pyrenoïdosa	1.6×10^3

Plancton mixed : (Eyman & Trabalka, 1980)

CF 239,240Pu = 2420 --- 5700

zooplankton : (Livingston and Bowen, 1976)

CF 239,240Pu = 380; CFAm = 230

(Eyman & Trabalka, 1980)

CF 239,240Pu = 350 --- 630

Macrophyta : (Emery et al., 1974 ; Emery and Klopfer, 1976)

Rorippa	CF 239,240Pu = 6200	CFAm = 7700
Typha (submergent)	CF 239,240Pu = 2100	3200
(emergent)	= 380	480
Scirpus (emergent)	= 42	41
Lagorosiphon major	6×10^2	2×10^2

Plants : (Livingston & Bowen, 1976)

Green algae (Lake Ontario)	CF 239,240Pu = 1480
Angiosperms (Lake Ontario)	= 1050
Plancton (Lake Ontario)	= 25, 50, 380

(depending on the year of sampling)

(C.W. Wayman et al., 1977), CF dry weight (depending on the sampling-place)

	CF 238Pu	CF 239,240Pu
Cladophora sp.	220- 3500	340-28800
	15800-103000	4100-16400
	14200- 59300	1800- 7800
Potamogeton sp.	30- 70	500; 620
	5870- 6430	90; 620
	700- 3500	
Myriophyllum sp.	1000, 1600	100,2700
Cattails	30 - 1.7×10^5	
Duckweed	400- 2900	

Invertebrates : (Emery et al., 1974 and Emery and Klopfer, 1976)

	CF 239,240Pu	CF_{Am}
Lymnaea	4100	5700
Physa	2100	2200
Daphnia	1600	2000
Hyolella	620	430
Libellula larvae	6700	6100

	Aeschna larvae	1600	1400
	Chironomidae	150	150
	Coleoptera	64	61
	(C.N. Murray et al., 1979)		
	Lymnaea stagnalis	8×10^2	5×10^2
	Biomphalania glabrata	1×10^3	4×10^2
	(M. Metayer-Piret et al., 1981) Concentrations expressed as 10^{-8} Ci/kg dry matter per Ci released		
		CF Pu	CF Am
	Erpobdella octoculata	43 ± 8	47 ± 13
	Glossiphonia sp.	147 ± 34	35 ± 21
	Hirudinea sp.	340 ± 34	235 ± 34
	Gastropod eggs	124 ± 106	1507 ± 146
	Oligochaeta	595 ± 146	849 ± 206
	Asellus aquaticus	340 ± 18	312 ± 22
	(Livingston & Bowen, 1976)		
	Pelecypode-Unionidae (Lake Ontario)		
	CF 239Pu soft parts	890	
	shells	1800	
Vertebrates :	(C.N. Murray et al., 1979)		
	Lebistes reticulates	CF Pu = 4×10^2	CFAm = 33
	Xiphophorus helleri	3×10^2	76
	(Emery et al., 1974 and Emery and Klopfer, 1976)		
	Carrassius	CF 239,240Pu = 290	CFAm = 26
	Anas	2	2
	(Garten et al., 1981)		
	Rana	17	34
	(Eyman & Trabalka, 1980)		
	Large mouth bass	CF 239,240Pu = 0.04	
	Blue gill	3	
	Carrasius	3	
	Shad	4	
	Alewife	25 to 176	
	Smelt	1 to 235	
	Perch	1 à 2	
	Chub	37	
	White fish	14	
	Chinook	4 to 7	
	Lake trout	1	

Different CF for Am, obtained in laboratory conditions in different surface waters.

Thiels et al., 1984 ; Thiels & Vanderborght, 1980
Bierkens et al., 1981 ; Bierkens, 1984
Vangenechten, 1984.

Lymnaea stagnalis :	Prinsenpark	= 130
	Reivennen	= 230
	Lago di Varese	= 160
	Ronde Put	= 140
	Lago Maggiore	= 200
	Lago di Monate	= 260
Lymnaea stagnalis :	Effect of pH	
	pH 4	= 1300 ± 200 SE
	pH 8	= 420 ± 80 SE
Astacus leptodactylus :	Prinsenpark (slightly basic)	= 40 - 150
Warche river	after papermill	= 20 - 70
	before papermill	= 0.5 - 20

	CF in "Prinsenpark" (slightly basic pH 7.8)	CF in "Reivennen" (acid, pH 4.2)
Lymnaea peregra	250 - 325	26 - 68
Planorbis corneus	100 - 235	27 - 56
Anisus planorbis	350 - 450	38 - 66
Gasterosteus aculeatus	0.5 - 3.4	0.4 - 1.3
Gammarus pulex	700	500
Corixa punctata	0.5 - 2	1 - 2
Tubifex sp.	8.1 ± 1.2 (95%)	3.3 ± 0.3 (95%)
Unio pictorum	4.0 ± 0.8 (95%)	3.2 ± 0.8 (95%)
Tadpoles Rana esculenta	400	3000

References

J. Bierkens, J.H.D. Vangenechten, S. Van Puymbroeck and O.L.J. Vanderborght, Bioavailability of the transuranic 241Am in different types of freshwater for 4 freshwater animals, Proc. Congress Impact Radiological Impact of Nuclear Power Plants and other Nuclear Installations on Man and his Environment, Lausanne 1981, p 216-223.

J. Bierkens, (abstract) Biological availability of the transuranic ^{241}Am in different types of freshwater for Astacus leptodactylus Eschscholtz (Turkish Crayfish), Colloquium on the Toxicity of Radionuclides, BLG 563, SCK/CEN, Brussels, 1984, p. 128.

R.M. Emery, D.C. Klopfer, Ecological distribution and fate of plutonium and americium in a processing waste pond on the Hanford Reservation, Report BNWL-2001 Pt.2 (1977), 4.15 - 4.17.

R.M. Emery, D.C. Klopfer, W.C. Weimer, The ecological behavior of plutonium and americium in a freshwater ecosystem. I. Limnological characterisation and isotopic distribution. Report BNWL-1867, 1974, 73 p.

L.D. Eyman and J.R. Trabalka, Patterns of transuranic uptake by aquatic organisms : consequences and implications, in Hanson, W.C. (ed.), Transuranic elements in the environment, Department of Energy, Oak Ridge, TN (USA), 1980, 612-624, DOE/TIC-22800.

T.C. Garten, J.R. Trabalka, M.A. Bogle, Comparative foodchain behaviour and distribution of actinide elements in and around a contaminated fresh-water pond, Oak Ridge National Laboratory, Report CONF-810722-5, 1981, 19 p.

H.D. Livingston and V.T. Bowen, Contrasts between the marine and freshwater biological interactions of plutonium and americium, US Energy Res. Develop. Agency, Health and Safety Lab. Report HASL-315, 1976, I-157, I-173.

M. Metayer-Piret, K. Hofkens, J. Colard, R. Kirchmann and L. Foulquier, Radioecological study of a river receiving radioactivity liquid wastes containing actinides, First results, Proc. of Techn. Comm. Techniques for identifying transuranic speciation in aquatic environment, IAEA, Vienna, 1981, p. 157-170.

C.N. Murray, A. Avogadro and G. Lazzari, The distribution of actinides in a freshwater microcosm ; comparison of simulated input sources ; Presented at II International Radioecology Symposium, Cadarache, 1979, 24 p.

M.H. Smith et al. Annual report of ecological research at the Savannah River Ecology Laboratory, SREL-6, Institute of Ecology, University of Georgia, 1976, 88 p.

G.M. Thiels, C.N. Murray, M. Hoppenheit, O.L.J. Vanderborght, Effect of the physico-chemistry of surface waters at neutral pH on the ^{241}Am uptake by the snail Lymnaea stagnalis L., Proc. Seminar on the Environmental Transfer to Man of Radionuclides Released from Nuclear Installations, C.E.C., Luxembourg 1984, p. 263-271.

G.M. Thiels, O.L.J. Vanderborght, Distribution of ^{241}Am in the freshwater snail Lymnaea stagnalis L. and water acidity, Health Physics, 39, 1980, 679-682.

C.W. Wayman, G.E. Bartelt and J.J. Alberts, Distribution of ^{238}Pu and 239,240Pu in aquatic macrophytes from a midwestern watershed, Energy Research and Development Admin., Las Vegas Nev. (USA), Nevada Operating Office, 1977, 505-516, NVO-178.

J.H.D. Vangenechten (abstract) Bioaccumulation of ^{241}Am in different freshwater types in tadpoles (Rana esculenta) and waterbugs (Corixa punctata), Colloquium on the Toxicity of Radionuclides, BLG 563 SCK/CEN, Brussels, 1984, p. 126-127.

THE ROLE OF NATURAL DISSOLVED ORGANIC COMPOUNDS IN DETERMINING THE CONCENTRATIONS OF AMERICIUM IN NATURAL WATERS*

D. M. NELSON AND K. A. ORLANDINI

Argonne National Laboratory
Environmental Research Division
Argonne, IL 60439
U.S.A.

ABSTRACT

Concentrations of ^{241}Am, both in solution and bound to suspended particulate matter, have been measured in several North American lakes. Dissolved concentrations vary from 0.4 µBq/L to 85 µBq/L. The ^{241}Am in these lakes originated solely from global fallout and hence entered all lakes in the same physiocochemical form. The observed differences in solubility behavior must, therefore, be attributable to chemical and/or hydrological differences among the lakes. Concentrations of dissolved ^{241}Am are highly correlated with the corresponding concentrations of ^{239,240}Pu(III,IV), suggesting that a common factor is responsible for maintaining both in solution. The K_D values for ^{241}Am and ^{239,240}Pu(III,IV) are highly correlated with the concentrations of dissolved organic carbon (DOC) in the waters, suggesting that the common factor is the formation of soluble complexes with natural DOC for both elements. This hypothesis was tested in a series of laboratory experiments in which the DOC from several of the lakes was isolated by ultrafiltration. Plots of K_D, as a function of DOC concentration, show K_D to be very high (~10^6) at low DOC concentrations. Above critical concentrations (a few mg/L DOC) the K_D values begin a progressive decrease with increasing DOC. We conclude that in most surface waters, the dissolved ^{241}Am concentration is regulated by an adsorption/desorption equilibrium with the sediments (and suspended solids) and the value of K_D that characterizes this equilibrium is largely determined by the concentration of natural DOC in the water.

---

*This study was supported by the US DOE Office of Health and Environmental Research.

1.0 INTRODUCTION

Americium-241 is one of the most abundant of the long-lived radionuclides produced in reactors, and as such its environmental behavior has been the topic of many studies. Most of these studies have been short-duration laboratory experiments investigating the sorption of americium tracer to various sediments or soils.[1-4] These studies generally show americium to be strongly sorbed, comparable to plutonium, with systemmatic variations resulting from changes in pH, equilibration time, and solid-to-liquid ratio. Sorption is conveniently expressed as a distribution ratio (K_D), where:

$$K_D = Bq \cdot Kg^{-1} \text{ solids}/Bq \cdot L^{-1} \text{ liquid.}$$

Values of K_D range from a few thousand to a few million, depending on the study consulted. Such a wide variation suggests that either K_D is highly sensitive to water or adsorber type or the method used to evaluate K_D may have a substantial impact on the value found. Few measurements of the actual K_D of americium found in natural systems have been reported. Hence, it is difficult to assess the reliability of the laboratory K_D measurements. Also, the few measurements of ambient K_D that do exist are from areas around waste discharges, where the possibility of variable source-term effects must be considered.[5-7]

We have now investigated the behavior of americium in a diverse group of North American lakes that had all been contaminated by a single source -- global fallout from nuclear weapons testing. The lakes sampled are those for which Wahlgren and Orlandini[8] have previously reported data for $^{239,240}Pu$, ^{232}Th, and ^{238}U. Thus, the sorption behavior of ^{241}Am can be compared directly with that of these other actinide elements over a wide range of environmental conditions.

2.0 METHODS

Samples (~50 L) were collected from each lake during 1977, 1978, or 1980 at a water depth of 5 m. For lakes less than 10 m deep, samples were collected at mid-depth. All lakes except Great Slave were sampled during isothermal conditions; Great Slave was sampled shortly after stratification had begun. Immediately after collection, the samples were filtered sequentially through preweighted 3.0 and 0.45 μm membrane filters (Millipore SSWP and HAWP) with the filtrates collected in polyethylene carboys. Filtrates were acidified with 100 mL HCl and spiked with ^{243}Am, ^{242}Pu, ^{230}Th, and ^{232}U as chemical yield monitors.

In the laboratory, the filters were air dried, weighed, and ashed at 500°C. The ash was weighed, spiked with the yield monitors, treated with combinations of HCl, HF, and HNO_3, and ultimately taken up in 8 M HNO_3. Filtrates were evaporated to minimal volumes, treated with HNO_3, and also taken up in 8 M HNO_3. Plutonium and thorium were removed from these 8 M HNO_3 solutions by passage through anion exchange columns (AG1x8 resin). The feed stocks were evaporated to dryness, taken up in 9 M HCl and again passed through anion exchange columns to remove uranium. These feed stocks were again evaporated, taken up in 1 M HCl, then passed through cation exchange columns (AG50x8 resin) to isolate ^{241}Am and the rare earth elements. Americium was eluted from these columns with 6 M HCl. The eluates were evaporated to dryness, taken up in 10 mL of electrolyte

(1.0 M NH_4Cl plus 0.01 M $H_2C_2O_4$) and electro-deposited onto stainless steel discs. The discs were counted for up to several weeks on low-background silicon surface barrier detectors. Occasionally, samples were found to contain enough rare earth contamination to degrade the alpha spectra. In those cases, the americium was separated from the rare earths using the thiocyanate procedure described by Moore.[9]

Concentrations of dissolved organic carbon (DOC) were measured on aliquots of filtered water using a Barnstead Photo-chem dissolved organic carbon analyzer. Alkalinity was measured by titration to the methyl orange end point. The pH was measured in the field using a meter and glass electrode. Conductivity was measured with a YSI model 33 S-C-T meter.

The influence of DOC concentration on the sorption of americium to natural sediments was investigated in the laboratory using the method of Nelson et al.[10] In this method, natural, colloidal-sized organic matter is concentrated from lake water using a stirred ultrafiltration cell (Amicon model 2000). The material retained by the 1000 molecular weight cutoff membrane (Amicon UM2) is then recombined with the ultrafiltered water in various proportions to generate a series of waters differing primarily in their concentrations of colloidal organic matter. Each of these waters is then spiked with ^{145}Pm (as an analog of americium), a portion of a natural fine-grained sediment is added to each, and the suspensions are stirred for one week. In these experiments, a sediment from an abandoned section of the Miami-Erie Canal (Ohio) was used at a concentration of 37 mg/L. After equilibration, each water was filtered through a 0.45 μm membrane, and the concentrations of ^{145}Pm in both filter and filtrate were determined by gamma-ray spectroscopy.

3.0 RESULTS AND DISCUSSION

Some general limnological characteristics of the lakes included in this study are presented in Table 1. These lakes were selected to represent a variety of types, and a substantial range in each of the variables is evident.

The results of the americium measurements are presented in Table 2, with the lakes listed in order of increasing dissolved ^{241}Am concentration. Concentrations of dissolved ^{241}Am are extremely variable, differing by a factor of greater than 200. In contrast, the concentrations found in suspended particulate matter are much less variable, with no significant correlation found between the dissolved and particulate concentrations. The nearly constant concentrations on suspended particles are reasonable if it is assumed that: (1) they reflect the concentrations on surficial sediments; (2) inputs of fallout ^{241}Pu, the parent of ^{241}Am, are reasonably constant over the area of the study; and (3) the inputs are mixed with and diluted by nearly equal amounts of sediment, per area, in each lake. The variation in dissolved ^{241}Am concentrations can then be viewed as a consequence of differences in water chemistry among the lakes.

Distribution ratios calculated from these data cover the range found in laboratory experiments, with the lowest K_D values being found in those samples having the highest dissolved americium concentrations. Interestingly, the highest K_D values, ~ 2×10^6, are almost identical to those found in coastal seawater around Windscale.[5]

The concentrations of dissolved ^{241}Am reported here are highly correlated with the concentrations of dissolved $^{239,240}Pu(III,IV)$

TABLE 1 Limnological characteristics of the study lakes.[a]

Lake[b]	DOC (mg/L)	pH	ALK[c] (mg/L)	COND (μS/cm)	SS[d] (mg/L)
Lake Michigan, Mich.	1.6	8.0	113	230	1.0 + 0.1
Lake Katherine, Man.	-	8.5	180	240	1.1 + 0.3
ELA 302 S	<5	6.3	4	-	2.2 + 0.3
GSL (Main Basin)	5.0	8.2	80	155	1.3 + 0.3
Clear Lake, Man.	-	8.0	190	300	1.1 + 0.2
ELA 223	4.6	5.6	~1	22	2.6 + 0.5
Last Mountain, Sask.	-	8.4	302	3000	5.1
GSL (McLeod Bay)	2.5	7.5	14	30	0.3 + 0.1
GSL (Christie Bay)	6.8	8.7	70	135	0.5 + 0.1
ELS 305	5.1	6.9	8	20	0.9 + 0.1
Volo Bog, Ill.	15	5.5	33	65	8.2
Okeefenokee Swamp, Ga.	34	3.9	<1	45	1.0
ELA 239	8.3	6.3	9	22	2.3 + 0.3
ELA 661	15	5.6	3	15	2.8 + 0.8
Little Manito, Sask.	115	>8.7	770	72000	43

(a) ALK = alkalinity; COND = conductivity; SS = suspended solids.
(b) ELA = Experimental Lakes Area, Ont.; GSL = Great Slave Lake, N.W.T.
(c) Expressed as $CaCO_3$.
(d) Dry weight, the two numbers indicate the material retained by the 3.0 and 0.45 μm filters, respectively; if only a single number is present it is the total (i.e., no 3.0 μm filter was used).

TABLE 2 Measured americium concentrations in the study lakes.

Lake	^{241}Am concentration: Dissolved[a] μBq/L	^{241}Am concentration: Particulate[b] Bq/kg	K_D L/kg	Fraction dissolved
Lake Michigan	0.41 ± 0.04	0.83 ± 0.09	2.0 x 10^6	0.30
Lake Katherine	0.7 ± 0.1	0.90 ± 0.13	1.3 x 10^6	0.33
ELA 302 S	1.1 ± 0.4	1.18 ± 0.09	1.1 x 10^6	0.26
GSL (Main Basin)	1.3 ± 0.2	0.18 ± 0.04	0.14 x 10^6	0.80
Clear Lake	1.5 ± 0.4	3.26 ± 0.25	2.2 x 10^6	0.25
ELA 223	1.8 ± 0.4	1.07 ± 0.11	0.59 x 10^6	0.33
Last Mountain	2.2 ± 0.4	0.22 ± 0.04	0.10 x 10^6	0.66
GSL (McLeod Bay)	2.8 ± 0.3	1.11 ± 0.16	0.40 x 10^6	0.85
GSL (Christie Bay)	3.5 ± 0.3	0.75 ± 0.11	0.21 x 10^6	0.88
ELA 305	4.4 ± 0.7	1.13 ± 0.19	0.26 x 10^6	0.79
Volo Bog	5.2 ± 0.7	0.37 ± 0.04	0.071 x 10^6	0.63
Okeefenokee Swamp	20 ± 3	1.15 ± 0.18	0.057 x 10^6	0.95
ELA 239	22 ± 2	0.98 ± 0.09	0.044 x 10^6	0.89
ELA 661	34 ± 3	1.14 ± 0.11	0.034 x 10^6	0.88
Little Manito	85 ± 11	0.13 ± 0.02	0.0015 x 10^6	0.94

(a)That passing a 0.45-μm filter.
(b)From analysis of the 3.0-μm filters.

reported by Wahlgren and Orlandini[8] for the same lakes (Figure 1). The line in Figure 1 corresponds to a ratio of ^{241}Am to $^{239,240}Pu$ of 0.25, approximately that expected from the ingrowth of ^{241}Am (from decay of ^{241}Pu) through 1977.[11] This strong correlation suggests that a single mechanism is responsible for maintaining both americium and reduced plutonium in solution in these lakes. Furthermore, that mechanism is almost equally effective for each element.

Wahlgren and Orlandini[8] observed a strong inverse correlation between the concentrations of DOC in these lakes and the K_D values both for reduced plutonium and for thorium. This same relationship is observed for americium. Figure 2 compares the K_D values of ^{241}Am, $^{239,240}Pu(III,IV)$, and ^{232}Th. Although there is substantial scatter in the data, the general trend is unmistakable - a systematic decrease in K_D with increasing DOC. The fact that a similar trend is observed for the fallout radionuclides as for the natural thorium strongly suggests that source effects are negligible and that K_D values are responding solely to variations in water chemistry.

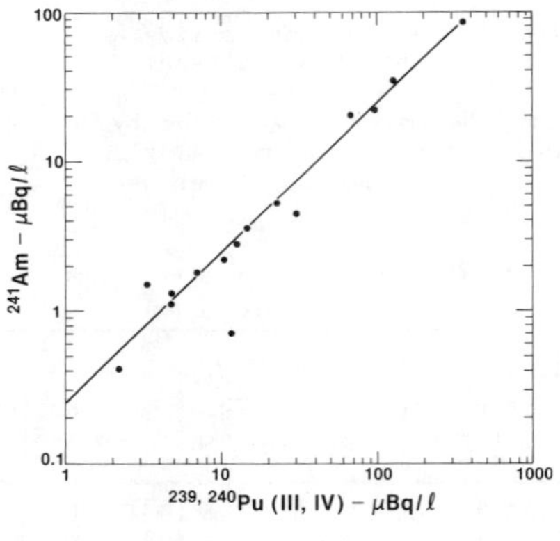

FIGURE 1. Relationship between the concentrations of ^{241}Am and $^{239,240}Pu(III,IV)$ in filtered water in a series of lakes.

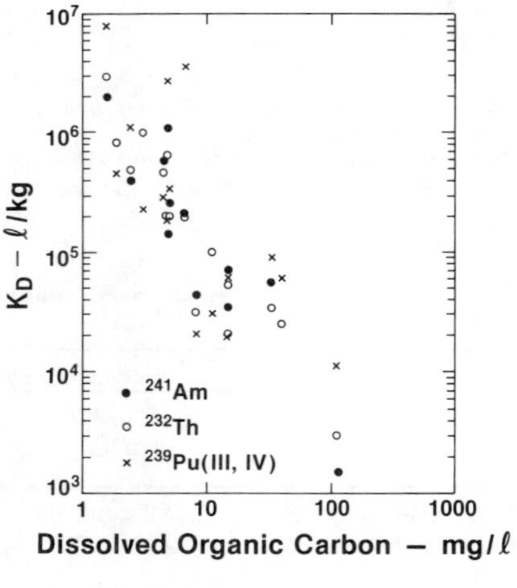

FIGURE 2. Comparison of the the K_D values for ^{241}Am, ^{232}Th, and $^{239,240}Pu(III,IV)$ as a function of dissolved organic carbon concentration.

Much of the variability observed in Figure 2 results from differences in the nature of the DOC and the suspended particulate matter in the various lakes. A much more regular variation in K_D values is observed in laboratory tracer experiments using a standard sediment and various concentrations of colloidal-sized DOC isolated from a single lake. Figure 3 compares the results of experiments conducted using DOC from Volo Bog and Lake Michigan according to the methods described in

Nelson et al.[10] Promethium-145 was used in these tracer experiments to avoid use of high concentrations of americium in our laboratory. Since the formation constants for most complexes of americium and promethium are almost identical, we feel their behavior in these experiments would not be detectably different.

At low DOC concentrations, the K_D values are very high ($>10^6$), and essentially constant. This suggests that in the absence of DOC, americium would display a uniformly high K_D in most water types. At

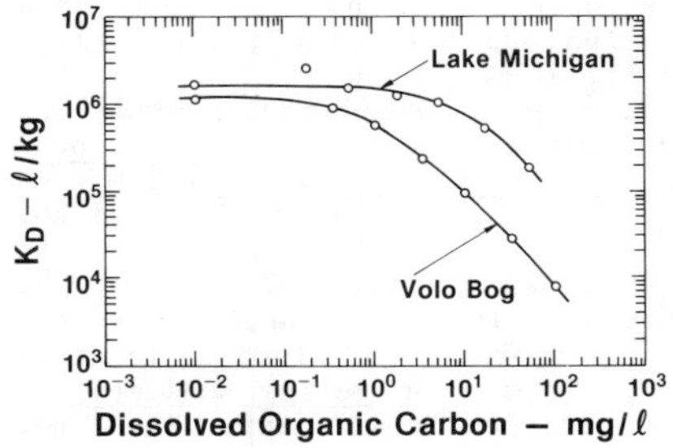

FIGURE 3. Dependence of the K_D of ^{145}Pm on the concentration of dissolved organic carbon in two sets of laboratory equilibration experiments.

higher DOC concentrations, the K_D values in both experiments decrease dramatically, with the decrease occurring at lower concentrations in the experiment using DOC from Volo Bog. Thus, while the general mechanism of K_D reduction at elevated DOC concentrations is probably similar in each case, i.e., formation of a non-adsorbed complex ion, there is a factor of ten difference in the effectiveness of the DOC from the two waters. The ambient K_D measured for ^{241}Am in each water is almost identical to the K_D found at the same DOC concentration in the tracer experiments. The short duration (~1 week) tracer equilibration had apparently reproduced to a significant degree the most important features regulating sorption in the natural environment.

The dependence of K_D on the concentration of DOC seen here for Am is very similar to that observed by Nelson et al.[10] for reduced plutonium. In that study, this dependence was shown to be consistent with a simple equilibrium model in which the metal atoms were distributed between the colloidal organic matter and the filterable particulate matter according to straightforward mass action principles. The results of this study support the proposition that a similar mechanism operates on americium and that the K_D of americium in most natural waters is determined, primarily, by the concentration of natural DOC in the water.

4.0 REFERENCES

1. C.N. Murray, and R. Fukai, Adsorption-desorption characteristics of plutonium and americium in the estuarine environment, in Impacts of nuclear releases into the aquatic environment (International Atomic Energy Agency, 1975), pp. 179-192.
2. H. Nishita, A. Wallace, E.M. Romney, and R. K. Schulz, Effect of soil type on the extractability of ^{237}Np, ^{239}Pu, ^{241}Am, and ^{244}Cm as a function of pH, Soil Science, 132, pp. 25-34 (1981).
3. A.L. Sanchez, W.R. Schell, and T. H. Sibley, Distribution coefficients for radionuclides in aquatic environments: adsorption and desorption studies of plutonium and americium (U.S. Nuclear Regulatory Commission, NUREG/CR-1852, 1981), 64 p.
4. T.H. Sibley, Comparative adsorption of selected radionuclides on different types of suspended particulates, in International Symposium on the Behavior of Long-Lived Radionuclides in the Marine Environment (Commission of the European Communities, EOR 9214EN, 1984), pp. 189-199.
5. R.J. Pentreath, D.F. Jeffries, M.B. Lovett, and D.M. Nelson, The behavior of transuranic and other long-lived radionuclides in the Irish Sea and its relevance to the deep sea disposal of radioactive wastes, in Proc. of the Third NEA Seminar on Marine Radioecology (Organization for Economic Cooperation and Development, 1980), pp. 203-221.
6. R.M. Emery, D.C. Klopfer, T.R. Garland, and W.C. Weimer, The ecological behavior of plutonium and americium in a fresh water pond, in Radioecology and Energy Resources, C.E. Cushing ed. (Dowden, Hutchinson and Ross, Stroudsburg, PA, 1975), pp. 74-85.
7. V.E. Noshkin, K.M. Wong, T.A. Jokela, J.L. Brunk, and R.J. Eagle, Comparative behavior of plutonium and americium in the equatorial Pacific, in Fourth International Ocean Disposal Symposium (Wiley-Interscience series Wastes in the Ocean, in press).
8. M.A. Wahlgren and K.A. Orlandini, Comparison of the geochemical behavior of plutonium, thorium, and uranium in selected North American Lakes, in Environmental Migration of Long-lived Radionuclides (International Atomic Energy Agency, 1982), pp. 757-774.
9. F.L. Moore, Improved extraction method for isolation of trivalent actinide-lanthanide elements from nitrate solutions, Anal. Chem., 38, pp. 511 (1966).
10. D.M. Nelson, W.R. Penrose, J.O. Kartunnen, and P. Mehlhoff, Effects of dissolved organic carbon on the adsorption properties of plutonium in natural waters, Environ. Sci. Technol., 19, pp. 127-131 (1985).
11. P.W. Krey, E.P. Hardy, C. Pachucki, F. Rourke, J. Coluzza. and W.K. Benson, Mass isotopic composition of global fall-out plutonium in soils, in Transuranium nuclides in the Environment (International Atomic Energy Agency, 1976), pp. 671-678.

STUDY OF THE PHYSICOCHEMICAL FORM OF COBALT IN THE LOIRE RIVER.

**Ph. PICAT * - F. BOURDEAU ** -J.P. THIRION*
M. SIGALA* - J.M. QUINAULT* - M. ARNAUD* - Y. CARTIER ***

*CEA, CEN Cadarache, B.P 1, 13115 SAINT PAUL LEZ DURANCE ; ** EDF, 28-30 avenue de Wagram - 75382 Paris - FRANCE

ABSTRACT

Cobalt isotopes 58 and 60 are basically found in dissolved form in the liquid effluents from the PWR plants studied. When introduced in Co** form into Loire River water, cobalt progressively evolves into particulate forms (suspensions and precipitates); radioactive equilibrium is never reached in less than three days, regardless of the water sample used. A realistic model of radionuclide behavior in waterways must therefore take account of the kinetics of solute transfer to solid phases if it is not to provide erroneous estimates of the actual hazards involved, given the short time period (generally a few hours) between the effluent discharge into the river and human utilization of the water.

1 - INTRODUCTION

Based on a detailed analysis of several accidents that occurred in a number of facilities, LEVENSON[12] concludes that the accident hazards calculated from existing models are overestimated by a factor of 10 or more. An EEC study[13] emphasizes the lack of available knowledge concerning the physicochemical forms of the radionuclides dispersed in an accident situation. Unfortunately, it is difficult to predict the behavior of a pollutant in the environment without knowing the form(s) in which it is present; the form(s) may also evolve differently, depending on the environmental characteristics.

The ultimate objective of environmental radionuclide transfer studies is to develop a realistic model of the natural environment capable of predicting the behavior of the pollutants introduced. The vast number of parameters and mechanisms involved mean that the model must be above all a simplified but coherent representation of the actual system[14] in order to constitute a decision aid for national authorities in the area of nuclear protection and safety. A simplified representation of a complex reality implies adequate knowledge of the environment/nuclear plant system and of the major processes involved. Field studies and laboratory simulations are thus indispensable in developing models, and only a continuous three-way confrontation between field, laboratory and theoretical results can provide realistic estimates.

2- OBJECTIVES AND METHODS

2.1. Field Studies

Investigations of the nuclear site itself and the surrounding environment are intended to determine the characteristics of the source term (effluent types, physicochemical forms of the radionuclides and of related stable elements) and to obtain full scale measurements of waste dispersion in the air and water under the specific conditions prevailing at each site.

2.2. Laboratory Simulations

Simulations allow the process that determine radionuclide behavior in the environment to be studied under controlled, although necessarily simplified, conditions. The tests are designed to quantify the role of each significant mechanism with regard to the actual values of the physical, chemical and biological parameters of the waste/environment system. The objective is to obtain the simplest possible presentation (formula or chart) together with an estimate of the errors and variability of the results so that they may be easily and correctly incorporated in the general model for the transfer phenomenon studied.

2.3. Modeling

One of the primary objectives, and not the least dificult, is to obtain the most accurate possible representation of an often complex reality. This implies not only a selection of a limited number of important processes which alone will be used to simulate the transfer, but also a number of carefully chosen simplifications in the way they are presented, based to a large extent on laboratory test results (§ 2.2.). An effort is also made to maximize the accuracy of the values for the most important parameters and to assess their influence on the final result by means of sensitivity tests.

3 - COBALT AND THE LOIRE RIVER : JUSTIFICATION OF THE CHOICES

3.1. Cobalt

Radioactive cobalt isotopes often account for a significant fraction of the total non-tritium radioactivity present in the liquid effluents from PWR and BWR plants, which constituted 87% of the total world nuclear power generating capacity in 1983.[5]

Cobalt is relatively abundant in the earth's crust (mean Co content : 25 mg/kg^{-1}) and widely used in industry. Cobalt is also a transition element with complex chemistry[6-7]. Its behavior in the environment may vary considerably depending on the existing physicochemical conditions : i.e. the initial waste form and the properties of the receiving environment.

3.2. The Loire River

The Loire river basin is a major nuclear power producing region : twelve reactors on three sites generate a total of 9 GWe, with four additional PWRs now under construction .

The Loire is also a major river in France : physically, it is 1000 km long and drains 115 000 km², with a flow rate of 870m³/sec; its water is pumped both directly and from the alluvial water-sheet to supply potable water to riverside cities. Its waters are used for agriculture and industry.

As a result of serious eutrophication, the Loire teems with plankton that cause wide variation in the water pH and oxygen content.[8] These two parameters significantly effect the behavior of polluants in aquatic media.

4 - RESULTS

4.1. Physicochemical Forms of Cobalt Release into the Environment

The results of a one-year study of liquid wastes from the Fessenheim plant show that the activity of ^{58}Co and ^{60}Co is essentialy related to dissolved chemical species (Tables I). This observation also applies to other radionuclides in the effluents (^{131}I, ^{54}Mn, ^{137}Cs), although not without exception (e.g. ^{51}Cr).

The dominance of the dissolved forms* in the wastes is related to the fact that radioactive liquids** from a PWR plant are successively filtered to 0.5, 5 and/or 25 microns, depending on their origin.

TABLE I. Physicochemical forms of radionuclides in liquid wastes from the Fessenheim power plant (from August 1, 1979 to July 31, 1980) as measured by gamma-Ge spectrometry on averagbe monthly aliquots[(a)]

RADIO-ISOTOPES (half-life)	PERCENTAGE OF TOTAL ACTIVITY RELATED TO DISSOLVED FORMS[(b)] average	(extremes)	NUMBER OF MONTHLY ALIQUOTS IN WHICH ISOTOPE WAS FOUND IN 1 YEAR	ISOTOPE CONTRIBUTION TO TOTAL NON-TRITIUM GAMMA ACTIVITY RELEASED 1 YEAR %
^{58}Co (71d)	86	(84-96)	12/12	56.5
^{131}I (8 d)	97	(93-100)	9/12	15.2
^{60}Co (52 y)	76	(34-95)	12/12	14.5
^{54}Mn (280 d)	68	(0-95)	12/12	5.5
^{137}Cs (30 y)	88	(0-99)	10/12	1.8
^{51}Cr (28 d)	6	(0-57)	9/12	1.6[(c)]

(a) Aliquot : mixture of 200cl from eache of the 20 tanks discharged on average each month ; aliquots were stored at 4°C in darkness prior to conditioning and measurement.

(b) Dissolved forms : forms measured after 0.45 micron filtration. Total activity : dissolved activity + Particulate activity (retained on the 0.45 micron filter).

(c) The remaining 5% are due to the presence of the following radionuclides (in decreasing order of importance) : ^{59}Fe, ^{134}Cs, ^{144}Ce, ^{110m}Ag, ^{124}Sb, ^{65}Zn, ^{125}Sb, ^{57}Co, ^{7}Be, ^{95}Zr, ^{95}Nb, ^{136}Cs.

----

* activity present in the effluent after filtration to 0.45 micron.

** primary and secondary circuit samples, chemical and utility effluents.

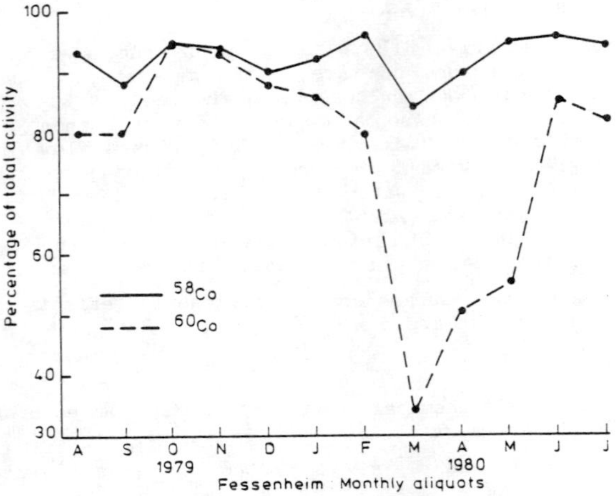

Figure 1 Abundance (%) of dissolved forms for cobalt isotopes in liquid wastes from the Fessenheim plant (1 August 1979 to 31 July 1980)

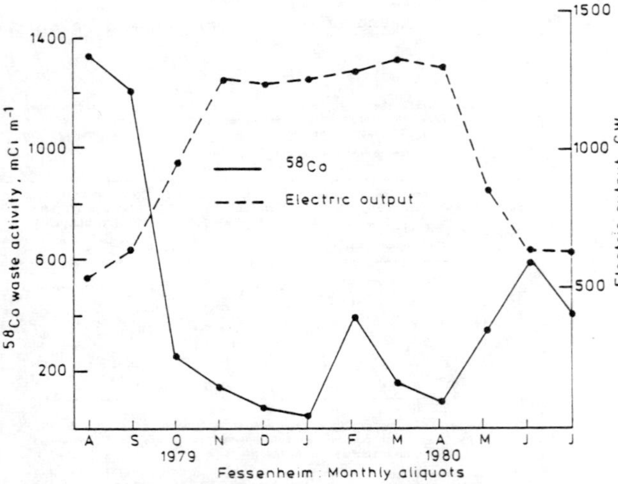

Figure 2 Cobalt-58 waste activity variations v electric output in liquid wastes from the Fessenheim plant (1 August 1979 to 31 July 1980)

For certain radionuclides, however, these filtrations do not prevent particulate forms* from predominating : this is true, for examples, of ^{51}Cr, which is explained by the presence of this effluent in valence III^{+} and by the formation of relatively insoluble compounds such as $Cr(OH)_3$. Another interesting example concerns the two cobalt isotopes : for certain monthly aliquots the partition between the dissolved and particulate forms was different (Figure I).

In this case, the nuclear reactions that produce these isotopes from nickel [^{58}Ni (n,p) ^{58}Co] or cobalt [^{59}Co (n, γ) ^{60}Co] could also account for this difference. The presence or absence of various stable chemical species in certain monthly aliqots could modify the solubility of one of the stable parent elements (e.g. nickel) without affecting the second (cobalt) by formation of ions or complex molecules. From a chemical standpoint, these effluents are dissolved boric acid solutions** containing certain dissolved chemical species*** at concentrations generally below 10 ppm, some of which are capable of forming complex molecules with the radioactive ions present. Moreover, detergents and decontaminants were not determined, althoug these product properties with respect to radionuclides are well known.

An analysis of the dissolved radionuclide chemical forms (identification of simple or complex anions, cations and uncharged molecules) did not provide reliable data. The separation methods used (e.g. ion exchange) gave highly variable results, for example regarding the partition between dissolved cationic and anionic forms depending on the effluent studied and the operating conditions. The chemical properties of the environment are sufficiently well known, but more information is required on stable species (ions, molecules, waste concentration levels) in order to better define the conditions under which methods may be implemented to obtain reliable results.

Moreover, from a chemical standpoint, the discharged effluents are aqueous solutions with practically no matter in suspension, with pH values ranging from 5.5 to 8.5. The highest monthly activity release values for ^{58}Co, for example, correspond to periods when unit 1 or 2 at the Fessenheim plant was shut down. This finding may be considered in the light of the well known rise in the corrosion product activity concentration of the primary coolant fluid during a PWR reflueling shutdown period.[9] The evolution of the activity breakdown between dissolved and particulate forms for cobalt isotopes does not, however, appear to correlate with the shutdown periods of the two Fessenheim reactors (compare Figures 1 and 2). The possibility that the radionuclide physicochemical forms might evolve during the monthly aliquot preparation ans storage period (1 month) led us to take samples of individual effluent batches just before they were released into the environement, and to condition and measure them immediately. The results obtained at Fessenheim and Dampierre on 6 'fresh" effluent discharge (Table II) seem to confirm the observations of monthly aliqots (Table I), notably for the predominance of dissolved forms.

----

* Activity retained on a 0.45-micron filter.
** Boron content ranging from 20 to 240 ppm.
*** Ca^{2+}, K^{+}, Mg^{2+}, SO_4^{+}, Cl^{-}, CO_3H, NO_3^{-}, PO_4-

TABLE II Physicochemical forms of radionuclides present in "fresh" effluents from the Fessenheim and Dampierre plants

PLANT (sampling) date)	^{58}Co	^{131}I	^{60}Co	^{54}Mn	^{137}Cs	OTHER MAJOR ISOTOPES[a]
Fessenheim (02-02-80)	83[b] (10)[c]	98 (57)	77 (11)	87 (9)	99 (5)	
Fessenheim (02-03-80)	73 (38)	98 (1)	17 (40)	44 (13)	100 (1)	
Fessenheim (05-03-80)	64 (34)	98 (1)	13 (46)	31 (12)	95 (< 1)	
Fessenheim (03-06-80)	97 (72)	99 (7)	78 (14)	68 (3)	96 (1)	
Dampierre (16-12-83)	97 (39)	100 (1)	92 (8)	98 (7)	96 (8)	
Dampierre (21-03-84)	80 (42)	100 (1)	78 (10)	(4)	99 (11)	^{124}Sb ^{110m}Ag ^{134}Cs

(a) Isotopes accounting for more than 5% of the total non-tritium activity in the waste discharge.

(b) Percentage of total activity present in dissolved form for a given isotope.

(c) Contribution of isotope (all forms) to total non-tritium gamma activity in the waste discharge.

4.2. Effects of Discharge Conditions on Radionuclide Physicochemical Forms

French regulations require significant dilution of radioactive liquid wastes from PWR plants before they are released into the environment. For example, at Dampierre[10] the minimum dilution before release into the Loire River is a factor of 4000 : after an initial dilution by a factor of at least 500 in non-radioactive water (generally cooling tower drain water supplied from the Loire), waste releases are prohibited when the river flow rate is less than $60 m^3/sec$. One liter of liquid waste thus rapidly mixes with at least $4\ m^3$ of Loire River water. The chemical nature of the aqueous solution in which the radionuclides are initially present (§ 4.1.) is thus completely modified by the dilution and by the characteristics of the Loire water (§ 4.3.1). As dilution favorizes the dissociation of complexes, some dissolved radioactive complexes present may also be broken down if the Loire does not include the chemical species responsible for the chelation.[11]

4.3. Loire Water Effects on the Evolution of Physicochemical Forms of Cobalt

4.3.1. Physicochemical Characteristics of Loire River water.

Three points are worth emphasizing : the diversity of dissolved chemical species liable to associate with the radionuclides or to be in competition with them in various reactions (Table III); the presence of suspended matter (SM) with well known decontaminant ab- and adsorptive properties ; and the high pH values (> 9) frequently observed in the Loire, together with the wide pH fluctuations over the hydrological cycle (Figure 3) or on a day to day basis during periods of intense photosynthetic activity (Figure 4).
These pH variations are due to the photosynthetic activity of phytoplankton (primarily diatoms, chlorophyceae) that teem in the Loire as a result of major phosphate discharges throughout the river basin, and to the often high temperatures, turbidity and low calcium content of the river water.[12]

TABLE III Physicochemical characteristics of Loire River water measured downstream from Orleans[(a)] between January 1971 and December 1980[21]

PARAMETER	NUMBER OF MEASUREMENTS	MAXIMUM PERCENTILE VALUES[(b)]					OVERALL AVERAGE
		10%	25%	50%	75%	95%	
Temperature (°C)	132	4	7	12	18	22.4	12.3
pH	132	7.2	7.4	7.8	8.1	9.2	7.9
Conductivity (µS)	132	156	175	210	264	372	228
Dissolved O_2 (mg/l)	132	8.1	9.4	10.4	11.5	13.8	10.3
Total SM (mg/l)	132	12	20	31	46	97	39
Cl- (mg/l)	132	8	10	12	16	27	14.5
SO4-- (mg/l)	132	10	15	20	25	35	20.6
HCO3- (mg/l)	131	70	79	90	114	167	101
PO4--- (mg/l)	130	0.1	0.24	0.36	0.53	2.75	0.74
Ca++ (mg/l)	132	22	26	29	35	48	31
Mg++ (mg/l)	132	3.4	3.7	4.4	5.2	6.3	4.6
K+ (mg/l)	132	2.6	3	3.5	4.1	7.6	4.1
Na+ (mg/l)	132	7.9	9	11	14.5	25.4	13.4

(a) at La Chapelle Saint Mesmin

(b) i.e. the actual values are less than or equal to the indicated value in the specified percentage of cases.

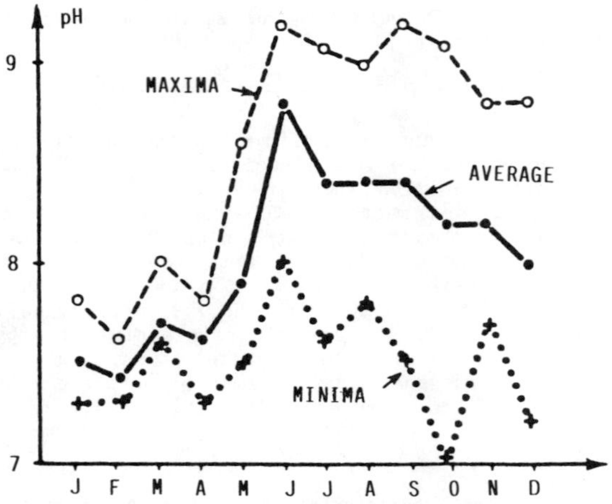

FIGURE 3 : pH VARIATIONS IN THE LOIRE DOWNSTREAM FROM DAMPIERRE IN 1983[21]

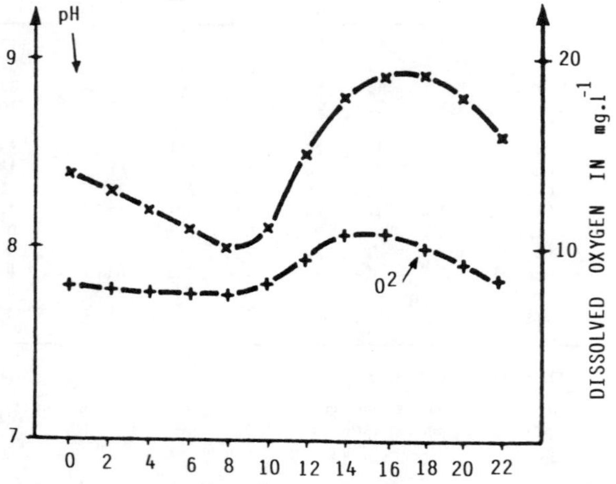

FIGURE 4 : pH AND DISSOLVED OXYGEN DOWSTREAM FROM DAMPIERRE ON SEPTEMBER 15, 1984[21]

4.3.2. Cobalt Monitoring in the Loire

The waste discharge from the Dampierre plant provided a possibility for monitoring the behavior of cobalt isotopes 58 and 60 in the Loire. However, two prerequisites must be ensured to obtain interpretable results from downstream sampling, irrespective of the river flow rate :

(1) the sampling station must be located in the homogeneous mixture zone, otherwise the samples might not even include effluent content (figure 6);

(2) samples must not be taken before the waste induced concentration reaches steady-state conditions, i.e. a stable contaminant content in the water (figure 7).

These conditions are determined for each sampling station and for each flow rate by releasing tracers into the river. It was considered preferable to approach the study of cobalt in the Loire by laboratory simulations, which are easier to conduct and to analyze.

4.3.3. Laboratory simulation of Cobalt Behavior

4.3.3.1. **Loire River Sampling Techniques**

Two methods are used. Large samples (approx. $1m^3$) may by be taken, and the matter in suspension separated and concentrated by 0.45-micron tangential filtration : the concentrates can be used to study the properties of the suspended matter by experimentally simulating different SM concentrations in filtered Loire water samples such experiments are known as "reconstituted water" tests. Or the Loire water samples may be used directly without prior SM filtrate separation ("raw water" tests). Water samples are transported and stored at 4°C in darkness before the experiments.

4.3.3.2. **Experimental Protocol and Equipment for Radioactive Testing**

A plexiglas beaker containing five liters of "raw" or "reconstituted" Loire River water is placed in a thermostatically-controlled tank regulated to maintain the water temperature at the value recorded when the sample was taken. A magnetic stirrer keeps the particulate matter in suspension ; a pH meter and a automatic base and acid microtitrater are used to maintain a preselected pH value throughout the seven-day test period (figure 7). After an initial 24-hour phase during which the Loire water is restored to equilibrium (temperature and dissolved gas content) and the natural pH variation is observed, the contamination is introduced and the pH regulating system is starded. Cobalt-60 is introduced in the form of a microvolume of $CoCl_2$ in a hight specific activity (0.9mCi/g) 0.1 N HC1 solution after dilution in 200 ml of distilled water to obtain an initial activity concentration in the Loire water sample of about 5µCi/1. This procedure not only limits the amount of stable cobalt and acidity introduced into the water sample, but also ensures that the activity introduced into the 5-liter sample is quickly mixed. The stable cobalt content of the Loire water as determined by neutron activation is quite low; 0.09-0.15 µg/1 for dissolved cobalt (0.45-micron filtration) and 0.26-0.32 µg/1 for total cobalt : dissolved + adsorbed on suspended matter (obtained by acidification of one liter of Loire water with 5 ml of highly pure concentrated HNO_3).

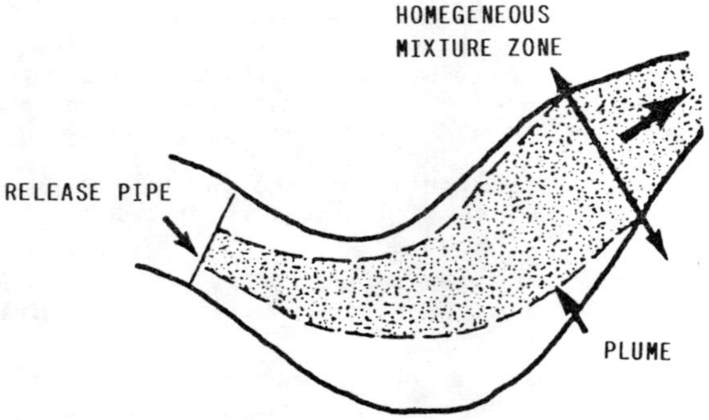

FIGURE 5 : SELECTING A DOWNSTREAM SAMPLING SITE

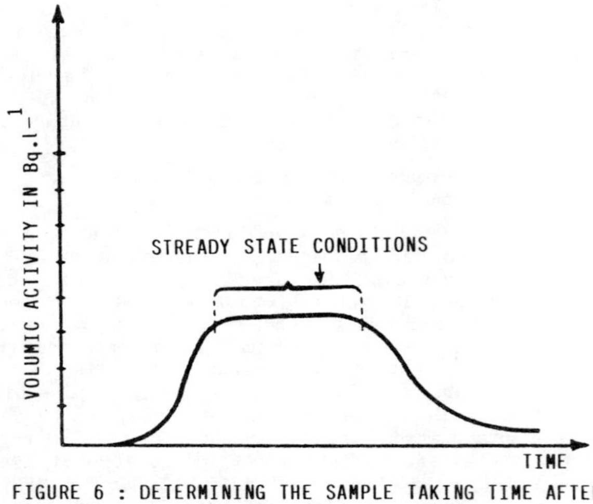

FIGURE 6 : DETERMINING THE SAMPLE TAKING TIME AFTER WASTE RELEASE

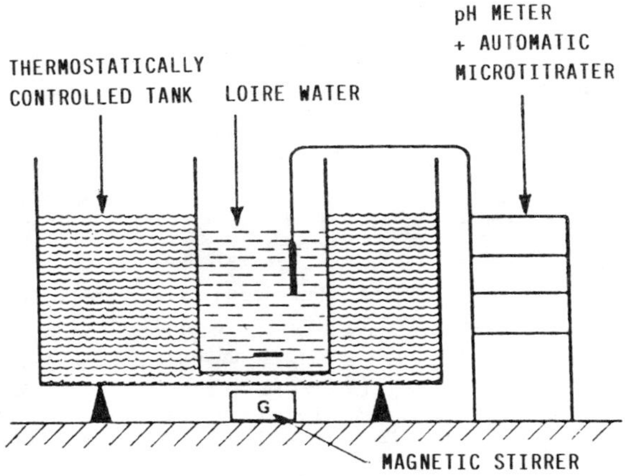

FIGURE 7 : EXPERIMENTAL SETUP FOR RADIOACTIVE TESTING

FIGURE 8 : TRANSMISSION ELECTRON MICROSCOPIE VIEW OF LOIRE SUSPENDED MATTER SAMPLED IN AUGUST 1984 :
note quantity of silicified dintomaceous skeletons.

After contact times ranging from 2 minutes to 7 days, 50 ml aliquots are sampled and filtered to 0.45 micron to follow the evolution of the remaining dissolved activity. On completion of the test, a final balance is prepared by total filtration of the remaining volume and measurement of the activity retained on the suspended matter or deposited on the beaker.

All of the tests are conducted without any treatment of the suspensions, whose effect on modifying radionuclide sorption properties is well known.[14] The water sample is submitted to a complete chemical analysis before and after the experiments.

4.3.3.3. **Results**

Figure 9 shows the results obtained for a series of tests with ^{60}Co at three stations located along the waterway for three different flow conditions (low, average and high water level). It may be seen that in each case the ^{60}Co, initially introduced in dissolved form, is progressively fixed by the matter in suspension. The cobalt fixation differs from one station to another at a given flowrate, but less than for different flowrates at any given station.

These tests were conducted without any pH regulation. However the role of pH on metal fixation by suspended matter is ofen mentionned,[15] especially for cobalt[3]. Another series of tests was conducted at constant pH with water samples taken in summer (figure 10) and in winter (figure 11).

The results obtained confirm that even with a perfectly stable pH, and regardless of the type of Loire water, the cobalt initially introduced in dissolved Co^{++} form evolves into particulate forms and equilibrium is never reached before 3 days.

This transfer of the dissolved Co^{++} ion to solid particulate forms is due to the presence of suspended matter, but may also result from cobalt fixation, for example on calcium phosphate or carbonate precipitates (curve A in figure 10) : LAUDELOUT calculated the concentration of complexes likely to form in Loire water, based on models developed by MOREL[16] : this work shows that precipitated calcium salts may form (Table IV) especially as a result of pH increases related to photosynthetic activity of phytoplankton. A detailed examination of the suspensions sampled in August 1984 shows that they consist primarily of diatoms with a smaller mineral phase fraction (figure 8). However, these suspensions develop a large specific surface area ($10 m^2/g$) because of the highly porous nature of the silicaceous diatom skeletons.

Tests conducted in winter (figure 11) at three stabilized pH values (7.3, 7.8 § 8.3) did not reveal any very marked effect of this parameter on the partition, between dissolved and particulate cobalt forms. This study will be completed by other tests using water sampled in summer : the higher pH values and the presence of other stable ions in solution at this time of year could modify the cobalt forms and favorize the formation of precipitates (Table IV).

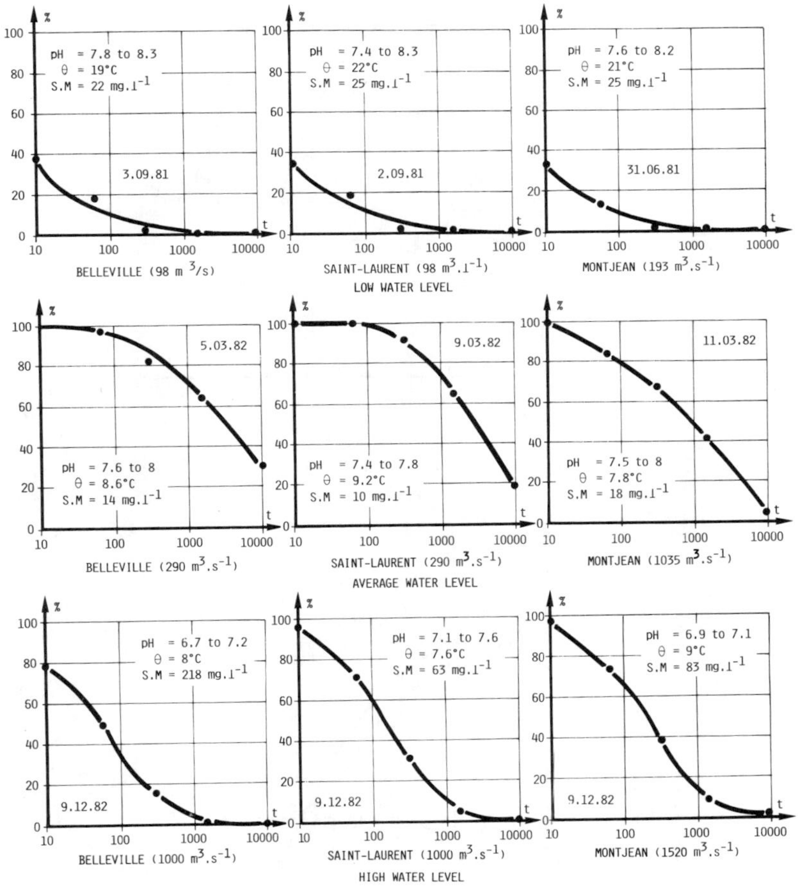

FIGURE 9 : PERCENTAGE (%) OF ^{60}Co ACTIVITY IN SOLUTION AS A FUNCTION OF TIME (t in min.)

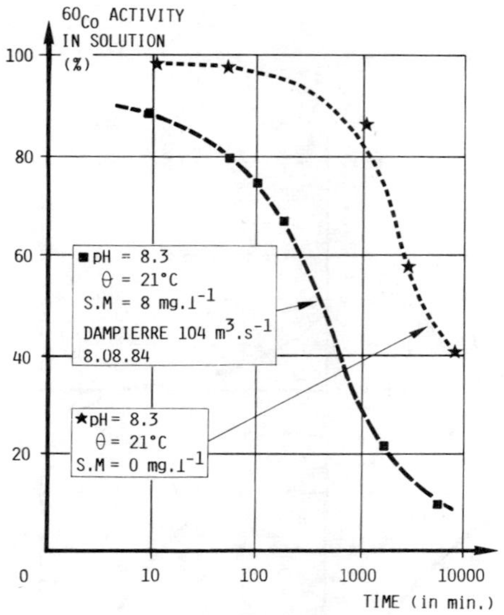

FIGURE 10 :
^{60}Co EVOLUTION IN LOIRE WATER SAMPLES COLLECTED IN SUMMER

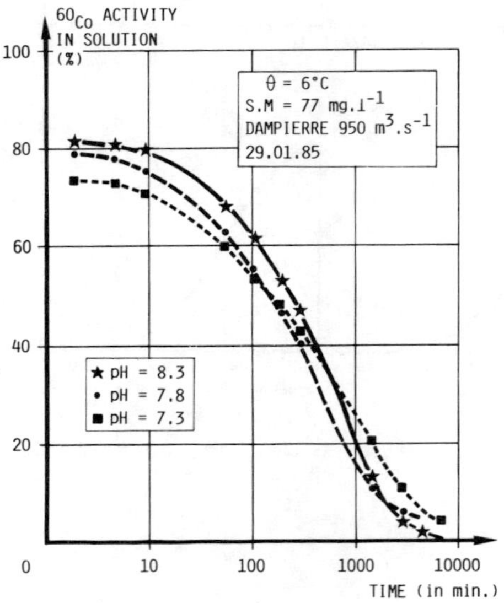

FIGURE 11 :
^{60}Co EVOLUTION IN LOIRE WATER SAMPLES COLLECTED IN WINTER

TABLE IV Distribution of certain cations and ligands in Loire River water percentage of each form relative to total element molarity (from LAUDELOUT)

SAMPLING DATE (pH)	COBALT		CALCIUM				PHOSPHATES		
	FREE Co^{++}	BONDED (CO_3^-)	FREE Ca^{++}	BONDED (CO_3^-)	BONDED (SO_4^{--})	BONDED $(PO_4)^{---}$	BONDED (Ca^{++})	BONDED (Mg^{++})	BONDED (H^+)
18.10.76 (7.2)	53.1	44.7	92.7	1.0	2.7	3.4*	1.2 87.8	0.6	10.3
12.04.76 (8.3)	18.6	80.2	95.4	3.0	1.4		1.9 92.2	0.9	5.0
14.09.76 (8.9)	5.8	92.4	25.6	2.3 70.9	1.4		12.2 26.4	16	44.5

* Underlined results correspond to precipitating forms.

5 - CONCLUSION

A study of cobalt introduced in dissolved Co^{++} form into the water of the Loire River has revealed a number of prerequisites for realistic predictions of radionuclide behavior in a waterway

(1) adequate knowledge of the water characteristics (stable species in solution and in suspension ; other pollution forms affecting the river basin by eutrophication);

(2) allowance for the kinetics of solute transfer from a liquid to a solid, which is a classic theme in chemical engineering.[17]

The failure to reach radioactive equilibrium is also observed with other radionuclides introduced into the Loire or the Rhône[18] in ion form (e.g. Cs^+, Mn^{++}, Cr^{+++}). Despite their apparent simplicity equilibrium constants such as Kd should not be used too routinely without some precautions in models.

Major parameters and processes must be selected to obtain a simple model that permits decisions to be made easily- and reliably. Especially in accident situation the kinetics of environmental transfer must be taken into account : the percentage of activity in solution, which is always higher before equilibrium in surface water, corresponds in fact to the most highly mobile waste fraction that most easily reaches humans through drinking water or by irrigation through the food chain.

****

The authors would like to thank the personnel of the Fessenheim and Dampierre plants, without whose cooperation the effluent study would not have been possible, as well as the following laboratories for their analytical support : LMEI Osay (M. PHILIPPOT: neutron activation and gamma-Ge spectrometry), BRGM Orléans (Mr BOULMIER: sedimentological analysis), SCEV Paris (M. MONTEIL: water analysis).

REFERENCES

1 M. Levenson and F. Rahn, "Realistic Estimates oif the Consequences of Nuclear Accidents", Nucl. Tech. 53, 99 (1981)
2 A.E.N., International Comparison Study of the Consequences of Nuclear Accidents, (OECD, Paris, 1984)
3 IAEA, "The Role of Sediments in the Tranport and Accumulation of Radioactive Pollutants in Waterways", Final Report of the Second Research Coordination Meeting Held in Budapest : 28 May 1 June 1984, 642.T2-RC-263-3/CL2.2 (1985)
4 F. Luykx and J. Fraser, Radioactive Effluents from Nuclear Power Stations and Nuclear Fuel Reprocessing Plants in the European Community : Discharge Data 1976-1980, Radiological Aspect, (CEC, Luxembourg, 1983)
5 "Power Reactors 1984", Nucl. Eng. Int., No 10 suppl, 2 (1984)
6 Trace Metals in the Environment, edited by I.C. Smith and B.L. Carson (Ann Arbor Sc. Publishers, Ann Arbor, 1979) vol 6 "Cobalt: An Appraisal of Environmental Exposure"
7 R. Bittel, Discussion bibliographique sur le comportement physico-chimique et la radioécologie du cobalt dans les systèmes hydrobiologiques (CEA-BIB-130, Saclay, 1968)
8 P. Crouzet, "L'Eutrophisation de la Loire", Water Supply, 1, 131 (1983)
9 R. Van Brabant and P. de Regge, "Study of the Corrosion Products in the Primary System of PWR Plants as the Source of Radiation Field Buildup", (Report BLG 552, Mol, 1982)
10 Journal Officiel de la République Française, "Autorisation de rejet d'effluents radioactifs liquides par la centrale de Dampierre en Burly (tranches 1 à 4)" (Journaux Officiels, Paris, 1979), 05.07.1979, pp 5741-5742
11 G. Charlot, Les réactions chimiques en solution, (Masson, Paris, 1969) 6° édition, p 51
12 Inventaire du degré de pollution des eaux superficielles (rivières et canaux) : Campagne 1981, (Ministère de l'Environnement, Paris, 1983) vol IV Bassin LOIRE-BRETAGNE
13 Centre de Production Nucléaire de Dampierre en Burly, Rapport de Surveillance de l'Environnement pour l'année 1983, Annexe n° 2 : "Résultats des mesures physicochimiques en Loire", (EDF, Direction de la Production et du Transport, Paris, 1984)
14 E.K. Duursma, Problems of Sediment Sampling and Conservation for Radionuclide Accumulation Studies, IAEA TEC-DOC-302, 127 (1984)
15 A.C.M. Bourg Un modèle d'adsorption des métaux traces par les matières en suspension et les sédiments des systèmes aquatiques naturelles, C.R. Acad. Sc, Paris, 294, 1091 (1982) Série II
16 F. Morel, R.F. MacDuff and J.J. Morgan, "Interactions and Chemostasis in Aquatic Chemical Systems : The Role of pH, pE, Solubility and Complexation", Trace Metal and Metal Organic Interactions in natural waters, edited by C. Singer (Ann - Arbor Sciences pub 1973 p 157-268)
17 D. Schweich, Les Lois physicochimiques d'interaction entre un fluide et un solide, IAEA TEC-DOC-302 (IAEA, Vienna, 1984) p71
18 M.C. Ferrer, Etude expérimentale du comportement de huit radionuclides artificiels dans le cours terminal du Rhône : Fixation sur les matières en suspension, étude des formes physicochimiques, (Thèse, Univ. Bordeaux I, 1983)

DISCUSSION

CREMERS:

Two points or comments - first, it should be worthwhile to try and simulate the scenario as proposed (coprecipitation of Co with $CaCO_3$) on a laboratory scale.

Second, if this is indeed the mechanism responsible for the removal of Co, then, in view of the rather larger variation in pH, would this not be a transient condition?

PICAT:

I think, following your proposal, it is interesting to check the role of $CaCO_3$ precipitation in the disappearance of cobalt from the soluble phase and the probably transient aspect of this coprecipitation effect considering the variations of pH derived in the Loire river.

As you propose, a laboratory experiment may be carried out to check these two points with a filtered Loire river water and a Carbonate solution, with appropriate pH regulation.

INTUBATION OF DIFFERENT CHEMICAL FORMS OF AMERICIUM-241 ON THE CRAYFISH *ASTACUS LEPTODACTYLUS*.

J. BIERKENS*; J.H.D. VANGENECHTEN+; S. VAN PUYMBROECK+ and O.L.J. VANDERBORGHT+*.

(*) Radionuclides Metabolism Laboratory
SCK-CEN, Belgian Nuclear Centre
B-2400 Mol, Belgium
(+) University of Antwerp
Biology Department
B-2610 Wilrijk, Belgium

ABSTRACT

The availability of ^{241}Am for the crayfish *Astacus leptodactylus* has been investigated following an intragastric injection of three chemical forms of ^{241}Am (^{241}Am-humate, non-complexed ^{241}Am and *in vivo* assimilated ^{241}Am in the hepatopancreas of the pond snail *Lymnaea stagnalis*) at two different pH's. The results show that both parameters can affect the whole-body retention of ^{241}Am. The effect of the speciation of ^{241}Am however is only seen at pH 4, whereas a difference in pH only influences the retention of ^{241}Am for animals intubated with ^{241}Am-humate.

The highest concentrations of ^{241}Am are found in the hepatopancreas representing 21 to 69 % of the whole-body radioactivity at the time of dissection. The radioactivity on the carapaces at that time account for 15 to 75 %. In the tail muscle between 0.1 and 3.8 % of the ingested ^{241}Am is found, which means that depending on the chemical form and the pH of the intubated ^{241}Am a difference in bioavailability of more than one order of a magnitude can occur.

1 INTRODUCTION

Environmental conditioning of ^{241}Am is thought to determine to a considerable extend the amount of ^{241}Am taken up by freshwater animals. Previously we reported on the effect of a few environmental parameters on the adsorption-desorption of ^{241}Am onto the exoskeleton of Astacus leptodactylus[1]. In the present paper attention is given to the availability of ^{241}Am following intragastric injection (intubation). We tried to assess the relation between the administered chemical form of ^{241}Am and its assimilation by the animal. Therefore the organ- and intracellular distribution of ^{241}Am, its direct release to the water and its elimination through defaecation were registered. We intubated three different forms of ^{241}Am: non-complexed ^{241}Am, ^{241}Am complexed to humic acids and in vivo ^{241}Am-labelled hepatopancreas homogenate of the freshwater snail Lymnaea stagnalis. Each of these forms was administered at pH 4 and pH 7.

2 MATERIALS AND METHODS

The experiments were carried out at constant temperature (18° C) and a 12 hr/12 hr light/dark cycle in separate plastic containers. Three intubates were administered to the crayfishes by intragastric injection: ^{241}Am-humate (^{241}Am added to a 10 mg/l humic acid solution[2]), non-complexed ^{241}Am (^{241}Am added to demineralised water) and ^{241}Am-labelled hepatopancreas homogenate of the pondsnail Lymnaea stagnalis. For this homogenate the top fractions of the hepatopancreas of in vivo contaminated pondsnails were homogenized. The dose of the intubates varied between 10 and 20 kBq of ^{241}Am except for the experiments with snail hepatopancreas homogenate. In this experiment each crayfish received 0.1 kBq of ^{241}Am. During the experiment the animals were starved. To determine whole-body retention as a function of time the crayfish were counted periodically on a γ-spectrometer (Quartz & Silize, Scintibloc, Type 25 with a NaI-(Tl) and a counting efficiency of 10 %). Meanwhile the radioactivity in the water was measured on two ml aliquots of water and the faeces were picked up with a forceps and monitored for ^{241}Am. After each measurement, the animals were replaced in renewed tap water. From animals intubated with non-complexed ^{241}Am

(pH=7) the intracellular distribution of the ^{241}Am in the hepatopancreas was determined. The radioactivity measurements in the water, on the feaces , in the organs and in the organel fractions were made with an automatic γ-spectrometer (Packard Autogamma Scintillation Spectrometer, Model 5986 with a counting efficiency of 22 %). Retention of ^{241}Am is expressed as the percent and as a concentration factor (CF) of the ingested radioactivity. The CF is defined as:

$$CF = \frac{(\text{residual radioactivity in organ}) \times g^{-1}}{(\text{ingested radioactivity}) \times g^{-1}}$$

3. RESULTS

The retention curves for the whole-animal ^{241}Am radioactivity are given in Figure 1. The initial loss of ^{241}Am over the first 5 days of the experiment represents a direct elimination of the intubate to the water together with a gut clearance of unassimilated ^{241}Am. After this period the radioactivity measured in the water is very low (around the detection limit) whereas the release through defeacation lasts throughout the experiments.

The setup of the experiments allows us to compare the different results both regarding an effect of the chemical form and the pH of the intubated ^{241}am. An effect of the speciation of ^{241}Am on the whole-body retention is only seen at pH 4. After 25 days significantly more ^{241}Am is retained by animals intubated with ^{241}Am-humate. Also when intubated with non-complexed ^{241}Am the animals retained more than in the experiment with ^{241}Am-labelled hepatopancreas homogenate. At pH 7 no significant effect of speciation on the ^{241}Am retention is noticed within a 30 days period. A difference between pH 4 and pH 7 on the other hand is only for seen for the ^{241}Am-humate intubated animals.

In Table 1 the organ distribution of ^{241}Am is shown as a CF of the ingested radioactivity. In all experiments the largest concentrations of residual ^{241}Am are found in the hepatopancreas whereas the lowest concentrations are calculated for the tail muscle. In the haemolymph the concentrations were at or below the limit of detection. Taking in account the weight of the hepatopancreas, it represents between 21 and 69 % of the residual radioactivity. On the carapaces between 15 and 75 %

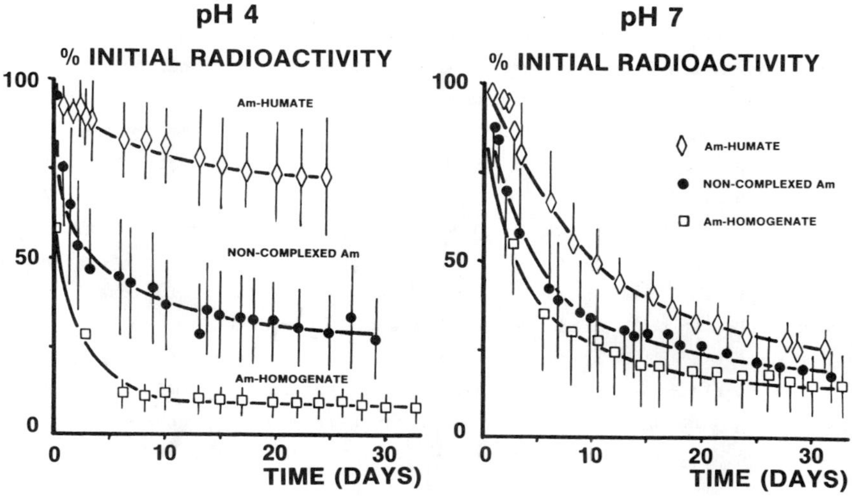

FIGURE 1 Whole-body retention patterns for Astacus leptodactylus intragastrically injected with different chemical forms of ^{241}Am at pH 4 (Fig. 1a) and pH 7 (Fig. 1b). The retention is expressed as the percent of the ingested radioactivity. Presented are the mean values with their 95 % confidence limits.

	NON COMPLEXED		HUMATE		HOMOGENATE	
	pH 4	pH 7	pH 4	pH 7	pH 4	pH 7
	CF x 10^{-4} (n = 8)	CF x 10^{-4} (n = 5)	CF x 10^{-4} (n = 7)	CF x 10^{-4} (n = 9)	CF x 10^{-4} (n = 4)	CF x 10^{-4} (n = 6)
STOMACH	32 ± 7	5 ± 1	310 ± 113	12 ± 3	15 ± 8	9 ± 6
GUT	91 ± 44	12 ± 2	720 ± 151	24 ± 13	41 ± 31	6 ± 11
HEPATO-PANCREAS	232 ± 49	25 ± 6	809 ± 273	40 ± 11	153 ± 60	58 ± 24
TAIL MUSCLE	2 ± 1	0.2 ± 0.1	42 ± 18	0.4 ± 0.1	5 ± 3	0.7 ± 0.2
GILLS	5 ± 1	0.9 ± 0.2	178 ± 70	3 ± 1	13 ± 7	6 ± 1
CARAPACE	5 ± 2	2 ± 1	61 ± 18	1 ± 1	34 ± 11	13 ± 2

TABLE 1 Concentration factors of the organs of Astacus leptodactylus intragastrically injected with different chemical forms of ^{241}Am at pH 4 and pH 7. Given are (CF ± Standard Error of the Mean) x 10^{-4}.

of the residual ^{241}Am is found. It is observed that a lower percentage of ^{241}Am in the hepatopancreas often coincides with a higher one on the carapaces. The tail muscle accounts for 1.3 to 5.3 % of the total body burden at the time of dissection. When we consider the ^{241}Am in the tail muscle as really assimilated by the animals, we can calculate that at least between 0.1 and 3.8 % of the ingested radioactivity is available for crayfish. At pH 4 the radioactivity retained in the tail muscle expressed as a percent of the ingested ^{241}Am is 0.4% for the non-complexed ^{241}Am, 3.8 % for the ^{241}Am-humate and 0.1% for the ^{241}Am homogenate. At pH 7 these percentage are 0.6, 0.7 and 0.2 %.

Compared to the animals intubated with neutral solutions, significantly more ^{241}Am is retained in all organs of animals intubated with acid (pH 4) ^{241}Am-humate, and in nearly all organs for those intubated with acid non-complexed ^{241}Am. The pH of the non-complexed ^{241}Am however had no significant effect on the ^{241}Am retention in the gut and the (secondarily) absorption on the carapaces.

The effect of the speciation of ^{241}Am on the concentration factors in the organs on the other hand is only seen at pH 4 between the animals intubated with non-complexed ^{241}Am and ^{241}Am-humate. We also calculate an significant effect of the speciation of ^{241}Am on the concentrations of ^{241}Am associated with the carapaces.

Furthermore there exists positive correlation between the hepatopancreas and the tail muscle (r=0.9854), between the stomach and the tail muscle (r=0.9803) and between the stomach and the hepatopancreas (r=0.9995).

The ^{241}Am accumulated by the hepatopancreas seems to be for the greater part associated with the lysosomal-mitochondrial fraction (41 %). It is likely that the lysosomes take most of this percentage for their account. From one until 15 days post administration no differences in the proportional subcellular distribution of ^{241}Am in the hepatopancreas were noticed.

4 DISCUSSION

The intubation experiments on the crayfish *Astacus leptodactylus* show a dependency of the whole-body retention of ^{241}Am on both the chemical form and the pH of the intubated ^{241}Am. At pH 4 (Figure 1a) significantly different whole-body retentions of ^{241}Am are seen for the three chemical forms of ^{241}Am used in our experiments. On the other hand we observe an increased retention of ^{241}Am for the animals intubated with ^{241}Am humate at low pH when compared with a neutral pH.

Comparable data on the availability of transuranics for crustaceans are scarce. A single worm-crab food chain experiment on the crab Cancer pagurus showed that depending on the valence of the ingested ^{237}Pu (+6 or +4), 12 to 41 % was retained in the tissues at the time of dissection[3]. In the present experiments the radioactivity in the tissues (whole-body retention minus radiaoctivity on carapaces) amounts between 3 on 33% of the ingested ^{241}Am. Some experimental work is also made on fish [4,5,6,7,8]. Eyman[4] and Trabalka point out that for the channel catfish Ictalurus punctatus chelation can either enhance or reduce the uptake of ingested ^{237}Pu compared to $^{237}Pu(OH)$. In the present experiments complexation of ^{241}Am with humic acids or with hepatopancreas homogenate increases respectively decreases the whole-body retention compared to non-complexed ^{241}Am. The effect disappears at neutral pH.

The major route of elimination seems to be defaecation. Whether there also occurs an excretion of ^{241}Am through the gills or the antennal glands remains uncertain because, except for the first days of the experiments, the amounts of ^{241}Am in the water were at the detection limit.

From our experiments we also notice the important contribution of the hepatopancreas (21 to 69%) in the whole-body retention of ^{241}Am for A. leptodactylus. This is comparable with findings in Cancer pagurus and Carcinus maenas in which 42-85 % of the ingested ^{237}Pu was found in the hepatopancreas at the time of dissection[3]. Within the cells of the hepatopancreas the ^{241}Am is mainly associated with the lysosomal-mitochondrial fraction. The same percent of ^{241}Am (46%) is observed in the lysosomal-mitochrondrial fraction of the hepatopancreas of Lymnaea stagnalis[9].

A speciation effect on the concentrations of ^{241}Am in the organs is especially seen at pH 4 between the organs of the animals intubated with ^{241}Am-humate and non-complexed ^{241}Am. At pH 7 the most important differences are seen between the carapaces. The carapaces concentrate the ^{241}Am less than the hepatopancreas but as they represent about 50 % of the total weight of the animal their contribution to the total body burden of ^{241}Am may reach 75 %. A low percentage of ^{241}Am often coincides with high percentages of ^{241}Am on the carapaces, which may indicate that amounts of non-assimilated ^{241}Am are secondarily adsorbed onto the carapace.

Important regarding ecological modelling is that between 0.1 and 3.8% of the ingested ^{241}Am is found in the tail muscle, which is the edible part of the crayfish. This means that depending on the chemical form of ^{241}Am and the pH of the intubated ^{241}Am a difference in bioavailability of more than an order of one magnitude can be seen. Furthermore, the present percentages are comparable to those of other crustaceans[3] but seem to be higher than the assimilation percentages reported for mammals[10]

This work was performed with a specialisation fellowship of the I.W.O.N.L. (Instituut tot Aanmoediging van het Wetenschappelijk Onderzoek in Nijverheid en Landbouw) and partly supported by CEC-contract nr BIO-329-B.

5. REFERENCES.

1. J. Bierkens, J.H.D. Vangenechten, S. Van Puymbroeck and O.L.J. Vanderborght, accepted for publication in Health Phys.
2. J.H.D. Vangenechten, S. Van Puymbroeck, J. Bierkens, L. Van Keer and O.L.J. Vanderborght, Proc. of Seminar on the Environmental Transfer to Man of Radionuclides Released from Nuclear Installations, Vol I, p. 174, (1984).
3. S.W. Fowler and J.C. Guary, Nature, 266, 827, (1977).
4 L.D. Eyman and T.R. Trabalka, Health Phys., 32, 475, (1977).
5. R.J. Pentreath, Mar. Biol., 48, 327, (1978).
6. R.J. Pentreath, Mar. Biol., 48, 337, (1978).
7. D.R.P. Leonard and R.J. Pentreath, Mar. Biol., 63, 67, (1981).
8. F.P. Carvalho, S.W. Fowler and J. La Rosa, Mar. Biol., 77, 59, (1983).
9. G.M.R. Thiels, Ph.D. dissertation, Univ. of Antwerp, Antwerp, p.122, (1982).
10. J.C. Nenot and J.W. Stather, in The toxicity of plutonium, americium and curium, C.E.C. (Pergamon Press, Oxford, 1979).

DISCUSSION

CREMERS:

It seems somewhat surprising that the effect of humic acid is absent in conditions which are favourable for the Am-HA complex formation (pH = 7) whereas at pH7, where the formation of Am-HA is much less favourable, there appears to be an effect. Any comment?

BIERKENS:

It was rather surprising for us too, but probably two variables do play in this. The inherent chemical environment in the stomach may alter the initial speciation of ^{241}Am in the intubate. Secondly we do not know the assimilation mechanism involved in the uptake of Am in the hepatopancreas cells: whether the ionic or complex form is favoured in uptake.

BONOTTO:

Would it be possible to separate mitochondria from lysosomes and to determine their ^{241}Am content?

BIERKENS:

It is, and we certainly will do in the near future. Moreover, we are planning to identify the molecule with which ^{241}Am is associated. This could influence the speciation of ^{241}Am.

ELECTROCHEMICAL STUDIES OF METAL SPECIATION IN MARINE AND ESTUARINE CONDITIONS

C.M.G. VAN DEN BERG

Department of Oceanography, University of Liverpool
Liverpool, L69 3BX, England

ABSTRACT

The degree of speciation of trace metals detected depends to some extent on the analytical method used. Possible changes in organic-metal interactions may be due to complex dissociation by competition with added chelators. Results obtained with cathodic stripping voltammetry are discussed in terms of interactions of dissolved organic ligands, and colloidal particles with dissolved metal ions. Estuarine data for copper, uranium, vanadium and iron are presented.

1. INTRODUCTION

Concentrations of radioactive elements are determined relatively easily because of their radioactivity. It allows for a certain degree of selectivity over other non-radioactive elements, or even over non-radioactive nuclides of the same element. However, radioactivity is non-specific in terms of the possible speciation of an element over various dissolved, or colloidal fractions, as the radiation is released no matter in what fraction it occurs. Radio-counting on its own is therefore a poor analytical tool to study speciation. This aspect could be investigated perhaps by using chemical separation techniques to preconcentrate certain metal fractions as a result of their different chemical reactivities. By their nature these separatory techniques tend to disturb solution equilibria in an unpredictory manner and it is therefore unlikely that a precise estimate of the solution composition is obtained. For this reason it will be argued in this paper, that perhaps other analytical tools should be looked at to investigate metal speciation to predict or explain the geochemical pathways of radionuclides. The metals copper, zinc, uranium, vanadium and iron will be used as examples in this discussion. The technique used in our laboratory to study speciation of these metals is cathodic stripping voltammetry (CSV) and a brief explanation of this technique is given first.

2. METHOD

2.1 Cathodic stripping voltammetry (CSV)

The technique of CSV is based on the reductive measurement of metal ions preceded by adsorptive collection of surface active complexes of these metals on a hanging mercury drop electrode (HMDE). Copper, iron, uranium and vanadium, for instance, can be determined in presence of catechol as added surface active chelating agent[1-14]. Adsorption takes place at a potential more positive than the reduction potentials of these metals (they are therefore collected in their original, oxidized form) whereafter measurement is by reductive voltammetry of the adsorbed complexes. Optimal conditions for the simultaneous determination of these four elements are a catechol concentration of 5×10^{-4}M, a solution pH 7.3, adsorption time of 1 min from a stirred solution, adsorption potential of -0.05 V (against Ag/sat.AgCl, KCl reference elctrode) and a scan using the differential pulse modulation. The solution pH is buffered at pH 7.3. A typical measurement in sea-water is shown in Fig. 1.

Zinc is determined by CSV in presence of amino pyrrolidine dithiocarbamate (APDC) after adsorption at -0.8 V.[5] The solution pH is buffered at pH 7.3 as before, or at a higher pH of 8.5.

3. RESULTS

3.1 Effects of complexation by natural organic ligands.

Natural organic complexing material forms complexes with the determined metal ions in competition with the added surface active chelator. The CSV peaks are then suppressed to an extent which depends on the respective ligand concentrations and the strength of the complexes formed. The effect of complexation of copper by natural organic material in sea-water on the peak height of copper at various concentrations of catechol is shown schematically in Fig. 2. Comparison of measurements in sea-water (SW) with those where the

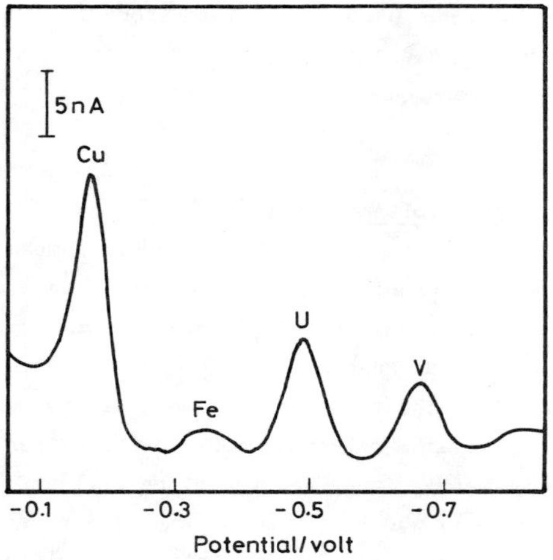

Fig. 1 Measurement of copper, iron, uranium and vanadium in sea-water using CSV. Metal concentrations are respectively 3nM, 5 nM, 15 nM and 9 nM.

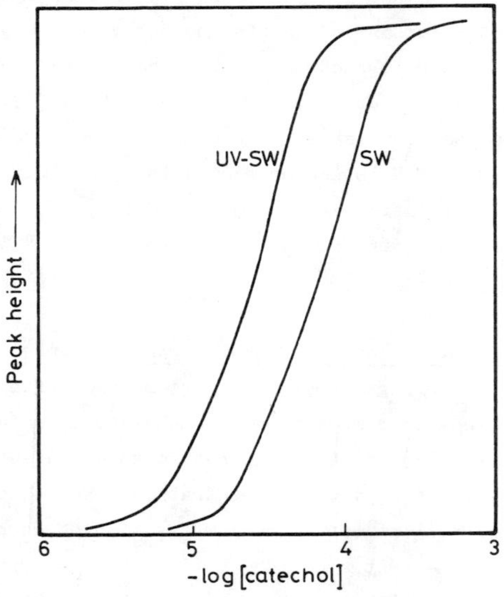

Fig 2. Effect of natural organic complexing material on the complexation of copper by catechol (represented by the peak height) in sea-water before (SW) and after (UV-SW) irradiation.

dissolved organic material has been removed by UV-irradiation (UV-SW) shows that the peak-height is suppressed at catechol concentrations less than 10^{-3}M by competition with natural organic ligands in the sample. There is not a particular catechol concentration where it would be justified to claim that the free metal ion concentration is determined. In a sample of the Atlantic for instance the apparent, dissolved copper concentration was found to be 0.8 nM in presence of 8 x 10^{-5}M catechol, 1.3 nM with 3 x 10^{-4}M catechol and 2.3 nM with 8.3 x 10^{-4}M catechol. [1] The detected metal concentration therefore varies with the amount of chelating agent added, the free (not complexed by organic ligands) metal concentration being of the order of 10^{-11}M as calculated from complexing capacity considerations. This finding is of importance to relativate metal speciation studies (e.g. of zinc[6], or, more recently, selenium[7]) based on extraction of a single metal fraction from a mixture of fractions with which it may be in equilibrium. A new equilibrium is attained quite rapidly (in a matter of seconds in case of copper[8]) when a high concentration of chelator is added to the solution as the natural organic complex only needs to dissociate to release metal ions to the competing ligand. Such studies may give valuable results but the estimates are likely to be low and possible complex dissociation should be checked.

The degree of complex dissociation depends on the relative magnitudes of the products of K'_{ML} (the conditional stability constant for the complex of metal M with ligand L) and C_L (the total concentration of L) for the natural ligand and the added chelator. It is possible to obtain an estimate of the ligand concentration present in a sample (also called the metal complexing capacity) by means of a stepwise titration of the ligand with metal ions, while measuring the detectable metal ion concentration using CSV. [8, 9] The result of such an experiment is shown in Fig. 3. A catechol concentration of about 3 x 10^{-5}M had been selected; the added chelator concentration should be below that needed to obtain the total metal concentration as shown before in Fig. 2. The straight line in Fig. 3 is the response in absence of natural organic ligands; the curved line is a best fit to the data which revealed the presence of two ligands at concentrations of 1.1 and 3.3 x 10^{-8}M, and with conditional

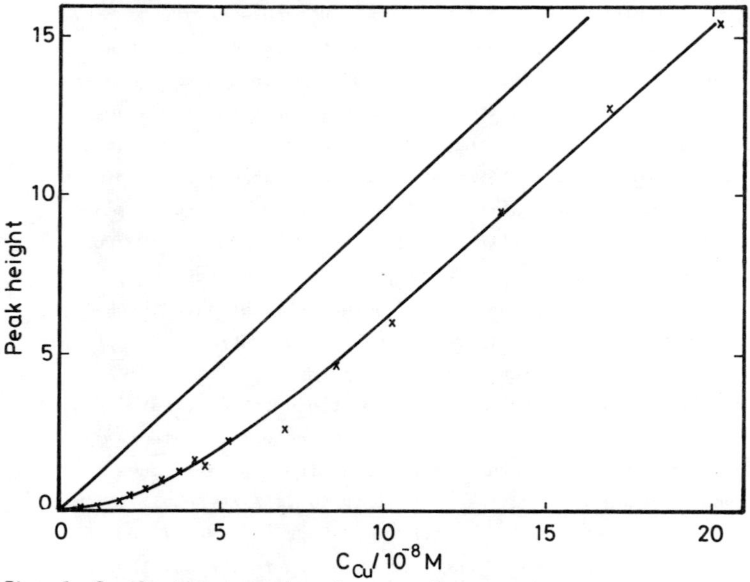

Fig. 3. Complexing capacity of copper in sea-water derived from the Atlantic, as determined using CSV. Data from van den Berg[8].

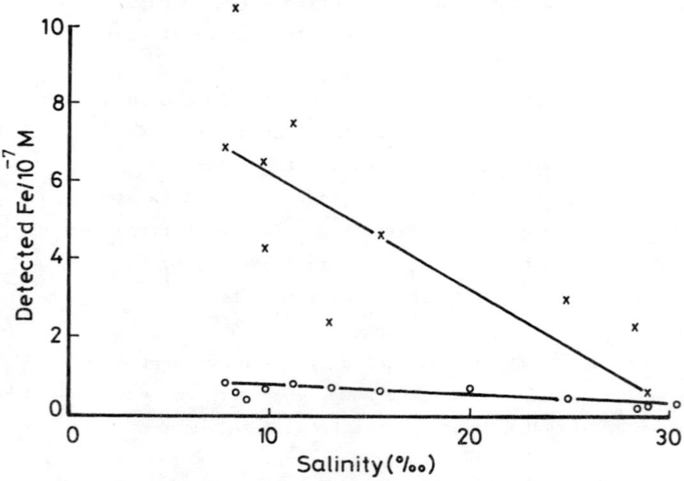

Fig. 4. Labile concentrations of iron in filtered samples of the Ribble estuary, before (oo) and after (xx) acidification and UV-irradiation. Measurements by CSV. Data from van den Berg et al.[14]

stability constants, log K'_{ML}, of 12.2 and 10.3 respectively.

3.2 Organic metal interaction in sea-water: copper and zinc

Results for copper and zinc are presented here only because these metals have been investigated relatively thoroughly.

Metal complexation studies similar to that described above have now shown that up to about 98% of copper [9-12] and perhaps about 25-40% of zinc, [9,11,13] is complexed by dissolved organic material in sea-water from coastal or oceanic origin. Ligand concentrations found vary with the technique used, but are of the order of 5-50 nM. Ligand concentrations tend to be greater in estuarine waters and decrease with increasing salinity. [12-14] Even higher ligand concentrations, up to micromolar levels, have been found in interstitial waters.[13]

3.3 Interactions of metals with colloids in estuaries.

The time it takes to perform a measurement by CSV after addition of chelating agent is only a few minutes. Only dissolved components are therefore measured by this technique. In fact tests have shown that long equilibration times are needed, on the order of an hour, for particulate, or colloidal, material to react. CSV is therefore a convenient tool to study speciation over dissolved and colloidal components. In Fig. 4. the CSV-detectable iron concentration has been plotted in filtered (0.45 um) samples from the Ribble estuary, [14] before and after UV-irradiation and acidification. It appeared that only a small fraction, about 5×10^{-8} M of dissolved (filtrable) iron was CSV-labile, whereas the total dissolved iron concentration varied between 10^{-7} and 10^{-6} M.

Variation of the dissolved ligand concentration (C_L) and the concentrations of copper and vanadium in samples from the Tamar estuary is shown in Fig. 5 as function of the salinity. There is no evidence of strong removal of copper from solution in this estuary, whereas it is considered to be a geochemically reactive metal in its free ionic form. Perhaps this apparently conservative behaviour of copper is caused by the fact that it occurs for 98% organically complexed in these waters. [14] Vanadium on the other hand behaves generally conservative in oceanic conditions, [15] but variation in the concentration found in samples from the Tamar estuary (Fig. 5) could perhaps be explained by interaction with suspended particles.[14]

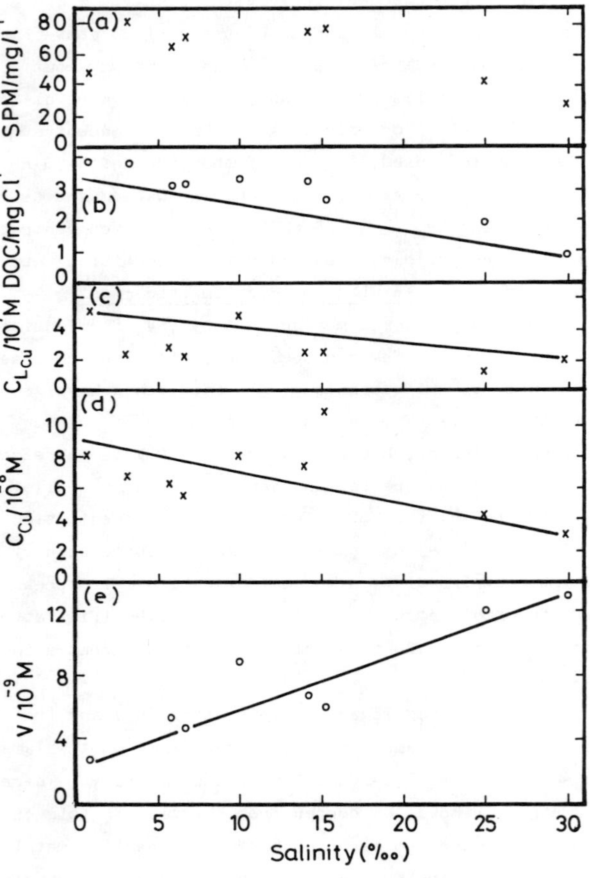

Fig. 5 Solid particulate matter (a), dissolved organic carbon (b), complexing ligand concentration (c), dissolved copper (d)[14] and dissolved vanadium (e) in samples of the Tamar estuary.

Colloidal speciation of vanadium, uranium and iron can be observed in the results from an experiment in which sea-water is gradually mixed with synthetic freshwater, as shown in Fig. 6. The metal concentrations at the freshwater end were 1.8 um Fe, 2×10^{-8}M U and 1.4×10^{-8}M V. Any organics present in the solutions were oxidized by UV-irradiation prior to mixing. The pH varied from 7.5 at the freshwater end to 7.8 at the sea-water end. The mixed solutions were allowed to equilibrate overnight by shaking in a waterbath at 20°C. The labile iron concentrations were between 2 and 3 nM at salinities greater than 18×10^{-3}, higher at lower salinities. Some dissolution of colloidal iron may have contributed to the measurement especially at the lower salinity end where the colloidal iron concentration was up to 1.8 uM. Nevertheless it is clear that only a very minor fraction of filtrable iron is in fact dissolved. The colloidal iron fraction removed dissolved vanadium and uranium from solution at the low salinity end. These metals were released back into solution upon mixing with sea-water: uranium is almost completely re-dissolved at salinities greater than 5×10^{-3}, whereas vanadium was released more gradually over the entire salinity.

4. DISCUSSION AND CONCLUSIONS

Measurements by CSV of the labile concentration of copper in sea-water at various added levels of catechol showed that natural organic complexes dissociate quickly when a competing chelator is added. This effect should be taken into account when organic-metal interactions are studied using extraction techniques. Measurement of the metal fraction complexed by the added chelator (by CSV for instance) and at various added metal concentrations can then be used to determine complexing capacities and conditional stability constants.

Uranium has been found to behave conservatively in estuarine conditions, [16,17] although some indication for uranium removal has been observed in one estuary.[18] Mixing experiments discribed here indicate that uranium is strongly adsorbed by colloidal iron hydroxides at salinities less than 5×10^{-3}. Vanadium is adsorbed by colloidal iron hydroxide over a much greater salinity range up to about 25×10^{-3}. Both elements can perhaps be remobilized if colloidal iron hydroxide is transported from low saline to sea-water conditions.

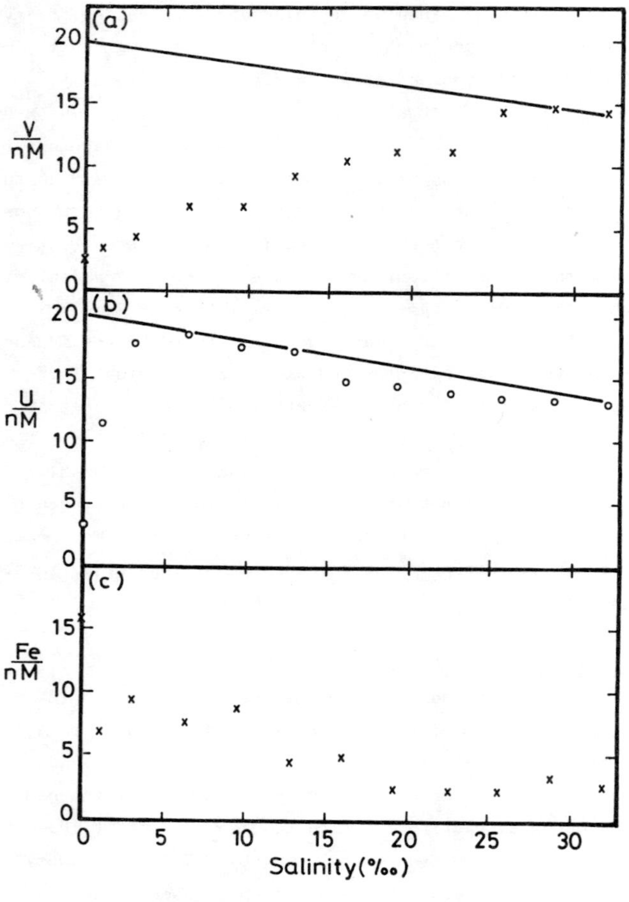

Fig. 6. Mixing of synthetic freshwater (1mM HCO_3, 1.8 μM Fe, 2 x 10^{-8} M U, 2 x 10^{-8} M V) with sea-water. Straight lines indicate total metal concentrations, data points are CSV-labile concentrations (van den Berg, unpublished results).

5. REFERENCES

1. C.M.G. van den Berg, Anal.Chim.Acta, 164, 195(1984).
2. C.M.G. van den Berg, and Z. Q. Huang, Anal.Chim.Acta, 164, 209 (1984).
3. C.M.G. van den Berg and Z.Q. Huang, Anal.Chem, 56, 2383 (1984).
4. C.M.G. van den Berg and Z.Q. Huang, J.Electroanal.Chem, 177, 269 (1984).
5. C.M.G. van den Berg, Talanta, 31, 1069 (1984).
6. J.F. Slowey and D.W. Hood, Geochim. Cosmochim. Acta, 35, 121 (1971)
7. K.Takayanagi and G.T.F. Wong, Geochim.Cosmochim.Acta, 49, 539 (1985)
8. C.M.G. van den Berg, Mar.Chem., 15, 1 (1984)
9. C.M.G. van den Berg, Mar.Chem., in press (1985)
10. C.M.G. van den Berg, Mar.Chem, 14, 201 (1985)
11. K.Hirose, Y.Dokiya and Y.Sugimura, Mar.Chem, 11, 343 (1982)
12. C.J.M. Kramer and J.C. Duinker, in Complexation of Trace Metals in Natural Waters., edited by C.J.M. Kramer and J.C. Duinker, Nyhof/Junk publishers, p.217-228 (1984).
13. C.M.G. van den Berg and S.Dharmvanij, Limnol.Oceanogr., 29, 1025 (1984).
14. C.M.G. van den Berg, P.J.M. Buckley, Z.Q. Huang and M.Nimmo, Est.Coast.Shelf.Sci, submitted (1985).
15. A.W. Morris, Deep-Sea Research, 22, 49 (1975)
16. J.M. Martin, M.Meybeck and M.Pusset, Neth.J.Sea Res., 12, 338 (1978).
17. D.V. Borole, S.Krishnaswami and B.L.K. Somayajulu, Est.Coast.Mar.Sci, 5, 743 (1977)
18. M.Maedar and H.L. Windom, Mar.Chem., 11, 427 (1982)

DISCUSSION

SIBLEY:

Do you think cathodic stripping voltammetry can be applied to the transuranic elements?

VAN DEN BERG:

The technique of cathodic stripping voltammetry can measure changes in oxidation states, preceded by some form of adsorptive collection. It is not necessary to reduce the metal to the metallic state. It is therefore quite possible to determine transuranic elements as well, but I have not tried it.

FISHER:

Do you see any correlation between the metal complexing capacity and the primary productivity in different water columns? Are the phytoplankton contributing significantly to the production of naturally occurring ligands in the sea?

VAN DEN BERG:

High complexing capacities are always found in surface waters. Furthermore all algae species I have tried excrete organic ligands. It is therefore most likely that algae produce the complexing capacities observed in the water column.

GUEGUENIAT:

What do you think of the discrepancies which are observed between polarography and manganese dioxide coprecipitation for the estimation of trace elements?

VAN DEN BERG:

By polarography the electrochemically labile metal concentration is measured, which is almost equivalent to the metal fraction not complexed by organic material. Addition of manganese dioxide, however, changes the equilibrium condition part of the complexed metal fraction dissociates and adsorbs onto the oxide. You measure then an enhanced "free" fraction. Perhaps this is not really a discrepancy, but a different type of measurement which gives you a different result.

THE EFFECT OF OXYGEN TENSION IN THE SEDIMENT ON THE BEHAVIOUR OF WASTE RADIONUCLIDES AT THE NEA ATLANTIC DUMPSITE

M.M. RUTGERS VAN DER LOEFF and D.A. WAIJERS

Netherlands Institute for Sea Research, P.O. Box 59, 1790AB Den Burg, Texel, The Netherlands

ABSTRACT

Predictions of the transport and fate of waste radionuclides at the NEA Atlantic dumpsite require a knowledge of the behaviour of these nuclides within the sediment. Since redox conditions are known to influence the mobility of many elements in deep-sea sediments, we have investigated the speciation of some trace elements in relation to the dissolved oxygen concentration in sediments from the dumpsite.

Dissolved oxygen in these sediments penetrates mostly between 50 and 100 cm, although at topographic highs and on hillsides oxygen penetrates more than 2 m into the sediment because of the special hydrodynamic and sedimentological conditions there. Remobilization at lowered redox potentials below the depth where oxygen reaches zero causes an upward diffusive transport of Mn and the manganese-associated trace metals Co and Ni. Whether this diagenetic mobilization influences other elements such as rare earth elements and actinides as well, remains to be investigated. Under normal sedimentological conditions this mobilization can not be expected to return radioactivity to the water column through an oxidized sediment layer of 50 cm. However, burial of radioactivity to dephts beyond the reach of deep burrowing organisms can be significantly delayed. Kd values (solid/dissolved partition coefficients) of redox sensitive elements vary over orders of magnitude and are inappropriate to model the behaviour of these elements in sediments with redox gradients.

1 INTRODUCTION

A considerable part of the radioactivity that is dumped at the NEA dumpsite for low-level radioactive waste will be adsorbed on sediment particles and then buried by bioturbation and sediment accumulation. The subsequent behaviour of the radionuclides will depend on their chemical properties, especially speciation changes, and on the geochemical behaviour of the solid phases to which the nuclides are bound. The relatively short history of waste dumpings precludes the study of speciation of actual waste radionuclides in dumpsite sediments. However, predictions of their behaviour can be based on a study of their stable analogues.

The speciation of many elements and the behaviour of the carrier phases are sensitive to changes in the redox potential in the sediment. The shallowest redox transition occurs at the depth in the sediment where oxygen becomes depleted. Below this depth, manganese oxide, nitrate and sulphate take over the role of electron acceptor in the mineralization of organic matter. Some elements such as V,U (1) and Pu (2) become less mobile in anaerobic layers. Others, such as Mn (3,4) and associated elements like Co (5,6) and Ni (6,7), are mobilized in the reduced zone and accumulate in the oxidized surface layer. A behaviour similar to Mn is inferred for other elements from their enrichment in the oxidized zone. Dissolved oxygen is therefore a key parameter in the description of the diagenetic behaviour of trace elements including radionuclides at the dumpsite.

The vertical distribution of dissolved oxygen in sediments at and near the dumpsite has been presented in an earlier report (8). The present paper relates these data to the partitioning of transition metals between the dissolved phase (pore water), and four different phases of the sediment that are distinguished by a selective extraction procedure. Transition metals were selected for this study because of their known dependence on redox conditions. The large range of observed distribution coefficients and the implications for the modelling of radionuclide behaviour are discussed.

We thank captain and crew of RV Tyro and our NIOZ colleagues for their support during the cruise. The project is financed by the Netherlands Energy Research Foundation, the Dutch government (Ministry of Environment and Ministry of Economic Affairs), contributions from the Swiss and Belgian governments and through a contract with the Commission of the European Communities (No. BIO-B-509-NL (N)).

2 METHODS

TABLE I Sequential extraction procedure

	name	extractant	repeat	extracted fraction	ref.
1	HAC	NaOAc (1M) + HAc, pH 5.0	*3	carbonate, sorbed,salt	9
2	HAM	$NH_2OH.HCl$(1M) + Na citrate (0.175M), pH 5.0	*3	Mn-oxyhydroxide	9
3	CH	$NH_2OH.HCl$(1M) + HAc (25% v/v)	1	Fe-oxyhydroxide	10
4	HCl	HCl (1M)	1	residual leachable	

Location of stations where sediments were sampled during a cruise with RV Tyro in 1982, and methods for dissolved oxygen analysis and pore water extraction are described elsewhere (8). 0.5 g aliquots of freeze-dried and homogenized sediment were sequentially extracted using the procedure outlined in Table I, modified from (9), and the extracts were analysed with atomic absorption spectrometry (AAS). Pore water samples, extracted on board, were acidified with 1 ml/l of 6 mol/l HCl suprapur and stored at 4 ^{0}C. Trace metal contents were determined by flameless AAS using direct injection for Mn and a preconcentration (11) for Ni and Co. Analytical blanks for these metals were .02, .02 and .003 µg/l, respectively.

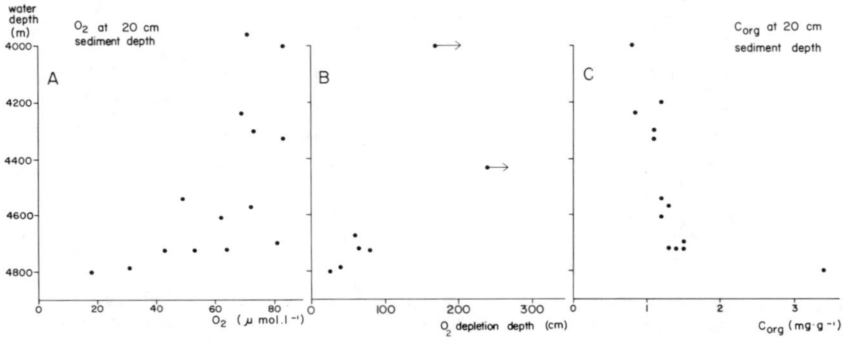

FIGURE 1, A:dissolved O_2 at 20 cm sediment depth (bottom water value: 245 µmol/l); B: O_2 depletion depth; C: C(org) in sediment at 20 cm sediment depth vs water depth in dumpsite area.

3 RESULTS AND DISCUSSION

3.1 Oxygen penetration in the sediment

As a result of the consumption of oxygen for the mineralization of organic matter, the dissolved oxygen concentration in the pore water decreases rapidly with increasing depth in the sediment (8). The oxygen concentration at 20 cm depth tends to be lower in the valleys than in the hill areas (Fig. 1.a). Also, the depth in the sediment where oxygen is depleted tends to be shallower in the valleys (Fig. 1.b). At topographic highs, anaerobic layers were in fact too deep to be reached by the available coring equipment (Fig.1.b). The decreased oxygen penetration at increased water depth contrasts with the usual trend of increasing oxygen penetration depth with increasing water depth that results from a decreasing supply of organic matter . The anomalous behaviour can be explained by winnowing of the sediment at local highs (12). This winnowing results in a relatively coarse sediment on the hills with large Foraminifera tests and abundant ice-rafted debris (Jaquet in 13). A difference in the sedimentation regime is also evidenced by the observation that the Benthic Nepheloid Layer is much more developed in the valleys than over the hills (14). The accumulation of fine particles in the valleys could cause the shallower oxygen penetration through a higher organic matter content (Fig. 1.c) and a lower permeability of the sediment.

3.2 Speciation of transition metals

3.2.1 Evaluation of the selective extraction method

The ability of the selective attack to distinguish between various operationally defined phases in the sediment is demonstrated in Fig. 2 for an oxidized boxcore. The HAC attack dissolves completely the calcite which makes up 85% (top) to 79% (bottom) of the sediment in this core. Beside Ca and Sr,

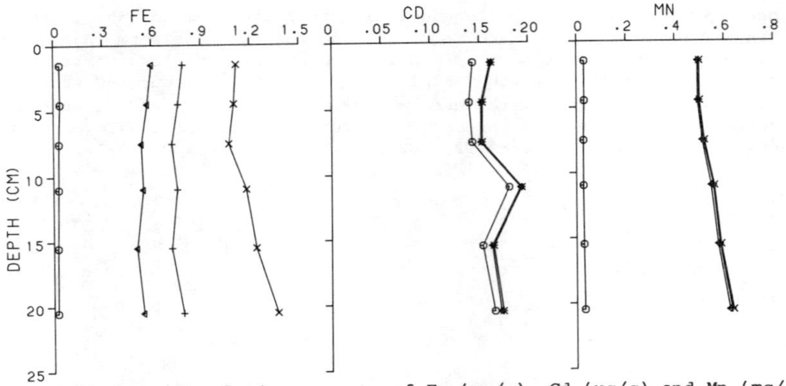

FIGURE 2, Cumulative amounts of Fe (mg/g), Cd (µg/g) and Mn (mg/g) extracted by HAC (O), HAM ( ), CH (+) and HCl (X) from an oxidized boxcore (station 6 in ref. 8).

this extract contains the major part of HCl-extractable Cd (Fig.2.b).

The HAM leach releases all Mn and the Co and Ni associated with it (Fig. 2.c). This leach is thus specific for the Mn-oxyhydroxides.

The CH leach (after Chester-Hughes, 10) releases additional amounts of Fe (Fig. 2.a), without attacking clay minerals too heavily, and is therefore meant to distinguish which metals are associated with Fe-oxyhydroxides (e.g. Zn and Cu) rather than with Mn.

The HCl leach is only included for reference. An analysis of the residual fraction has not been done.

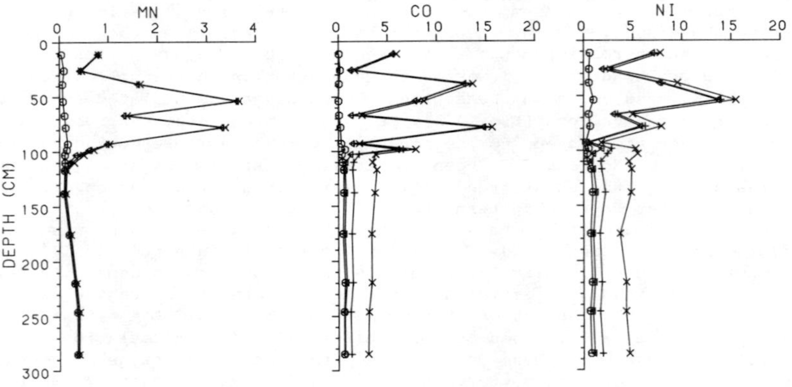

FIGURE 3, Mn (mg/g), Co and Ni (µg/g) speciation in core 11 (ref. 8) with oxygen depletion at ca 80 cm. For explanation of symbols see Fig.2.

3.2.2 Results of a core with redox transition

The selective extraction was performed on core 11 from the western valley (Fig. 3). Oxygen is depleted at a depth of approx. 80 cm in this core (8). Mn, Co and Ni are strongly enriched in the oxygen containing layer. These metals occur here predominantly in the HAM extract, corresponding to the Mn oxyhydroxide phase, whereas the much smaller amounts in the reduced layer

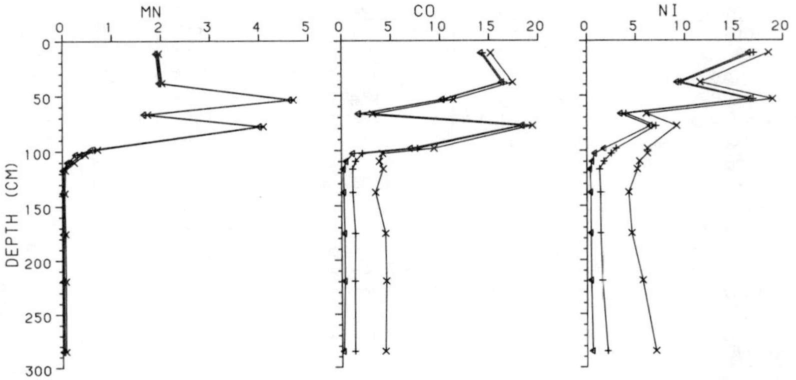

FIGURE 4, Cumulative amounts of Mn (mg/g), Co and Ni (µg/g) in HAM, CH and HCl leaches from core 11, expressed on a $CaCO_3$-free basis.

occur in the HAC extract (Mn, probably as mixed carbonate overgrowths on Foraminifera tests, 15) and in the CH and HCl extracts (Co and Ni).

The transition from the oxidized to the reduced layer becomes more apparent if the data are expressed on a $CaCO_3$-free basis (Fig. 4). This figure has further been improved by the omission of data from three depth horizons (26, 93 and 246 cm) where the sediment is diluted by much ice-rafted material, as evidenced by X-radiographs and the Mg content in the leaches (indicative of dolomite).

The gradients in solid-phase Mn, Co and Ni (Fig. 4) are sustained by reductive dissolution in the anaerobic layer and upward diffusion along strong gradients of dissolved metals (Fig. 5). This upward diffusion delays the burial of these metals beyond the reach of deep burrowing organisms. The opposing gradients imply that the solid/dissolved partition coefficients (Kd) vary over orders of magnitude between the two layers and are inappropriate to model the behaviour of such redox-sensitive elements in sediments with redox gradients.

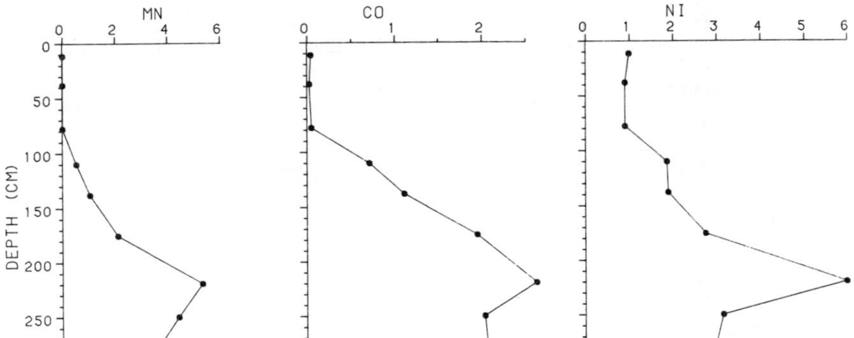

FIGURE 5, Mn (mg/l), Co and Ni (µg/l) in pore water of core 11.

3.3 Implications for radionuclide behaviour

Research on the formation of manganese nodules has revealed that many elements are associated with the oxyhydroxide phase in marine sediments. Wyttenbach (in 13) showed that 79% of Co, 50% of Ce and 25-40% of other REE, V, U and As in oxidized dumpsite sediments was present in the Fe/Mn-oxyhydroxide phase (corresponding to our phases 2+3+4). Upon reduction of the carrier phase, the associated elements are set free, and a transport process similar to the one established here for Mn, Co and Ni would be possible. In a core where no significant difference exists between total ($CaCO_3$-free) REE contents in the oxidized and the reduced layer, the speciation of REE was shown to change with depth (16). Analyses of selective leaches and of pore water samples for REE and other analogues of radionuclides could reveal diagenetic mobilization processes that are not apparent from analyses of bulk sediments alone.

5 REFERENCES

1. S. Colley, J. Thomson, T.R.S. Wilson and N.C. Higgs, Geochim. Cosmochim. Acta, 48: 1223-1235 (1984).
2. D.M. Nelson and M.B. Lovett, IAEA/OECA international symposium on the impact of radionuclide releases into the environment. IAEA,Vienna (1980).
3. S.E. Calvert and N.B. Price, Earth Planet. Sci. Lett., 16:245-249 (1972).
4. P.N. Froelich, G.P.Klinkhammer, M.L.Bender, N.A.Luedtke, G.R.Heath, D.Cullen and P.Dauphin, Geochim. Cosmochim. Acta, 43: 1075-1090 (1979).
5. D. Heggie and T. Lewis, Nature, 311: 453-455 (1984).
6. M. Hartmann, Chem. Geol. 26: 277-293 (1979).
7. G.P. Klinkhammer, Earth Planet. Sci. Lett. 49: 81-101 (1980).
8. M.M. Rutgers van der Loeff and M.S.S. Lavaleye, Neth. Inst. for Sea Res. int. rept. 1984-2 (1984).
9. M. Lyle, G.R. Heath and J.M. Robbins, Geochim. Cosmochim. Acta, 48: 1705-1715 (1984).
10. R. Chester and M.J. Hughes, Chem. Geol. 2: 249-262 (1967).
11. L.-G. Danielsson, B. Magnusson and S. Westerlund, Anal. Chim. Acta, 98: 47-57 (1979).
12. R.B. Kidd, in Interim Oceanographic description of the North-East Atlantic dumpsite for the disposal of low-level radioactive waste, edited by P.A. Gurbutt and R.R. Dickson (NEA/OECD, Paris), Chapter 4 (1983).
13. M.M. Rutgers van der Loeff, J.-M. Jaquet, P. Ruch and A. Wyttenbach, in Interim oceanographic description of the North-Eastern Atlantic dumpsite II, (NEA/OECD Paris), Chapter 4 (1985).
14. F. Nyffeler, A. Wyttenbach and J.-M. Jaquet, in Interim oceanographic description of the North-Eastern Atlantic dumpsite II, (NEA/OECD Paris), Chapter 2 (1985).
15. E.A. Boyle, Geochim. Cosmochim. Acta, 47: 1815-1819 (1983).
16. V. Marchig, P.Möller, H.Bäcker and P.Dulski, Mar. Geol. 62:85-104 (1985).

DISCUSSION

CREMERS:

What is the mineralogy of the sediments?

How do you visualize the scenario of radionuclides reaching deeper fractions of the sediment?

RUTGERS VAN DER LOEFF:

Apart from calcite, the sediment consists of the clay minerals smectite, illite, kaolinite and chlorite, especially in layers with ice-rafted material. Iron oxides make up about 7% of the non-carbonate phase, but I do not know the mineral form in which they occur.

The sediment accumulation rate is only 2 cm per 1000 yr at present. Sediment is also buried by bioturbation which is most rapid in the upper 6-8 cm, but is also effective to greater depth. Mobilization in the reduced zone will only affect the behaviour of radionuclides with relatively long half-lives.

BONOTTO:

Organisms living in deep sediments could interact with dumped radionuclides. Could you, please, briefly comment on the possible effects?

RUTGERS VAN DER LOEFF:

The infauna does certainly play a role in the transport of radionuclides. Bioturbation is the major mechanism of vertical transport of particle-bound radionuclides on timescales shorter than thousands of years. Benthic organisms ingest particles from the bottom water and in this way help to incorporate radionuclides into the sediment. Bacteria mediate the mineralization of organic matter, and thus mobilize radionuclides that were bound to this organic matter.

TOOLE:

You carried out 2 repeat leaches for the carbonate/salt and the manganese oxyhydroxide phases. Was there any additional removal of metals in the second and third leaches?

RUTGERS VAN DER LOEFF:

Three successive extractions with the acetate buffer were required to dissolved all calcite, and the associated metals, present in samples with the highest carbonate content (85%). The second and third extraction of the manganese oxyhydroxide phase also yielded additional amounts of metals. Further extractions of this phase did not yield significant additional amounts of Mn, Co and Ni, but yielded increasing amounts of iron.

CHEMICAL SPECIATION OF TRANSURANIUM NUCLIDES DISCHARGED INTO THE MARINE ENVIRONMENT

R. J. PENTREATH, B. R. HARVEY and M. B. LOVETT

Ministry of Agriculture, Fisheries and Food,
Directorate of Fisheries Research, Fisheries Laboratory,
Lowestoft, Suffolk NR33 OHT, England

ABSTRACT

Of the long-lived nuclides discharged into the marine environment, the most detailed studies have been made on the transuranium nuclides, although some work has also been carried out on technetium. These studies have included an evaluation of the chemical forms discharged from the British Nuclear Fuels plc (BNFL) reprocessing plant at Sellafield, Cumbria, their subsequent behaviour upon contact with sea water, their oxidation states in the waters around the UK coast-line, their adsorption onto particulate matter, their oxidation states in interstitial waters, and their accumulation by biological materials. From the evidence obtained to date it appears that Pu, in particular, undergoes changes in chemical form following discharge, and that the various oxidation states exhibit marked differences in their adsorbative properties for environmental materials. The most marked changes for both Pu and Np occur in interstitial waters, being dependent upon the Eh and pH conditions prevailing. There are also marked differences in the apparent biological availabilities of each element, which is presumed to be more than a reflection of the predominant oxidation states obtaining in ambient sea water and the sedimentary particles.

1 INTRODUCTION

The Ministry of Agriculture, Fisheries and Food (MAFF) is one of the joint authorizing departments for the discharge of low-level liquid radioactive wastes into United Kingdom coastal waters. Part of the research work which MAFF undertakes as support for its functions of inspection and monitoring is concerned with delineating the mechanisms by which radionuclides become dispersed in the marine environment and the pathways by which such radionuclides may lead to human exposure or may produce adverse environmental effects.

A basic requirement for obtaining an understanding of such mechanisms involves the determination of the chemical forms of nuclides at discharge and any subsequent changes which occur in the environment. The discharges from Sellafield provide a valuable opportunity to study the behaviour of long-lived nuclides in the marine environment. Early studies, for example, indicated the existence of both oxidized and reduced forms of Pu in Irish Sea water[1]. The relevance of this difference is that the distribution coefficient (K_d) between water and sedimentary material differs by some two orders of magnitude depending on the oxidation state. This paper describes the most recent studies relating to the speciation of transuranic nuclides discharged from Sellafield.

2 THE SELLAFIELD DISCHARGES

The low-level radioactive liquid effluents discharged arise from water used to purge the cooling ponds in which spent fuel elements are stored and from a variety of other processes which are operated throughout the plant. Pond-water is discharged continuously but other liquid wastes are routed through holding 'sea' tanks in which the predominantly acid liquors are neutralized with ammonia before being discharged.

Although the chemical nature of the Sellafield effluent has not been studied extensively, it is known that tiny particulate fragments $\ngtr$ 5 μm are present in the discharges[2]. These "hot particles" appear to persist in the environment for some months before dissolving[3]. Unlike pond-water effluent, which can be considered essentially oxidizing in character, the sea-tank liquors contain both the residues of organic complexing agents and some iron which arises largely from the use of ferrous sulphamate as a

TABLE I Distribution of transuranium nuclides between particulate and liquid phases of a 1982 Sellafield effluent sample (errors are ± 1σ propagated counting errors)

	Filtrate (0.22 μm), Bq l^{-1}		Particulate, Bq l^{-1}	
	Sea tank	Pond water	Sea tank	Pond water
^{237}Np	24.8 ± 1.8	0.15 ± 0.06	39.1 ± 3.3	0.020 ± 0.005
$^{239/240}Pu$	142 ± 15	78.6 ± 4.3	10 343 ± 350	824 ± 18
^{241}Am	54.2 ± 8.3	9.1 ± 1.3	3 611 ± 150	620 ± 38
$^{243/244}Cm$	Not detectable	1.5 ± 0.6	9.0 ± 6.8	34.3 ± 9.1

reducing agent to control the valency of plutonium, in particular, during fuel reprocessing. Even after neutralizing, the sea-tank effluent retains a mildly reducing character but, as would be expected, a floc composed of iron and other metal hydroxides separates to give an effluent which may contain suspended solids of up to about 50 mg l^{-1}. The distribution of transuranium nuclides between particulate and liquid phases of an effluent sample is given in Table I.

3 THE SPECIATION OF TRANSURANIUM NUCLIDES IN SEA WATER

Studies which have attempted to make approximate budget estimates of the distribution of transuranium nuclides in the Irish Sea have shown that only a small fraction of Pu and Am is retained in sea water, the majority being removed to settled sediments[2]. This is to be expected in view of the K_ds determined for sedimentary materials in suspension, which indicate that the values are of the order of 10^4 for Np, 10^5 for Pu and 10^6 for Am and Cm. Nevertheless, this still implies that a fraction of all of these nuclides exists in solution, a fraction which will be transported throughout the waters of the Irish Sea and beyond. With regard to Pu, Nelson and Lovett[1] demonstrated that more than 75% in filtered eastern Irish Sea water was in an oxidized form but their technique - which employed potassium dichromate as a holding oxidant - was unable to distinguish between Pu(V) (PuO_2^+) and Pu(VI) (PuO_2^{2+}). Since then studies with freshwater have shown that Pu exists predominantly as Pu(V), as indicated by the use of silicic acid or calcium carbonate as co-precipitants[4]. Studies have now been made with UK coastal waters, using $CaCO_3$ precipitation at pH 9 - by adding Na_2CO_3 and $NaHCO_3$ - which utilizes the indigenous Ca in sea water. Some examples are given in Table II. Samples were initially separated into oxidized and reduced forms by the method of Lovett and Nelson[5]. Replicate samples were taken to which ^{236}Pu (oxidized) and ^{242}Pu (reduced) tracers were added prior to calcium carbonate precipitation. All of the ^{242}Pu tracer (Pu III+IV) and at least 85% of the ^{236}Pu was co-precipitated. Replicate samples were also taken to which sufficient $KMnO_4$ was added - to give a 20 μM concentration - in order to oxidize the Pu(V) to Pu(VI). Again all of the added ^{242}Pu tracer was co-precipitated with calcium carbonate but only about 10% of the ^{236}Pu was co-precipitated, the major fraction - being in the PU(VI) oxidation state - was not removed from solution.

TABLE II Removal of Pu from sea water by $CaCO_3$ co-precipitation before and after oxidation with $KMnO_4$ (errors are ± 1σ propagated counting errors)

	Location (1984)	
	Sellafield area	South Scottish coast
$^{239/240}Pu$(III+IV) mBq l^{-1}	2.45 ± 0.08	0.55 ± 0.02
$^{239/240}Pu$(V+VI) mBq l^{-1}	21.30 ± 0.40	2.58 ± 0.06
$CaCO_3$ co-precipitation		
$^{239/240}Pu$(III+IV+V) mBq l^{-1}	21.16 ± 0.31	2.64 ± 0.06
$KMnO_4$ + $CaCO_3$ co-precipitation		
$^{239/240}Pu$(III+IV+VI) mBq l^{-1}	6.16 ± 0.15	0.93 ± 0.03

Studies have also been made on ^{237}Np, for which the development of ^{235}Np as a yield tracer has been a considerable advantage[6]. The two oxidation states, Np(V) and Np(IV), have been separated by using NdF_3 co-precipitation, ^{235}Np(V) being used to indicate the small amount of ^{237}Np carried down by the Np(IV) co-precipitate. The results are given in Table III, from which it can be seen that close to Sellafield some 1% may exist as Np(IV) and that a measurable amount of ^{237}Np still exists as Np(IV) as far away as the southern Scottish coast.

TABLE III ^{237}Np in replicate samples of filtered (0.22 μm) Irish Sea water, 1984 (errors are ± 1σ propagated counting errors)

	Location	
	Sellafield area	South Scottish coast
^{237}Np(V) mBq l^{-1}	8.99 ± 0.09 9.01 ± 0.10	0.368 ± 0.010 0.364 ± 0.010
^{237}Np(IV) mBq l^{-1}	0.11 ± 0.01 0.10 ± 0.01	0.0014 ± 0.0003 0.0024 ± 0.0002
% ^{237}Np(IV) of total ^{237}Np	1.21 ± 0.11 1.10 ± 0.11	0.38 ± 0.08 0.66 ± 0.06

The possibility that some oxidized ^{241}Am may exist in sea water cannot be ignored. The earliest separation of Cm from Am was achieved by oxidizing Am to the AmO_2^+ ion in alkaline carbonate solution[7]. A separation technique has now been developed which involves co-precipitation of ^{241}Am with 0.1 mg l^{-1} of ferric iron, as hydroxide, in the presence of $^{243}Am(III)$ as a yield tracer. After filtering off the $Fe(OH)_3$ precipitate, which carries $^{241}Am(III)$, the filtrate is acidified, $^{244}Cm(III)$ added, and reduction with 2 g l^{-1} of sodium sulphite is achieved over a period of some 16 hours. A further co-precipitation removes any oxidized ^{241}Am which has been reduced. The use of the tracers allows an estimate to be made of any carry-over of the reduced fraction into the oxidized fraction, but not vice versa. The technique was validated using Pu and found to give results comparable to those obtained with NdF_3. Some results are given in Table IV. In the eastern Irish Sea it has been shown that some 5 to 10% of the ^{241}Am in the filtrate does not co-precipitate with the $Fe(OH)_3$ + $^{243}Am(III)$, even following a second addition of 0.1 mg l^{-1} of ferric iron, but does co-precipitate with $^{244}Cm(III)$ after reduction with Na_2SO_3. It is therefore tentatively concluded that an oxidized Am species is present, presumably Am(V).

TABLE IV Oxidation state of $^{239/240}Pu$ and ^{241}Am in filtered sea water from various locations (errors are ± 1σ propagated counting errors)

Location	$^{239/240}Pu$ Reduced μBq l^{-1}	$^{239/240}Pu$ Oxidized μBq l^{-1}	% oxidized	^{241}Am Reduced μBq l^{-1}	^{241}Am Oxidized μBq l^{-1}	% oxidized	Suspended load mg l^{-1}
Sellafield area	4700±140	21300±400	81.9	9900±300	587±19	5.2	5.1
South Scottish coast	549±15	2580±60	82.5	289±11	52.0±2.5	15.2	4.4
Scilly Isles	4.8±0.3	3.6±0.5	42.7	1.6±0.2	0.11±0.05	6.5	0.3
Shetland Isles	11.6±0.7	14.9±0.7	56.2	5.2±0.4	0.15±0.06	2.8	1.0
Arctic Ocean	6.7±0.3	8.0±0.5	54.7	2.3±0.2	0.18±0.03	7.2	0.5

It is not easy to estimate the quantities of oxidized forms of Np, Pu and Am in sea water relative to the discharges. Estimates of the quantities of Pu removed from the Irish Sea in sea water - filtrate plus particulate - through the North Channel of the Irish Sea vary from 3 to 17% of the previous year's discharge, depending upon the method of computation[2]. The quantity of Pu in the filtrate will depend on the sediment load of the water but estimates vary from 30 to 62% within the Irish Sea and from 61 to 94% in the North Channel[2]. Taking averages of the two sets of values in Pentreath et al.[2] it would appear that some 4% (0.63 x 0.07 x 100) of the annual discharge is present as soluble Pu of which at least 75%, or 3% of the total, is likely to be Pu(V).

The possibility clearly exists that this small amount of Pu(V) could have arisen directly from the discharges. Attempts have been made to characterize the dominant oxidation states of Pu in the effluent samples using the techniques for Pu separation as described by Lovett and Nelson[5]. The initial results indicated that, in the pond-water effluent, the $^{239/240}Pu$ in the filtrate was primarily (> 95%) oxidized whereas in the sea-tank filtrate it was primarily (> 97%) reduced. More than 98% of the $^{239/240}Pu$ in particulate fractions from both sources was in the reduced form. It should be noted that during 1982 the pond-water effluents constituted some 10% of the total Pu discharges and that, as can be seen from Table I, less than 10% of the Pu in the pond water is in the filtrate. The direct discharge of oxidized Pu was thus very low during that year. In recent years (1978 to 1984) the Pu in pond water has varied from 2.5 to 10% of the total and thus, if the effluent samples are typical, the direct discharge of oxidized Pu is not more than about 1%, and may have been only 0.25%.

With regard to Np, again the major fraction (90%) is associated with the sea-tank discharges, of which some 40% was found to be in the filtrate and 60% in the particulate fractions (Table I). The Np in the filtrate was > 95% Np(V), and in the particulate about 12% Np(V), indicating a total of some 45% of Np(V) in the sea-tank effluent. As virtually all of the Am and Cm is associated with particulate material it was assumed at that time that these existed primarily in reduced forms.

In view of the likely errors associated with all of these calculations it cannot clearly be stated whether the small fraction of Pu(V) observed in the environment does or does not originate from the discharges. Estimates for the other transuranium nuclides in the environment would be even more difficult to make because the ^{241}Am originates both from direct discharges and from grow-in from ^{241}Pu, the ^{237}Np discharges vary considerably from month to month throughout a year, and the Cm nuclides are discharged in very low quantities. It has therefore been considered more profitable to make some experimental studies.

4 EXPERIMENTAL STUDIES

4.1 Sellafield Effluent

In order to determine the possible behaviour of the Sellafield effluent as a result of contact with sea water following discharge, 1 ml aliquots of unfiltered effluent were diluted into 10 l of filtered North Sea water contained in polythene carboys. After thorough mixing, sub-samples were taken at intervals over a period of many months. Each sub-sample was filtered (0.22 μm) and the relative amounts of different oxidation states determined; the Pu was simply separated into Pu(III+IV) and Pu(V+VI)[5] and the Am separated into oxidized and reduced forms as described above. The results are shown in Figures 1 and 2. With regard to the pond water, it is clear that both Pu and Am in the particulate fractions slowly dissolve such

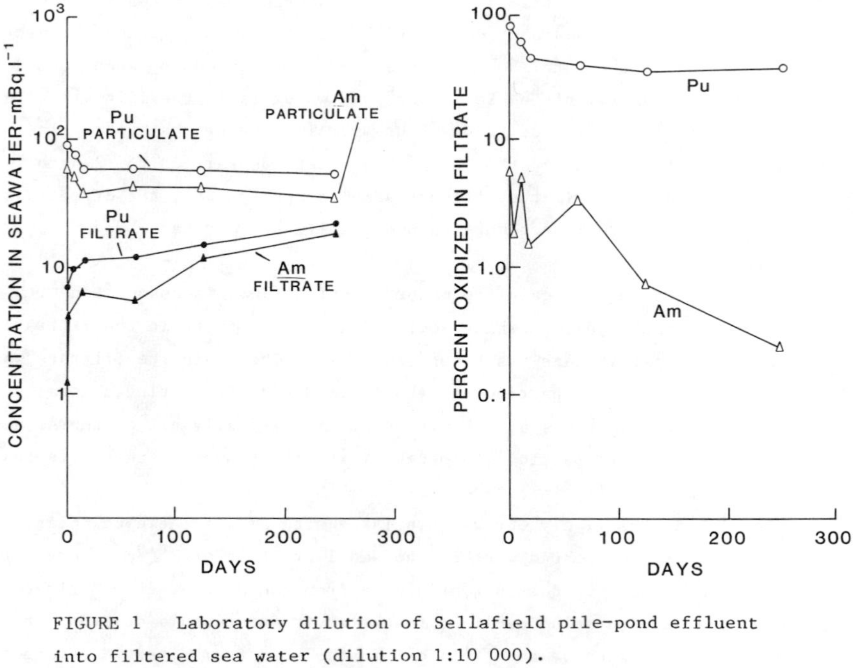

FIGURE 1 Laboratory dilution of Sellafield pile-pond effluent into filtered sea water (dilution 1:10 000).

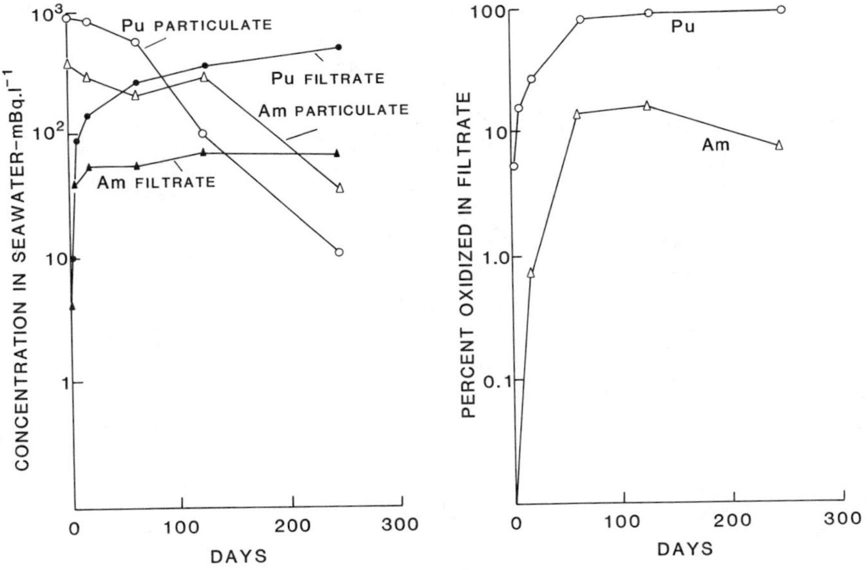

FIGURE 2 Laboratory dilution of Sellafield sea-tank effluent into filtered sea water (dilution 1:10 000).

that eventually some 20% of the $^{239/240}Pu$ and 10% of the ^{241}Am can be recovered in the filtrate. At the same time, the percentages of oxidized forms of both nuclides in the filtrate were continually reduced; thus the total of oxidized $^{239/240}Pu$ fell from 80 to 40% and that of ^{241}Am from 6 to 0.2%. In contrast, the sea-tank effluent behaved somewhat differently. First of all some 56% of the $^{239/240}Pu$ and 75% of the ^{241}Am were lost to the walls of the container, which did not occur with the pond-water effluent. Nevertheless, the concentrations of both nuclides in the recovered filtrate did increase with time and, at the end of the experiment, most of the filtrate $^{239/240}Pu$ and some 10% of the ^{241}Am was in an oxidized form (Figure 2). Experiments with ^{237}Np have yet to be made.

4.2 Sea Water

Experiments have also been made by adding tracers to filtered sea water. Both ^{237}Pu added in oxidized and reduced forms, and ^{236}Pu (reduced) and ^{239}Pu (oxidized) have been used in separate experiments. In each case the tracers were added to sea water held in polythene bottles at 10°C in the dark. The results (Figure 3) indicate that whereas some conversion of the

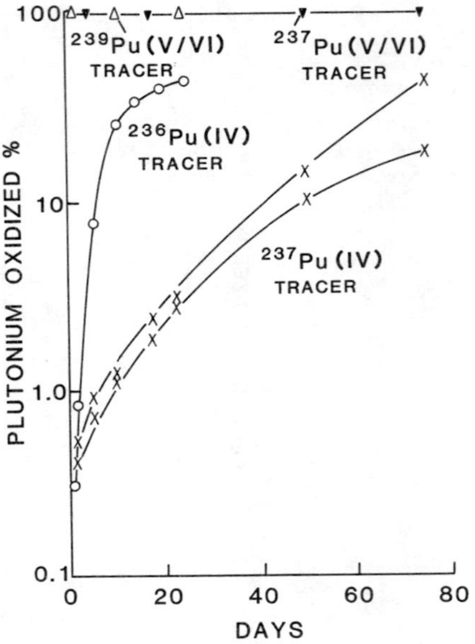

FIGURE 3 Oxidation of Pu IV tracers in filtered sea water (10°C in the dark).

reduced forms to oxidized forms does occur, there is little reduction of the oxidized forms. The reason for the differences in rate of change observed in the two experiments is not known.

Experiments have also been made using a 1 litre (Hanovia) photochemical reactor. Ultra-violet irradiation was carried out overnight (16 hours) with the pH of the sea water being adjusted from 0.7 to 8.0 by the addition of NaOH/HCl. The preliminary results, as indicated in Figure 4, show that the percentages of reduced forms remaining differ considerably from nuclide to nuclide, but at pH 8 the relative quantities of oxidized forms are not markedly different from those observed in the environment.

5 INTERSTITIAL WATERS

Redox conditions within interstitial waters are less oxidizing than those of overlying waters, so that the equilibrium between different oxidation states will be more variable. A virtual total absence of oxidized $^{239/240}$Pu in the interstitial waters of eastern Irish Sea sediments has

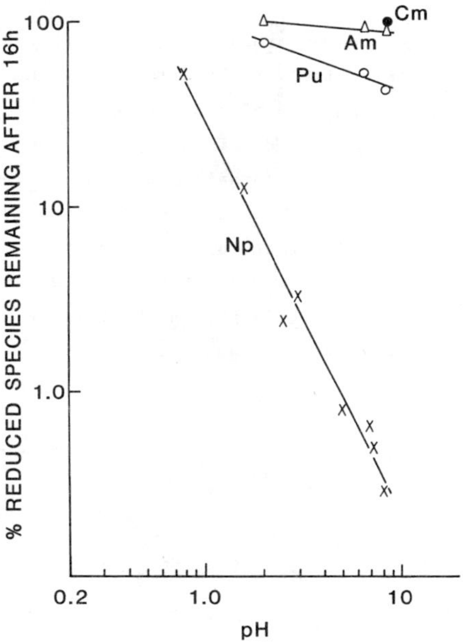

FIGURE 4 Photochemical oxidation of the reduced species of transuranic elements in sea water (16 hours ultra-violet irradiation, pH adjustment by NaOH and HCl only).

been recorded by Nelson and Lovett[8], and the estimated K_d values were therefore about 10^6. Substantial fractions of reduced ^{237}Np in the same area have also been recorded by Harvey and Kershaw[9], and the quantities prevailing have been correlated not only with pH and Eh but with the concentrations of reducing agents such as ferrous iron. All of these data have been published previously but it is worth noting here that preliminary studies have failed to identify the presence of any oxidized ^{241}Am in interstitial waters in the eastern Irish Sea.

One aspect which requires further study is that of complexation with fulvic and humic acids. Such complexation is thought most likely to occur with the (IV) oxidation state. Initial studies of the quantities associated with 0.5 NaOH extractions of Irish Sea sediments have indicated that, of the total quantities present, the order of association with humic acids is Np > Pu > Am[10].

6 BIOLOGICAL AVAILABILITY

The transuranium nuclides can be adsorbed to, and absorbed by, biological materials; both processes are likely to be affected by chemical speciation. Two examples of the likelihood of such processes occurring in the Irish Sea will suffice to illustrate the importance of such effects. In the first, concentrations of the transuranium nuclides in the benthic alga Ascophyllum nodosum, collected at two locations on two occasions, have been determined and their concentration factors (CF) relative to 0.22 μm filtered sea water derived as shown in Table V. It is clear that the order of accumulation (adsorption?) is generally that of Cm > Am >Pu > Np, but the the difference between the values of Np and the other three elements is more than two orders of magnitude.

TABLE V Concentration factors relative to filtrate (0.22 μm) sea water for the alga Ascophyllum nodosum

Location	^{237}Np	$^{239/240}Pu$	^{241}Am	$^{243/244}Cm$
St Bees, 1980	2.5 x 10	7.5×10^3	9.4×10^3	1.7×10^4
Balcarry, 1980	1.5 x 10	2.8×10^3	5.8×10^3	7.3×10^3
St Bees, 1982	5.6 x 10	3.0×10^3	6.2×10^3	2.3×10^3
Balcarry, 1982	0.8 x 10	1.7×10^3	4.4×10^3	7.0×10^3

A second example is that of the absorption of the transuranium nuclides in some tissues of the lobster (Homarus gammarus) as indicated in Figures 5 and 6 in which the quotients of radionuclide concentrations are given. The samples are pooled ones of animals collected close to Sellafield on four occasions during 1981. Although there is considerable variation between the sampling periods - which may not necessarily be simply due to seasonal effects because the discharges also vary from month to month - some underlying trends can still be seen. Thus, for example, comparing the concentrations of the transuranium nuclides in hepatopancreas (a storage organ) relative to those of either the gut and its contents or the gill, Pu appears to be the least assimilated. Having been absorbed into the hepatopancreas, however, there is a clear indication, as previously noted[10], of a translocation of Pu, and Np,into claw muscle compared with tail muscle and for the opposite to be the case for Am and Cm. Quite clearly such phenomena are related to the chemical forms prevailing both before, and subsequent to, their being absorbed.

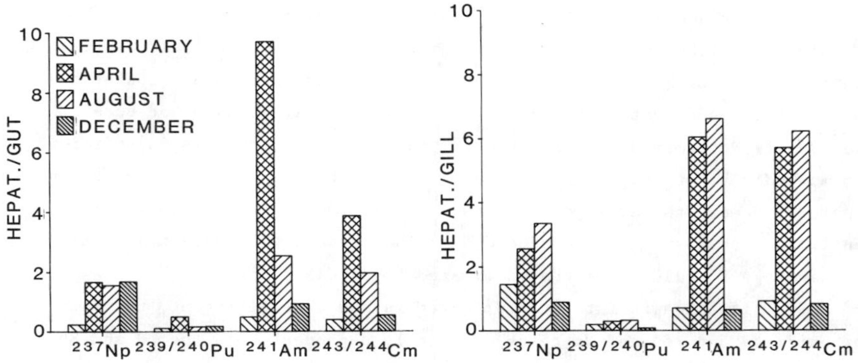

FIGURE 5 Relative absorption of transuranium nuclides by various tissue of Irish Sea lobsters (Homarus gammarus).

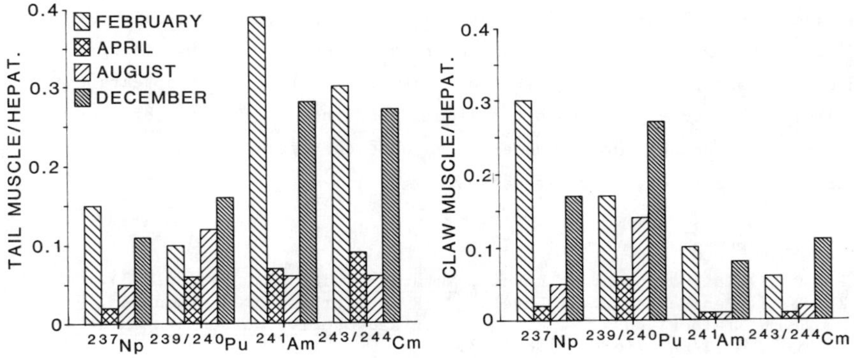

FIGURE 6 Redistribution of absorbed transuranium nuclides between tissues of Irish Sea lobsters (Homarus gammarus). HEPAT. = hepatopancreas.

7 DISCUSSION

It is evident from the rather limited data available that a thorough understanding of the environmental distribution of transuranium nuclides cannot be obtained without an understanding of the different chemical forms prevailing. One of the most important considerations is the extent to which the nuclides will be partitioned between the soluble and particulate fractions of the sea water, and thus the extent to which they will be removed to settled sediments. Where both oxidized and reduced forms exist, the average K_d value will clearly be a reflection of the separate contributions.

For Pu, the reduced and oxidized K_d values are approximately 10^6 and 10^4 respectively, the two forms being present in a ratio of at least 1:3 giving an average K_d of 2.6×10^5. The bulk of the Pu is therefore associated with settled sediments. For Am, although the K_d values have not been separately determined, the fraction of oxidized Am present is insufficient to markedly alter the average K_d value from that of the reduced form; in the environment the average value is some 2×10^6. For Np the opposite applies, with insufficient reduced Np being present to substantially alter the average K_d value from that dominated by oxidized Np.

The situation in interstitial water is somewhat different, the balance of oxidized and reduced Np being critically dependent upon the pH and Eh conditions prevailing, and Pu being dominated by the reduced form. There are other considerations, however, because solubility in the future could be effected by complexation with organic substances. This may also be important with regard to biological availability. From the two examples presented in this paper it is clear that simple relationships do not exist between the extent to which these nuclides are accumulated by biological materials and their forms in ambient sea water. Finally, it appears that whatever the forms present in the liquid effluents at the time of discharge, some changes are to be expected once the effluents come into contact with the receiving water masses.

8 REFERENCES

1. D. N. Nelson and M. B. Lovett, Nature, Lond., 276, 599, (1978).
2. R. J. Pentreath, M. B. Lovett, D. F. Jefferies, D. S. Woodhead, J. W. Talbot and N. T. Mitchell, in Radioactive Waste Management, Vol. 5 (IAEA, Vienna, 1984), 315.
3. E. Hamilton, Nature, Lond., 290, 690, (1981).
4. D. N. Nelson and K. A. Orlandini, Argonne National Laboratory Report, ANL-79-65, pt III, 57, 1979.
5. M. B. Lovett and D. N. Nelson, in Techniques for Identifying Transuranic Speciation in Aquatic Environments (IAEA, Vienna, 1981), 27.
6. B. R. Harvey and M. B. Lovett, Nucl. Instrum. Meth. Phys. Res., 223, pts II and III, 224, (1984).
7. L. R. Werner and I. Perlman, in Transuranium Elements, edited by G. Seabourg (National Nuclear Energy Series, Div. IV, Vol. 14B, pts 2 and 3, 1949), 1586.
8. D. N. Nelson and M. B. Lovett, in Impacts of Radionuclide Releases into the Marine Environment (IAEA, Vienna, 1981) 105.
9. B. R. Harvey and P. J. Kershaw, in International Symposium on the Behaviour of Long-lived Radionuclides in the Marine Environment, edited by A. Cigna and C. Myttenaere (CEC, Luxembourg, Eur 9214EN, 1984) 131.
10. R. J. Pentreath, in Impacts of Radionuclide Releases into the Marine Environment (IAEA, Vienna, 1981) 241.

DISCUSSION

BONOTTO:

My question concerns the uptake of radionuclides by *Ascophyllum nodosum*: were differences observed between young and old plants, or between young and old parts of the same plant, in the ability to absorb (or adsorb) radioactivity?

HARVEY:

No attempt was made to sample the *Ascophyllum* selectively, thus the concentration factor data given in table 5 refer to plant material of mixed age. Such differences do undoubtedly occur but the effect may well be somewhat obscured for those radionuclides showing strong adsorption to fine particulate matter which adheres to most marine plant surfaces.

EXPERIMENTAL STUDIES ON THE GEOCHEMICAL BEHAVIOUR OF 54-Mn CONSIDERING COASTAL AND DEEP SEA SEDIMENTS

P. GUEGUENIAT*, D. BOUST*, J.P. DUPONT** and G. APROSI***

* C.E.A. IPSN - Laboratoire de Radioécologie Marine
Centre de La Hague - B.P. 270 - F 50107 CHERBOURG

** Laboratoire de Géologie
B.P. 67 - F 76130 MONT SAINT AIGNAN

*** E.D.F. Direction Etudes et Recherches
6 Quai Watier - F 78400 CHATOU

ABSTRACT

In order to study the geochemical behaviour of 54-Mn in the marine environment (Mn^{2+}) 200 sediments gathered in deep sea and in coastal waters were contaminated experimentally. To correlate the various results, the oxidation processes occuring with or without sediments should be specified. Without sediments, in "blanks", the deposition rate of 54-Mn on the walls brings into play oxidation developing approximately according to a single order linear function. Consequently, it is characterized by a half-life (time for half 54-Mn to be retained) very similar to a residence time (T_R). In our water samples, T_R ranged from 12 to 150 days.
With oxidising sediments, oxidation processes could be very different :

1/ Oxidation was almost instantaneous whatever the experimental conditions. This was the case for Cape Verde samples only ; K_D ranged between 2 000 - 3 000 ml g after 24 hours contact.
2/ Oxidation occured as a first order reaction. The most oxidising sample, of this group, was collected in the Ionian sea for V/m = 100, 50 % of 54-Mn was oxidized after 6 days for T_R = 53 days (V =sea water volume ; m = sediment mass).
3/ Oxidation was autocatalytic, this was encountered for samples, only for V/m = 2 000.

In reducing sediments, K_D ranged between 10 and 400, equilibrium being reached after 3 days - contact, T_R having little effect.

During this experimental work, the geochemical behaviour of manganese will be dealt with using a radioactive tracer (54-Mn) in the divalent state and sediments collected on french littoral (160) in deep sea (30).
The latest data published - Grill (1982), Baliestrieri and Murray (1983), Wilson (1980), Emerson et al. (1982), Sung and Morgan (1981), Balzer (1982),Boulegue et al. (1978), Diem-Stumm (1984), Burdige and Keplay (1983), Murray et al. (1984), Tiping (1984), Kahlorn and Emerson (1984),Morgan and Stumm (1965) offer an excellent assessment of research findings on manganese in marine and estuary environments and testify to the interest constantly generated by this subject.
It is difficult to establish a priori any predictions on the behaviour of manganese based on the properties of a given environment, notably as concerns redox conditions. The oxidation of manganese was found to be governed by a very slow autocatalysis mechanism (Morgan and Stumm 1965) capable of being concealed by surface catalyses on mineral phases in suspension (Coughlin and Matsui 1976, Crerar and Barnes 1974, Hem 1980, Wilson 1980, Sung and Morgan 1981) or oxidation due to bacteria (Ehrlich 1983, Nelson 1978). The residence time in sea water vary considerably depending on the case : as we shall see, from a few days to some tens of years.

1 EXPERIMENTAL PARAMETERS

1.1. Radioactive tracer

The radioactive tracers(54-Mn, 51-Cr) we supplied by Fontenay aux Roses.

1.2.Sea water obtained from Cap de La Hague (Channel). The radionuclides were allowed to age in sea water for 24 hours before contact with the sediment

1.3. 200 cm^3 of sea water in 250 cm^3 plastic bottles. The sorption was determined from changes in the liquid phase.

1.4. Temperature 15° C.

1.5. Sediment dried (70° C). The sediments were introduced into contaminated sea water and allowed to settle without stirring.

1.6. List of sediments.

a/ 160 samples collected all over the french littoral in estuaries, bays and harbour wherefine material accumulates.
b/ Deep sea sediments : Cape Verde (10), Pacifique (6), Golfe de Gascogne (10), Ionian Sea (2).
c/ Exotic sediments
Exotic sediments with a high iron content ($>$ 8 %) were utilized
Iron is supposed to be an important parameter for the picking up of ^{54}Mn.

2. PRELIMINARY STUDIES : CHOICE OF REFERENCE SEDIMENTS

We have established an empirical diagram of the redox properties of 220 sediments, considering for each sample the experimental "residence time" (T_R) of Mn^{II} and Cr^{VI} in the presence of 2 grams of sediment. The case of chromium supplements usefully that of manganese because, in the hexavalent state, it requires reduction in order to be absorbed. The use of these diagrams helped us to select the sediments, notably as concerns those which will be either oxidizing or reducing. The diagram in Figure 1 is an example chosen from among several others.

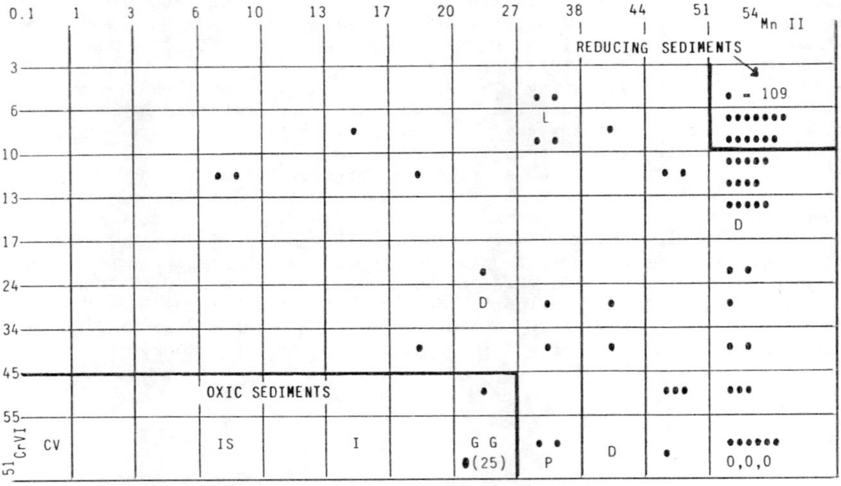

Figure n° 1 - Empirical diagram of redox properties of sediments : residence time of Mn II and Cr VI in the presence of 2 grams of sediment.
CV = Cape Verde, I = Inde, IS = Ionian sea, GG = Golfe de Gascogne, P = Pacifique, D = Dahomey, O = Orenoque, L = Labrador, • french coastal sediments.

The sediments, under the experimental conditions used, exhibit experimental residence times varying from 0.1 day to infinity for 54-Mn, from 3 days to infinity for 51-Cr.
We have selected 4 sediments which are placed in two opposite groups :

a/ High Cr T_R and low Mn T_R : oxidizing samples. The most interesting case in this regard is the sample from Cape Verde. We have also considered samples from ionian sea (1) and from Mont Saint Michel Bay (number 25).
b/ Low Cr T_R and high Mn T_R : reducing samples. The most representative case is sample 109 see which was subjected to Amoco Cadiz pollution.

3. KINETIC OF 54-Mn FIXATION ON OXIDIZING AND REDUCING SEDIMENTS

3.1. Properties of the sea water "Experimental residence time" in blank experiment

Series of sediments were contaminated by 54-Mn. Each serie included two "blanks" in order to study the evolution of deposits on the walls in the absence of sediments. When the same seawater is used, the adsorption evolvesin a comparable manner in the two bottles reserved for each experiment. On the other hand, the results can be very different from one water sample to another. Thus, for water taken in Goury on 16 February 1976, was 9.5 and 11.5 % of the 54-Mn was deposited on the walls after 30 days of contact, whereas for another water sample taken at the same place but at a different time (20 April 1982) these same wall deposits reached 84 and 71 %.

The adsorption on the walls is related to a reaction of the first order (see Figure 2), at least during the first 3 weeks. This offers the interesting possibility of characterizing each water sample by a residence time or half-life of 54-Mn which corresponds to an adsorption of half the 54-Mn following oxidation. For the series of experiments to be presented in this study, the residence times (T_R) of manganese was beetwen 12 days and 150 days.

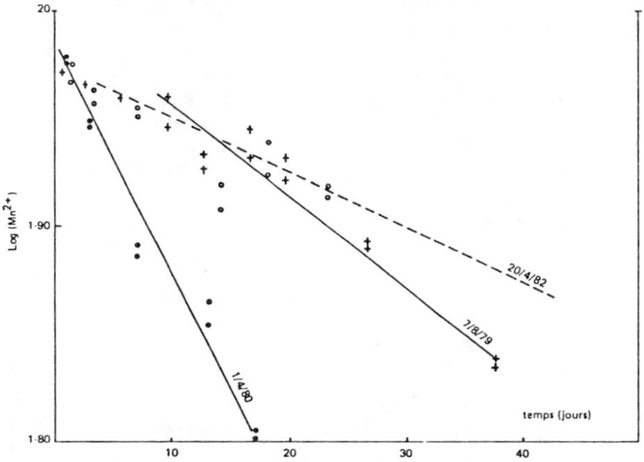

Figure n° 2
Effect of nature of sea water on $^{54}Mn^{2+}$ oxidation rates in blank experiment

3.2. Influence of the nature of sea water on sorption studies

We have considered 3 sets of experiments with samples of sea water A, B, C caracterised by T_R = 12 days (A), 130 days (B), 150 days (C). The 54-Mn K_D values for Cape Verde (n = 4 : KA 1, KA 2, KA 3, KA 4) and reducing sediments (n = 1 : reference 109) are given below :

	T_R = 12	T_R = 130	T_R = 150	
KA 1	1 380	820	400	oxic sediment t = 3 days
KA 2	5 900	1 800	840	
KA 3	3 080	1 350	480	
KA 4	9 300	3 500	1 440	
109	28	31	24	reductive sediment t = 3 days

For the reducing sediments K_D values are low and independant of T_R. For oxidising samples K_D are high but the values which are found are only relative ; they depend on the T_R value of the sea water. For example in the case of KA 4 K_D = 1 440 for T_R = 140 days, 3 500 for T_R = 130 days, 9 300 for T_R + 12 days.

3.3. Influence of the values of the ratio V/m (V = volume of sea water, m = sediments mass)

In working out the experimental process, we observed that for certain values of the ratio V/m, and for some oxidising samples, the manganese sorption reaction speeds up after a few days.

Sediment from Mont St Michel Bay (référence 25)

Thus, taking a sample from the bay of Mont St Michel (reference 25 of list) with contents of 500 mg, 5, 10, 25 and 50 grams per liter of seawater, it is observed that the results obtained with the lightest sample follows an altogether special fixation process. Whereas, in all the other cases, a state of equilibrium is reached after about 10 days, it is observed for m = 0.5 g/l that the fixation is accelerated after 3 days of contact. The result is that the percentage of manganese fixed under these conditions exceeds what is observed with sample weights 5, 10, 25 and 50 times greaer, provided a period of 12, 18, 26 and 30 days is reached (see Figure 3). For m = 0.5 g/l, the distribution coefficient (K_D) is highly variable : 360 after 3 days, then 2 000 after 30 days, whereas in the other cases this same K_D is relatively constant and clearly lower (20 to 70). When $m \geqslant 5$ g/l, the sorption is proportional to sediment concentration. For m = 1 g/l the uptake of $54\text{-}Mn^{2+}$ takes place by oxidation with a first order reaction (see Figure 4).

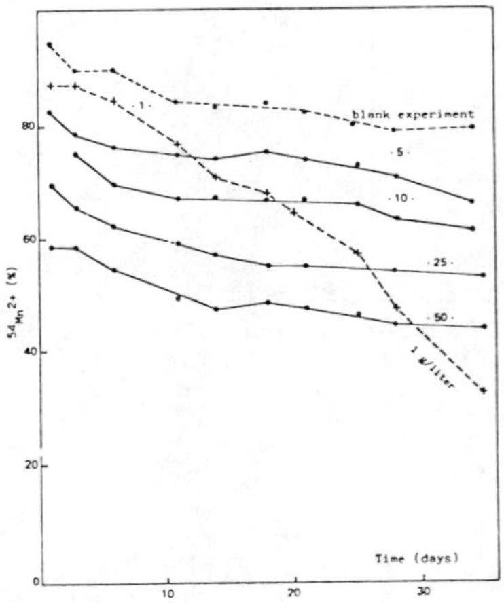

Figure n° 3 - Effect of sediment concentration (g/liter) on $^{54}Mn^{2+}$ uptake (oxic sediment) +---+ 1 g/liter (sediment n° 25)

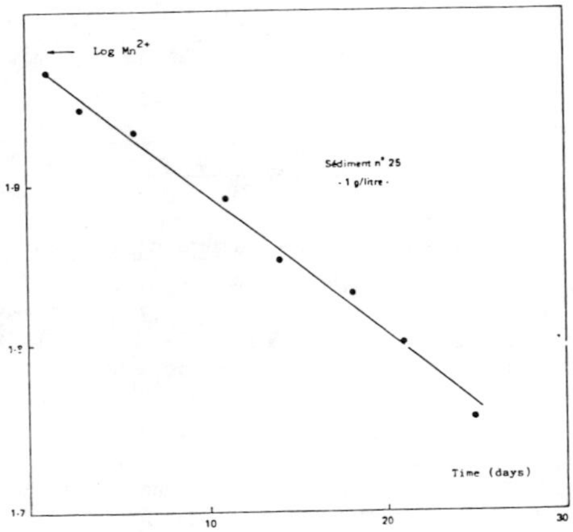

Figure n° 4 - Oxidation rate of $^{54}Mn^{2+}$ with sediment n° 25. Data plotted according to first order reaction.

Sediment from Ionian Sea

For m = 1 g/l, the sorption takes place by oxidation with a reaction which is no longer of the first order . It involves an autocatalysis mechanism similar to the one described by Wilson in the case of fresh water for manganese in the presence of $Fe_2O_3 n H_2O$ with a concentration of 10 mg/l . Fixation takes place such that the log $[(Mn^{2+})\ t_o/(Mn^{2+})\ t - 1]$ is proportional to the time. In this study, $(Mn^{2+})\ t_o$ and $(Mn^{2+})\ t$ represent respectively the concentrations of manganese in solution at the times t_o and t. Our situation approaches the one investigated by Wilson since he shows that, for a higher adsorbent concentration (125 mg/l), the autocatalysis phenomenon disappears. Figure 5 show the data illustrating this autocatalysis phnomenon for Ionian sea sediment.

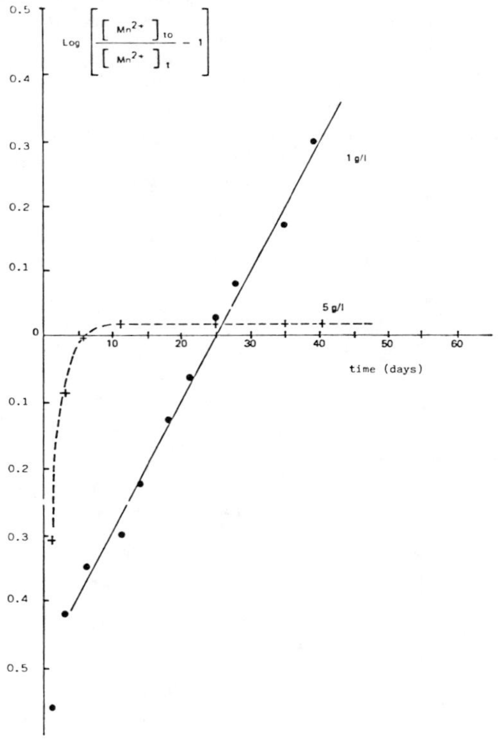

Figure n° 5 – Oxidation rate of $^{54}Mn^{2+}$ with sediment from Ionian sea (oxic sediment). Data plotted according to autocatalytic reaction.

Cape Verde Sample

Oxidation was quick (inside 72 hours) whatever the experimental conditions. The sediments from Cape Verde are exceptionally oxidising.

If we considere a reducing sediment (sample 109) the uptake of 54-Mn is proportional to sediment concentration (see figure 6), the equilibrium is reached after 3 days whatever the experimentals conditions.

CONCLUSION

The kinetics of Mn II oxygenation have been studied in the laboratory in the presence of coastal and deep sea sediments. The rate of Mn II removal was found to be dependent not only on the properties of samples studied but also of the nature of sea water (even when always collected at the same place) and the value of ratio V/m.

Two examples of oxic deep sea sediments (Cape Verde, Ionian Sea) may be choosen in this respect :

a) removal time scales for Mn II, of weeks for Ionian sea, of hours for Cape Verde, can be accounted for.

b) K_D values may vary by a factor ten when the sea water was changed. For example after 3 days, for the more oxic Cape Verde sample K_D = 1 400 when the experiment residence time of 54-Mn is 150 days, K_D = 9 300 when residence time of 54-Mn is 12 days.

c) the values of V/m may greatly affect the K_D values. For Ionian sea sediment K_D = 10 000 when V/m = 1 000, K_D = 1 600 when V/m = 200. For Cape Verde sediments K_D values are independent of V/m.

The main conclusion of this work, as reported by Sung and Morgan (1981), is that care should be exercised when extrapolating from laboratory data to natural systems. In particular K_D values are only relatives.

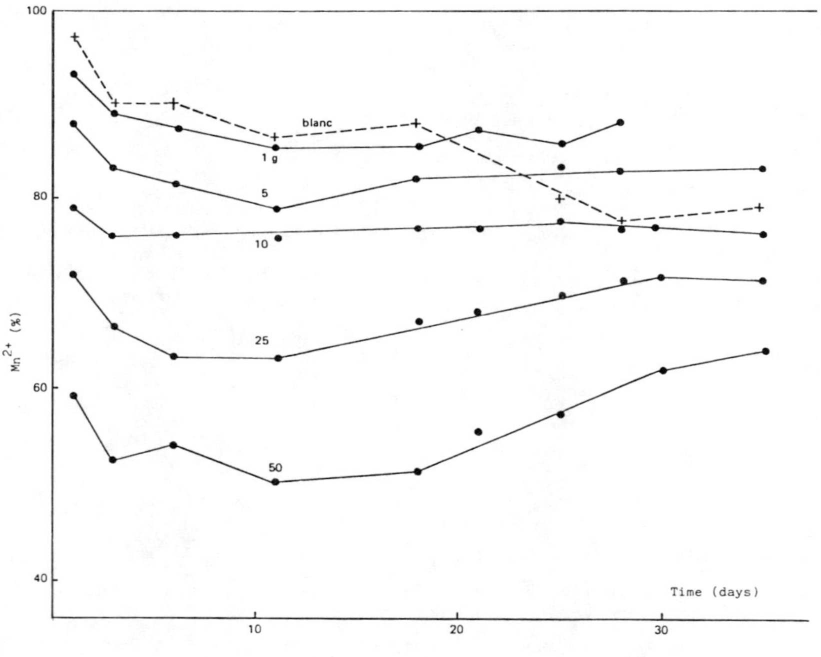

Figure n° 6 - Effect of 109 sediment concentration (g/liter) on $^{54}Mn^{2+}$ uptake (reductive sediment)

REFERENCES

W. Balzer, Geochim. Cosmochim. Acta, 46, 1153 (1982).
J. Boulegue, D. Renard, G. Michard and A.P. Boulad, Chem. Geol., 23, 41 (1978).
D. Burdige and P.E. Kepkai, Geochim. Cosmochim. Acta, 47, 1907 (1983).
R.W. Coughlin, I. Matsui, J. Catalysis, 41, 108 (1976).
D.A. Crerar, H.L. Barnes, Geochim. Cosmochim. Acta, 38, 279 (1974).
D. Diem and W. Stumm, Geochim. Cosmochim. Acta, 48, 1571 (1984).
S. Emerson, S. Kahlorn and L. Jacobs, Geochim. Cosmochim. Acta, 46, 1073 (1982).
E.V. Grill, Geochim. Cosmochim. Acta, 46, 2435 (1982).
J.D. Hem, Amer. Chem. Soc.(Eds M.C. Kavanough and J.O. Leckie) pp.56-72 (1980).
S. Kahlorn and S. Emerson, Geochim. Cosmochim. Acta, 48, 897 (1984).
J.J. Morgan and W. Stumm, J. Am. Water Works Assoc., 57, 107 51965).
J.W. Murray, L.S. Balistrieri and B. Paul, Geochim. Cosmochim. Acta, 48, 1237 (1984).
K.H. Nelson, Environmental Biogeochemistry and Geomicrobiology, 3, (ed. W.E. Krumbein) pp. 847-858. Ann Harbor Science.
W. Sung and J.J. Morgan, Geochim. Cosmochim. Acta, 45, 2377 (1981).
E. Tipping, Geochim. Cosmochim. Acta, 48, 1353 (1984).
D.E. Wilson, Geochim. Cosmochim. Acta, 44, 1311 (1980).

DISCUSSION

SIBLEY:

Do you have information on the chemical characteristics of the different seawaters you used for your experiments?

GUEGUENIAT:

Measurements of trace elements (Fe.Co.Zn.Ni-Cu.REE) were made but not for manganese. The method is based on coprecipitation of trace elements on manganese dioxide. For estimating the redox properties of the seawater used, it is precisely the method discussed in the work (estimation of residence time of ^{57}Mn in blank experiments which is used). For estimating the complexing capacities of sea water used additions of ^{60}Co were made and ionic species of cobalt were estimated in relation to the presence of organics ligands.

GEOCHEMICAL BEHAVIOUR OF Eu-152, Am-241 AND STABLE Eu IN OXIC ABYSSAL SEDIMENTS

D. BOUST* and J.L. JORON**

* IPSN-DERS-SERE - Laboratoire de Radioécologie Marine
C.E.A. Centre de La Hague, B.P. 270, 50107 CHERBOURG
** Groupe des Sciences de la Terre, Labo "Pierre Sue"
C.E.N. Saclay, 91191 GIF SUR YVETTE Cedex

ABSTRACT

The first results of sequential chemical leaching experiments carried out on natural sediments (stable Eu) and spiked subsamples (^{152}Eu, ^{241}Am), and related distribution coefficients are presented. Carbonate phases are shown to play a major role in the short term fixation of both radionuclides, while their long-term behaviour appears to be preferentially guided by Mn - Fe - rich mineral coatings. Natural distribution coefficients (10^6) are similar to those deduced from batch experiments (10^5 - 10^6). Nevertheless, the latter must be used very cautiously because of differences in the partition of stable Eu and ^{152}Eu between lithological phases.

1 INTRODUCTION

Since the eventuality of high level radioactive waste disposal into deep-sea sediments has occured, many works have been carried out in order to assess the feasibility of this concept (International Subseabed Disposal Program). Our laboratory has been especially involved in testing the efficiency of the sediment as a barrier. This implies a good understanding of physical and chemical processes occuring in the sedimentary column, including particles and interstitial waters. Special attention was drawn to vertical advection of pore fluids which are efficient vectors for transporting radionuclides from sediment to the overlying seawater.

Moreover, many in vitro sorption studies of radionuclides onto a variety of sediments allowed a rough classification of sediments in terms of binding capacities and of radionuclides in terms of relative mobility. Nevertheless, there is a big gap between in situ and in vitro conditions, and batch experiments suffers many drawbacks : (1) solid-liquid ratios are often much lower than in the natural environment ; (2) experimental medium (distilled water, natural or artificial seawater) is very different from in situ pore waters ; (3) those are short term experiments as compared with time scale of processes involved in long term behaviour of artificial radionuclides.

The interest of studying in situ behaviour of geochemical analogs of radionuclides which would be burried into abyssal sediments has been early pointed out. In this respect, rare earth elements are often cited as analogs of trivalent transuranics, especially Eu^{3+} as analog of Am^{3+}. Although this concept is now well accepted, as far as we know, no work has been yet done on the comparison between in situ and in vitro binding properties of abyssal sediments. In this paper, we give the first results of chemical leaching experiments carried out on natural sediments (stable Eu) and spiked subsamples (^{152}Eu, ^{241}Am), and related distribution coefficients.

2 MATERIAL AND METHODS

Abyssal sediments were collected from the CV2 site (20° N, 30° W) during the DIALANTE cruise, in May 1983. They consist of more or less marly nanno-foram oozes and have been demonstrated to be very oxic to 15 m depth [1]. Insterstitial waters were expressed by squeezing, at 2° C, under nitrogen.

Acidified 5 ml subsamples were analysed for Eu by neutron activation methods at the Osiris facility, C.E.N., Saclay, using a neutron flux of $2x\ 10^{14}$ n cm^{-2} S^{-1}. Moreover, chemical leaching methods applied to the sediments enabled us to distinguish four different carriers of Eu : carbonate (= biogenous) fraction, Mn - rich mineral coatings, Fe - rich mineral coatings and residual phase [2]. Additionnal parameters were also measured, including : volumetric calcimetry, clay mineralogy and specific areas using a modified methylene blue method [3]. Two sets of 10 sediment subsamples were then spiked with ^{152}Eu and ^{241}Am allowed to reach equilibrium in artificial filtered seawater (100 mg for 200 ml). After washing with an ammonium acetate buffer at pH 8.2, contaminated sediments followed the same chemical leaching steps as described above.

3 RESULTS AND DISCUSSION

3.1 Stable Eu in abyssal sediments

Eu concentrations in bulk sediment range between 0.67 - 1.62 $\mu g\ g^{-1}$, mainly depending on $CaCO_3$ content. Linear regressions (Eu versus $CaCO_3$) and chemical leaching lead to similar Eu concentrations for biogenous and lithogenous end-members : 0.2 $\pm$ 0.1 and 2.6 $\pm$ 0.3 $\mu g\ g^{-1}$ respectively. More than the half of lithogenous Eu (56 $\pm$ 9 %) is bound to mineral coatings (36 $\pm$ 5 % and 20 $\pm$ 4 % for Mn - rich and Fe - rich coatings, respectively) although they only account for 18 % weight of the lithogenous fraction. The profiles of Eu concentrations in the different phases (Fig. 1) show : (1) a decrease of Eu in residual phase with depth (2) an increase of Eu in mineral coatings with depth. This suggests post-dispositionnal exchanges between coatings and residual phase. Unfortunately, mineral lattice losses 0.03 $\mu g\ g^{-1}$ Eu per meter downwards while coatings gains 0.07 $\mu g\ g^{-1}$. So these processes imply a additionnal source of Eu which could be found in carbonate dissolution or hydrothermal influence.

Taking an average value of 0.7 pg cm^{-3} for dissolved Eu, distribution coefficients have been calculated between porewaters and each of the lithological phases we distinguished (Tab. I) ; these values emphasize the role of Mn-Fe rich coatings in Eu geochemistry.

TABLE I Distribution coefficients of Eu in various phases, assuming 60 % $CaCO_3$ content, Mn-rich and Fe-rich coatings 12 and 6 % of lithogenous phase, 50 % H_2O content.

$CaCO_3$ phase	Mn-rich coatings	Fe-rich coatings	Residual phase
0.3×10^6	12×10^6	10×10^6	2×10^6

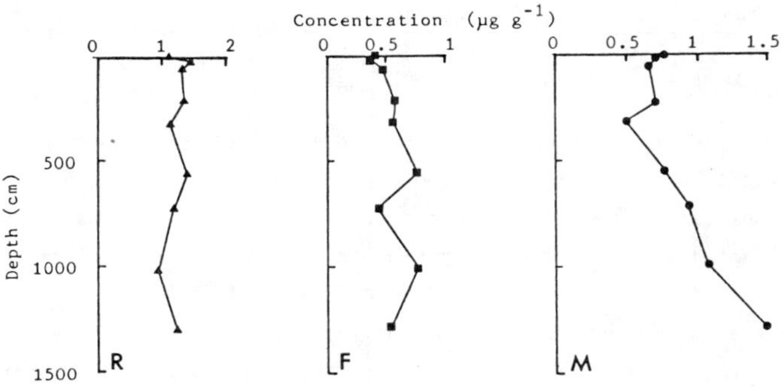

FIGURE 1 Profiles of Eu concentrations in residual phase (R), Fe-rich coatings (F) and Mn-rich coatings (M); values calculated on a carbonate - free basis.

3.2 <u>^{152}Eu and ^{241}Am in laboratory experiments</u>

The range of distribution coefficients (Kd) of ^{152}Eu and ^{241}Am for bulk sediments ($0.2 - 0.7 \times 10^6$ and $1.0 - 2.3 \times 10^6$, respectively) are not very different from Eu in situ data. The lower values obtained for ^{152}Eu as compared to ^{241}Am might be related at least partly, to the presence of unknown amounts of stable Eu in the solution. No relation has been found between Kd values and specific areas or $CaCO_3$ content. Insignificant amount of both radionuclides was leached by acetate buffer at pH 8.2. In contrast, carbonate fraction contains 90-100 % of ^{152}Eu and 60-70 % of ^{241}Am, thus confirming the great affinity of these elements for carbonaceous compounds [4]. When recalculated on a carbonate-free basis, the distributions of ^{152}Eu and ^{241}Am between the lithological phases are very similar (Tab. II) and suggest a downcore decreasing of binding capacities of residual phase and Fe-rich coatings, and subsequent increasing of both radionuclides bound to Mn-rich coatings.

TABLE II Partition of ^{152}Eu and ^{241}Am in lithological phases, in % of non-carbonaceous sediment.

	Mn-rich coatings	Fe-rich coatings	Residual phase
^{152}Eu	20 - 70	30 - 75	1 - 5
^{241}Am	5 - 70	20 - 75	10 - 20
Downcore trend	Increasing	Decreasing	Decreasing

4 CONCLUSIONS

- Distribution coefficients as determined from in situ measurements and batch experiments are in rather good agreement; but partition of ^{152}Eu between lithological phases is very different from stable Eu partition in natural sediments; thus, in vitro experiments cannot lead to a correct understanding of long-term behaviour of both ^{152}Eu and ^{241}Am.
- Carbonate phases appear to play a major role in the fixation of ^{152}Eu and ^{241}Am within a short time after release from canister (near-field). Because of their instability (below CCD), they cannot be ultimate host compounds for radionuclides of interest.
- Mineral coatings are shown to be very efficient scavengers because of (1) their increasing binding capacities at depth (Mn-rich coatings) and (2) their ability of continual accretion (Mn-Fe-rich coatings).

5 REFERENCES

1. D. Boust and A. Mauviel, Oceanol. Acta, 8, (in press, 1985).
2. D. Boust, J.L. Joron and P. Privé, presented at the 10e Réunion Annuelle des Sciences de la Terre, Bordeaux, April 1984.
3. D. Boust, P. Privé and N.L. Tran, Mar. Geol. (in press, 1985).
4. P.M. Shanbhag and J.W. Morse, Geochim. Cosmochim. Acta, 46, 241 (1982).

MARINE SPECIATION OF SOME EFFLUENT RADIONUCLIDES: INFERENCES FROM AN EMPIRICAL MODEL FOR TRANSPORT AND ESTUARINE DEPOSITION

J.E. CROSS and J.P. DAY

Department of Applied Chemistry, U.W.I.S.T., Cardiff
Department of Chemistry, University of Manchester, Manchester

ABSTRACT

A dynamic numerical model has used to relate the marine discharges of radionuclides from Sellafield to the sedimentary record of the nearby estuary at Ravenglass. Good agreement between observed core profiles and predictions from the model were obtained for the nuclides studied: 241-Am, 144-Ce, 134/137-Cs, 238/239-Pu, and 106-Ru. Differences in the nuclides' marine behaviour are characterised by the parameters of the model, and these differences are discussed in terms of possible nuclide speciation.

INTRODUCTION

Radionuclides, discharged to the Irish Sea from Sellafield, Cumbria, are incorporated to varying extents into the coastal, estuarine and sea-bed sediments of the area. In this paper, we report an attempt to model the various transport and depositional processes by which the released nuclides eventually appear in the sediment profiles of the estuary at Ravenglass, approximately 10 km south of Sellafield.

The main objective of this work is to pick out, through the use of the semi-empirical model, the most significant behavioural differences between the selected nuclides, and subsequently to relate the model-derived parameters to underlying physical and chemical factors determining each nuclide's behaviour.

EXPERIMENTAL PROCEDURES

Two cores (diameter, 5cm; length 35cm) were obtained from an inter-tidal mud-bank approximately 3km from the mouth of the estuary of the River Esk, at Newbiggin, Cumbria. The sample point is in an area previously studied and thought to be steadily accreting. The cores were frozen on collection, and later sectioned (1cm). The following nuclides were determined: 106-Ru, 134-Cs, 137-Cs, 144-Ce by gamma spectrometry, and 238-Pu, 239+240-Pu, and 241-Am, by alpha spectrometry following chemical separation.

THE ENVIRONMENTAL MODEL

The principle on which the model is based is that of inter-relating the radionuclide profiles observed in the sediment cores with the known history of nuclide discharges to the sea from Sellafield. The various environmental factors (sorption, transport, mixing, sedimentation, etc.) which determine the overall process of nuclide transfer, from discharge to deposition, in the environment are given parametric representation in a numerical, time-dependent simulation (the model). By continuous adjustment of the parameters for each nuclide, until a best fit is reached between the observed core profile and that calculated from the model, a set of environmental parameters are derived empirically for each nuclide.

An outline of the model is shown in Figure 1. The model's parameters are intended to represent physical processes which control nuclide movement in the marine environment. Five independent parameters are employed to describe the processes for each nuclide, together with a sixth parameter, defining rate of sedimentation, which is common to all nuclides. The steps in the model were simulated numerically, at 0.01 year intervals, with simultaneous adjustment for radioactive decay and ingrowth of 241-Am from 241-Pu. The calculation was programmed for a microcomputer

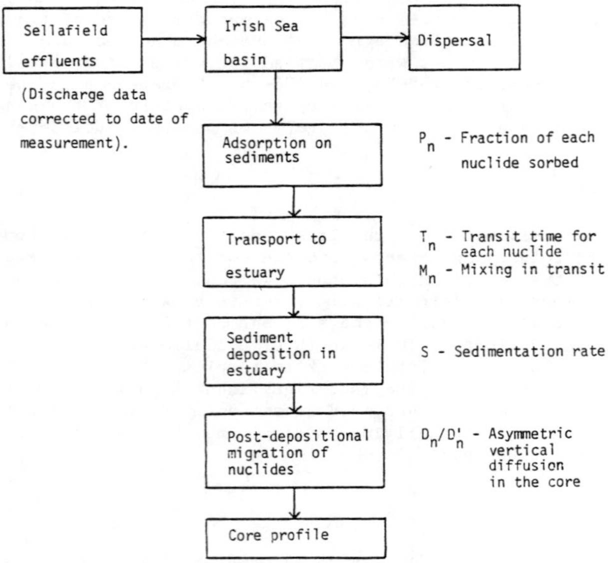

FIGURE 1 Outline of the model described in the text.

and adjustment of parameters was made interactively. The best fit was judged subjectively by comparison of plots of the observed core profiles for each nuclide against those calculated from the model.

The primary input to the model is the quantity of each nuclide discharged from Sellafield. The nuclides are assumed to equilibrate between solution and particulate phases and the partition parameter (P) defines the relative fraction of each nuclide eventually deposited in the sediment. Transport of each nuclide to the estuary is then assumed to take place by a linear movement of sediment in discrete "packets", characterised by an overall transit time (T) and fractional mixing (M) between adjacent packets. A constant rate of sediment deposition (S) is assumed (the calculations take account of the observed compaction of the core with increasing depth). Finally, upward or downward nuclide migration is adjusted by the diffusion parameters, D/D'.

RESULTS AND DISCUSSION

Calculations based on the model gave an acceptable fit with the observed core profiles for the seven nuclides studied (two examples are given in Figure 2). The degree of fit is sensitive to variation in all parameters, and the "best-fit" values are given in Table 1.

The sedimentation rate is the only physical quantity capable of independent verification. Although all the nuclide profiles are sensitive to the value of this parameter, the strongest determining factor in the calculation is the 137-Cs:134-Cs isotope ratio. The derived sedimentation rate approximates to a surface deposition of 2.0 cm/y, which is in good agreement with values previously derived for this area.

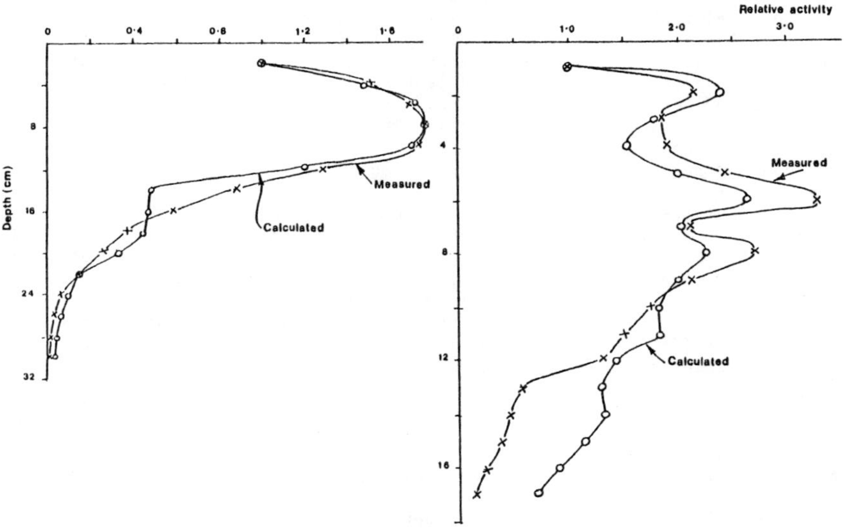

137-Cs 239-Pu

FIGURE 2 Calculated and observed core profiles for 239-Pu and 137-Cs.

TABLE I. Best fit values of the parameters for the numerical model.

Nuclide	Parameters				
	T (y)	P (%)	M (%)	D (%)	S(g $cm^{-2}y^{-1}$)
137-Cs	1.7	2	1.7	0.04	2.5
239-Pu	4.6	96	4.6	0	"
241-Am	4.1	96	9.2	0	"
144-Ce	0.1	94	0.05	0	"
106-Ru	0.9	65	0.54	0	"

Implications for Nuclide Speciation.

Plutonium and Americium.- The very large sediment uptake factors (96%), combined with the very long apparent transit times (over 4 y) is suggestive of these nuclides' immediate and nearly complete transfer to the solid phase, and transport almost entirely by movement of relatively immobile sea-bed sediments. Post-depositional migration is insignificant, indicating very little tendency for remobilisation from the solid phase. However, the degree of mixing in transit is considerably higher than for the other nuclides studied and a physical mixing of the sediments themselves, during transport, must be inferred.

Cerium.- The very high uptake (94%) coupled with a very short transit time suggests that cerium, on discharge, is removed irreversibly to a solid phase which, however, is highly mobile. On this interpretation, the cerium must deposit preferentially in the estuary, a process which might be related to altered chemical stability in estuarine water. The behaviour of cerium is in marked contrast to that of Am and Pu, although all three elements are almost totally removed from solution.

Caesium.- The very low overall uptake of caesium, coupled with the observation of significant post-depositional migration (indicating at least partial reversibility of sediment uptake), suggests that much of the transport process occurs in solution, with continuous re-equilibration between the aqueous and solid phases. The relatively long apparent transit time can then be interpreted as a measure of the average "age" of the caesium present in solution, and in equilibrium with the sediments, at the time of deposition.

Ruthenium.- This nuclide shows a behaviour intermediate between caesium and cerium. Sediment uptake is not complete. On one interpretation, sediment uptake might be partially reversible and, as with cerium, a degree of specific deposition in the estuary is suggested. Alternatively, a dual speciation is possible, in which one species might be supposed to show complete, irreversible removal to sediments (probably as ruthenium(III)-hydroxides), whilst the other might be supposed to remain in solution, possibly as complexes of nitrosylruthenium(II).

CONCLUSIONS

It has proved possible to simulate the behaviour of the nuclides of five, chemically diverse, elements using a simple model for marine transport and deposition. Predicted core profiles agree well with those determined experimentally, demonstrating the general utility of the model and, presumably, the absence of significant reworking or bio-turbation of the sediments studied. Chemical inferences from the model are in line with current theories of the marine speciation of these elements.

SPECIATION OF RADIONUCLIDES IN PLANTS

G. DESMET

Radiation Protection Programme, Commission of the European Communities, Brussels - Belgium

ABSTRACT

In living beings two different types of biochemical interactions with radionuclides may occur. Some radionuclides are tracers for indispensable nutrients and have been isotopically bound to and even assimilated in the biosphere. To these radionuclides belong all the isotopes of macro- and micronutrients necessary for the normal functioning of live.
A rather impressive amount of radionuclides, however, do not belong to this category such as e.g. U, Am, Np, Tc, Cs, Rb, etc.
Screening the literature back to 1972 a few remarkable facts were elicited concerning the fate of radionuclides in the biosphere. Tritium has well been investigated in animals but less well in plants ; carbon-14 metabolism is well-known in plants for obvious reasons ; strontium-90 and caesium-137 for many years have been used to trace calcium and potassium ; iron-59, zinc-65, manganese-54 and cobalt-60 are favorites in the studies on micronutrients in plants and animals.
Speciation of actinides in plants is rather tough, particularly due to their scarce uptake by plants. Elements like technetium on the other hand are readily accumulating in plants which led to an outburst of interest in its speciation. Careful examination by chromatography reveals indeed the existence of a large number of Tc-bio-organic molecules with likewise a broad range of molecular masses and chemical features. Technetium will be discussed therefore in more detail.

In routine dictionaries the word "speciation" is unfindable. One has to go as far as the Websters International Dictionary to find that speciation means : formation of biological species or the processes leading to this end, whether constituting a gradual divergence from related groups or occurring abruptly by combination or transformation of genomes.

Although it may come over a bit trifling to the audience, I had to mention this linguistic implication of that word "speciation" since I am convinced that we are not just going to study new species in the biosphere, which in our case means plants, but that we want specifying also these molecular species which already used to bind mineral isotopes since the Origin of Species. Bearing this in mind it may be necessary to find the reasons why, in view of the utilization in radiation protection, we ought to specify which chemical transformations may incur to radionuclides when assimilated by plants.

When plants are growing in places where radionuclides are available, selectively some species are absorbed, transferred and incorporated in the plant's biomolecules.

This is certainly true for elements like Tritium, Caesium, ^{65}Zinc, ^{54}Manganese, ^{60}Cobalt, ^{59}Iron, and some more of them belonging to the metal and transition elements of the table of Mendeleev. So, what does happen to them in the plant ?

First, I want to deal briefly with Carbon-14 and Tritium. Carbon-14, to my knowing need not really to be discussed from a biological point of view. Every biologist by now will be aware that in the late nineteen fifties there has been given a Nobel price to Melvin Calvin for his outstanding description of the CO_2 assimilation of green plants. This CO_2 assimilation results in an average amount of 500.000 ppm carbon in plants on a dry weight basis. All books of plant physiology describe very well how carbon behaves in plants, and how it criss-crosses through the plant's metabolism. This salient ease of incorporation of ^{14}C in plants urges therefore to keep on watching very carefully possible releases from nuclear power plants, especially in view of the gradually increasing CO_2 concentration in the atmosphere.

Tritium however is different. From the work of GUENOT and BELOT (1), one knows fairly accurately which percentage of tritium may be incorporated in plants. Though, it is worthwhile to give attention to some possible biochemical mechanisms that may lay at the basis of such an incorporation of tritium. KANAZAWA et al. (2), from the laboratory of Melvin Calvin, already in 1972 described the incorporation of tritiated water in the unicellular green alga Chlorella pyrenoidosa. In contrast to what happens with Carbon-14 CO_2 no single point of entry for tritium was discovered. This was not so quite unexpected, since several organelles and enzymatic systems in plants, H_2O is readily fixed, released and refined. Two important sites in the tricarboxylic acid cycle are shown here.

Fixation, release, refixation of H_2O in the TCA cycle

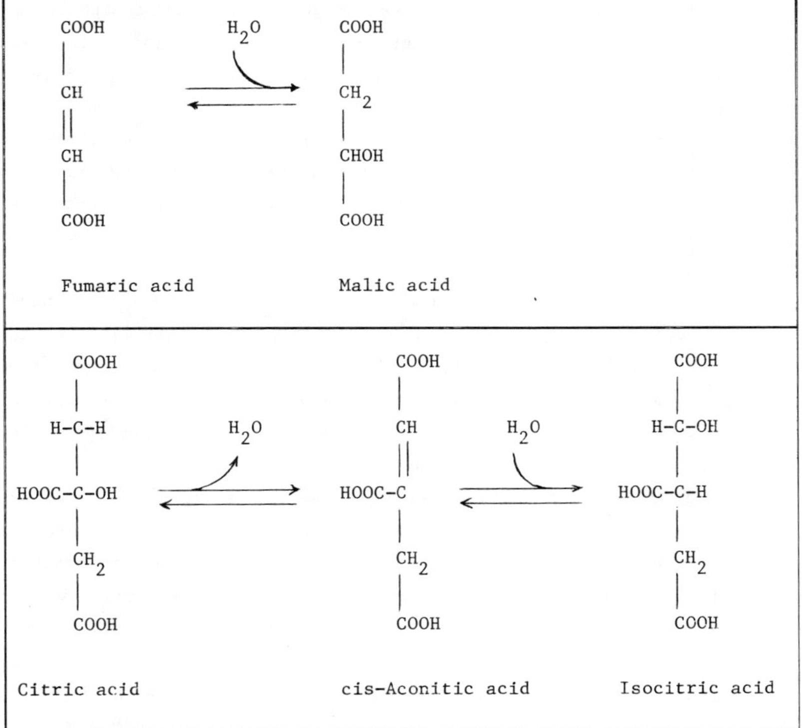

All such H_2O exchange reactions are virtual sites of entry of tritiated water. Usually the replacement of H_2O by such tritiated H_2O leads to some isotopic discrimination as mentioned by KANAZAWA (2). There was apparently no grave discrimination against tritium in the tricarboxylic acid cycle, but there was definitely one measurable in the lipid metabolism and in the purine and pyrimidine bases of nucleic acids.

This incorporation of tritiated water is certainly not the most intriguing aspect of tritium metabolism. Tritiated hydrogen incorporation in living cells poses obviously more problems. Although I am not a microbiologist, I was most anxious to find some basic information about this problem, since preponderantly HT will be set free from nuclear power installations, if any release would occur. In the leaves of higher plants there is no obvious site of entry for H_2 or HT. The oxidation pathways of H_2 in bacteroids particularly Rhizobium japonicum, an aerobic N_2 fixing bacterium, however have been neatly summarised by Günter EISBRENNER and Harold EVANS (3), and useful information can be drawn from their work. They showed schematically how close a relation exists between N_2 fixation and H_2 oxidation. A hydrogenase enzyme is responsible for the oxidation of H_2 into two electrons and two protons. The electrons can enter the electron transport chain further on at the level of ubiquinone having a standard potential of 0.10 volts. It is evident that in any organism possessing such a hydrogenase enzyme system, this H_2 oxidation may occur with the eventual production of tritiated water.

The conclusions of this enumeration are that there is enough substantial evidence for a biological oxidation of H_2 into H_2O. Once the H_2O is there, a lot of possibilities exists to guarantee its incorporation into organic molecules. KANAZAWA et al. (2) showed the conversion of tritiated water to occur indeed in higher plants, naturally making allowance for some isotopic discrimination.

An other group of important radionuclides are among the fission and activated corrosion products. They include Cs, Co, Zn, Mn, Fe, etc. I wish not so much to discuss some particular findings on the speciation or complexation of these radionuclides but rather to outline briefly what they could possibly undergo when taken up by plants. The reactions of Cs, quite similar to potassium are predominantly ionic, although some chelation can occur with some ligands, mostly however with very low stability constants. A quick look in the work of SILLEN and MARTELL (4) e.g. shows log stability constants of Cs with ATP of 0.9. ATP is abundantly generated in any living creature. Therefore even such an alkaline metal like Cs will never be fully present as a free cation. More substantial calculations concerning such relations are missing spitefully enough.

Elements like Co, Zn, Mn, Fe, may have been mentioned to form each specific complexes, also they are sequestered by a manifold of biological ligands of multifarious nature. An enumeration of most of the virtual chelators may be justified, such as there are carboxylic acids, amino acids, peptides and proteins, nucleotides and nucleic acids, this list not being exhaustive. All these chelators show their own metal-complex formation rate constants, and stability constants. It may deserve the effort to look at some of complexes into more detail. The following very superficial table list some of the possibilities (4).

Table 1 - List of the stability constants of some metals with a few organic ligands occurring in vivo.

	pK_1 values			
	Zn	Mn	Co	Fe
Oxalic acid	4.9	3.9	4.7	9.4
Citric acid	3.55	3.54	4.83	12.5
Glycine	5.03	3.21	5.23	10.0
ATP	4.85	4.30	4.66	?
ADP	4.28	4.10	3.68	?

Besides virtual binding with a large selection of these small chelators, a limited quantity of specific proteinic chelators are fairly well described in plants. Roughly it is possible to classify metal proteins systems under two headings, namely metalloproteins including metalloenzymes and metal-protein complexes. In the first group the metal ion and protein are firmly linked together so that the metal ion can be regarded as an integral part of the protein structure. The second group involves those proteins where the metal ion is reversibly bound with the ion. This reversibility of course is a matter of grade the second group gradually merging into the first group. The next table gives a concise classification of a few of these metal-proteins (5, 6, 7).

Table 2 - A few metalloproteins

Manganese		- Splitting of H_2O in photosynthesis - Photorespiration
Iron		
	non-haem	- Ferredoxin, (Fe-S proteins)
	haem	- Cytochromes
Cobalt		- Essential N_2 fixation
Zinc		- Carbonic acid anhydrase CO_2 $H_2O \rightleftharpoons H_2CO_3$ - DNA polymerase - RNA polymerase - Stability of ribosomes
Molybdenum		- NO_3^-, NO_2^- reduction

Among the small metal binding species, phytate (myo-inositol hexakisphosphate) has been given special attention by some scientists (8). They showed us that at least Pu phytate was absorbed significantly more by the gut of a rat than a Pu-nitrate for example. Actually, what is this mysterious phytate and where can we find it in plant tissues ? Some well made review articles, like the one by LOEWUS and LOEWUS (9), describe phytate being deposited in reserve tissues such as tubers and seeds, particularly in discrete regions (called globoids) of cellular organelles which are usually referred to as protein bodies. Formation of phytate and its subsequent breakdown are thought to be restricted to these subcellular regions, which results in its almost complete immobility. Plenty of information is available showing that much of the mineral reserves found in seeds accur as phytin, salt. The most commonly occurring cations in the phytin-rich globoïd crystals seem to be Mg and K, but in certain cases also Ca, Fe and Mn have been found with this phytin although in much smaller amounts.

Phytate therefore may be acting very well as a chelator of Pu in seeds and tubers. Considering however its very restricted localization, it is doubtful however whether it will play any role in complexing Pu or other radionuclides in the rest of the plant.

Good information about the complexation and thence enhanced solubility comes from the experiments of CATALDO, WILDUNG and GARLAND (10). They describe how chelation of Pu prior to its administration to plants results in several new Pu complexes in plants. Another modern topic today is the speciation of Tc in plants and some information is available now. Gel chromatography of plant homogenates reveals the existence of organic bound Tc in plants. An example is give here from work of my old laboratory in Wageningen, where still an intensive investigation is going on to counting and to characterizing, if possible, the number and the identity of the chemical species that are binding Tc in spinach plants (11, 12). Gel chromatography on Biogel-P6 and Biogel-P2 revealed the existence of both high molecular and low molecular Tc complexes, with however a variable percentage of TcO_4^- left unmetabolised and therefore present as the free anion. The weak point of either the study on Pu speciation in xylem exudates or the speciation of Tc in leaf homogenates lays in the fact that still no identity card can be stuck upon these compounds. This is spiteful and a real effort should be made to describe better their chemical properties. Indeed, if one of the objectives of our work is to predict whether and how much of the radionuclide in question arrives in a critical organ of animals and men, the chemical characteristics of the bioorganic isotope complex should be known as reliably as possible. The Pu-phytate example shows neatly how predictability has been improved knowing the chemical identity of the ligand.

CONCLUSIONS

- Many isotope-ligand interactions are possible in plant tissues ;
- Several isotope-ligand compounds have been discovered ;
- Overemphasizing the importance of some known biological chelators could lead to losing sight of the complexity of the physico-chemical possible speciation ;
- Very few isotope-ligand compounds have been characterized ;
- A considerable amount of information exists concerning sequestering of stable isotopes with organic ligands. It is amazing however to see how

little information there exists concerning their exchange rates and degrees with their radioisotopes ;

- For predictive aims at least some effort should be done to determine some physico-chemical and chemical properties of the sequestered radionuclides ;
- For predictive aims, basic research will probably be necessary to find out how plant bound radioisotopes are transferred from the outer to inner part of the human body eventually.

LITERATURE

1. J. GUENOT, Y. BELOT
Health Physics, 47, 849 (1984)

2. T. KANAZAWA, K. KANAZAWA, J.A. BASSHAM
Envir. Science and Technol., 6, 638 (1972)

3. G. EISBRENNER, H.J. EVANS
Ann. Rev. Plant Physiol., 34, 105 (1983)

4. L.G. SILLEN, A.E. MARTELL
Stability constants of metal-ion complexes
Special publication nr 17 (the Chemical Society, Burlington House, W.1, London) (1964)

5. M.N. HUGHES
The inorganic chemistry of biological processes
John WILEY and Sons, London, New York, Sydney, Toronto (1975)

6. D.J.D. NICHOLAS
in Trace Elements in Soil-Plant-Animal Systems, edited by D.J.D. NICHOLAS and A.R. EGAN (Academic Press Inc., New York, San Francisco, London) (1975), Chap. 4, pp. 181-195

7. D.T. CLARKSON, J.B. HANSON
Ann. Rev. Plant Physiol., 31, 239 (1984)

8. J.R. COOPER, J.D. HARRISON
Health Physics, 43, 912 (1982)

9. F.A. LOEWUS, M.W. LOEWUS
Ann. Rev. Plant Physiol., 34, 137 (1983)

10. T.R. GARLAND, D.A. CATALDO, R.E. WILDUNG
J. Agric. Food. Chem., 29, 915 (1981)

11. G.M. DESMET, J.F.F.M. LEMBRECHTS, H. OVERBEEK
J. Env. Exp. Bot., in press

THE INFLUENCE OF THE CHEMICAL FORM OF TECHNETIUM ON ITS UPTAKE BY PLANTS

L.R. VAN LOON
R.I.V.M., Lab of Radiation Research, p/a Association EURATOM-ITAL
Keyenbergseweg 6, NL-6704 PJ Wageningen

G.M. DESMET
Commission of the European Communities, Directorate General for Science Research and Development, Wetstraat 200, B-1049 Brussels

and

A. CREMERS
Laboratorium voor Kolloidchemie (K.U.Leuven), Kardinaal Mercierlaan 92, B-3030 Heverlee

ABSTRACT

Spinach plants, grown on a Steiner nutrient solution containing TcO_4^- at different concentrations, show a linear relationship between the concentration in the nutrient solution and the amount of Tc in the plant (concentration range 0 Bq/ml-58 Bq/ml). When Tc is added to the plants as a Tc-cysteine complex, less amounts of Tc are present in the plants. The Tc present in the plants is mainly due to the uptake of TcO_4^-, formed by reoxidation of the Tc-cysteine complex in the nutrient solution.
Plant tissue analysis together with a mathematical analysis of the uptake, show some evidences for TcO_4^- as the most important chemical form of Tc taken up by the plants.
In the case of anionic complexes, it's impossible to study only the uptake of the complex. Due to reoxidation of the complexed Tc, a mixture of TcO_4^- and the complex is present in the nutrient solution.
In the case of cationic complexes, the TcO_4^- can be removed from the nutrient solution by an anion exchange resin, so that only the complexed form of Tc is present in the nutrient solution. Its uptake by plants can be studied without interference of TcO_4^-. Uptake of Tc-complexes is possible, but the uptake rate (or transferfactor) is lowered by two order of magnitude as compared with TcO_4^-.

1 INTRODUCTION

The ionic form of an element, especially an element found in trace concentrations in the soil or nutrient solution, is very reactive with plant roots.[1,2]

Many research workers in the field of plant nutrition and soil chemistry now agree that the ionic form of an element is preferred by plants above other forms and that the uptake therefore is better related to the concentration of the ionic form than to the total concentration of an element.[3,4]

Every parameter that affects the activity (concentration) of the free ion in the soil or nutrient solution, will increase or decrease its uptake by plants or as Sposito mentions in his paper[1]: "The bio-availability of a chemical element in the soil solution may be result of competition among plant roots, soluble complexing compounds, and the solid phases in soil for the free ionic form of an element" (fig. 1).

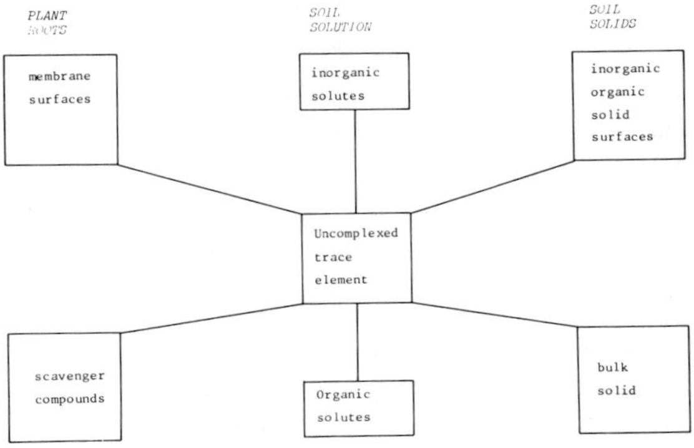

FIGURE 1 Schematic diagram of the competition among plant roots, soil solution ligands and soil solids for a trace element in the free ionic form.

Speciation of an element in the soil or nutrient solution is important to get the relationship between the activity (or concentration) of the free ionic form of an element and its uptake. It will enable us to determine transferfactors based on the intensity or concentration of an element in the soil solution. Such transferfactors are less dependent on e.g. soil type and time than the transferfactors based on the total amount of an element in the soil.

In this paper, the relationship between the TcO_4^- concentration in a nutrient solution and its uptake by plants will be discussed. A model system, in which Tc is converted from an unavailable (Tc-complexes) to an available (TcO_4^-) form during plant growth, is shown.

2 MATERIALS AND METHODS

Four experiments were carried out :

- Experiment 1,2 and 3.
  Spinach plants (Spinacea Oleracea L. cv Verbeterd Breedblad) were grown on a Steiner nutrient solution containing a constant concentration of TcO_4^- (exp. 1), a varying concentration of TcO_4^- (exp. 2) and the Tc-cysteine complex (exp. 3).
  Every five days, ten plants were harvested and the amount of Tc was determined by liquid scintillation counting after a wet destruction of the plant material (HNO_3 conc., H_2O_2 30 %, t = 80 °C). The concentration of TcO_4^- in the nutrient solution was monitored during the growth period.
- Experiment 4.
  Spinach plants were grown on a nutrient solution containing the Tc-tetren complex and a anion exchange resin (DOWEX AG 1 x-8, pre-equilibrated with the nutrient solution). After 15 days of growth, plants were harvested and homogenized by grinding the leaves in a buffer solution at pH = 7.5 (sucrose 0.5 M, TES 0.05 M, EDTA 0.5 mM,2-mercaptoethanol 1 mM).
  After filtering the homogenate through a 60 μm nylon cloth, it was centrifuged at 12,000 g during 20 minutes. The different Tc-compounds in the supernatant were separated by gel filtration (TSK HW 40F, L = 11.8 cm, Ø = 0.66 cm, V_t = 4.04 ml).

3 RESULTS AND DISCUSSION

3.1 Experiment 1

In case of constant TcO_4^- concentrations in the nutrient solution, there is a linear relationship between the amounts of Tc in the plants (Bq/g fresh weight) and the concentration of TcO_4^- in the nutrient solution (Bq/ml) (fig. 2).

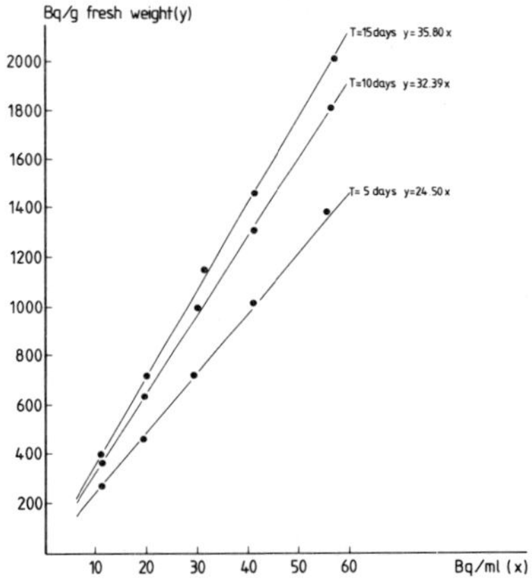

FIGURE 2 Concentration of Tc in spinach plants as a function of concentration of TcO_4^- in the nutrient solution.

The slope of the lines represents the transferfactor (defined as Bq/g fresh weight : Bq/ml nutrient solution). The transferfactor slightly increases as a function of time (table 1).

TABLE 1 Transferfactor of $^{99}TcO_4^-$ for spinach plants as a function of time.

Time	Function	T.C.
5	Y = 24.5X	24.5
10	Y = 32.4X	32.4
15	Y = 35.8X	35.8

Y = concentration of ^{99}Tc in the plants (Bq/g fresh weight)
X = concentration of TcO_4^- in the nutrient solution (Bq/ml)

The relationship enables us to predict concentrations of Tc in th plants when the TcO_4^- concentration in the nutrient solution is known.

3.2 Experiment 2

When the concentration of TcO_4^- is not constant during the growth period, it is much more difficult to evaluate the relationship between TcO_4^- in the nutrient solution and the amount of Tc in the plants. Especially when someone is trying to calculate a transferfactor, he is faced with these problems.

The question rises : "Which concentration do we have to use in the calculation ?" It is possible to calculate a weighed mean concentration of TcO_4^- :

$$\bar{C}_{wv} = \frac{{}_0\int^t C(t)e^{\ t}dt}{{}_0\int^t e^{\ t}dt}$$

Again there is a linear relationship between this mean concentration of TcO_4^- and the amount of Tc in the plants, or in other words : the same transferfactors are obtained as in experiments with constant concentrations (table 2). There is no reason why the transfer of TcO_4^- would have been changed.

The validity of the formula is due to the relationship between plant growth and uptake of TcO_4^-. Both phenomena can be approximated by exponential functions (fig. 3) :

growth function $W = W_0 \cdot e^{\mu t}$

W = fresh weight (g)
W_0 = fresh weight at $t = 0$
μ = relative growth rate $\frac{dW}{Wdt}$

uptake function $N = A \cdot e^{bt}$

N = amount of Tc in the plant (Bq)
A = amount of Tc at $t = 0$
b = relative uptake rate $\frac{dN}{Ndt}$

TABLE 2 Expected and measured concentrations of ^{99}Tc in plants grown on altering TcO_4^- concentrations.

	Day	$\bar{C}_{wv}$	C_{pl} (exp.)	C_{pl}	T.C.
1*	5	12.26	297 ± 4	290 ± 61	23.7
	10	19.02	631 ± 11	661 ± 70	34.8
	15	26.08	933 ± 24	908 ± 133	34.7
2*	5	7.78	184 ± 3	163 ± 27	21.0
	10	17.88	595 ± 11	543 ± 58	30.4
	15	29.46	1054 ± 27	998 ± 156	33.9
3*	5	28.66	711 ± 9	606 ± 59	21.1
	10	17.54	585 ± 11	490 ± 67	27.9
	15	10.31	369 ± 9	332 ± 23	32.2

* 1 = Tc-cysteine
2 = Increasing TcO_4^- concentration
3 = Decreasing TcO_4^- concentration

$\bar{C}_{wv} = \dfrac{\int_0^t C(t)e^{\mu t}dt}{\int_0^t e^{\mu t}dt}$ = weighed mean concentration of TcO_4^- in the nutrient solution (Bq/ml)

C_{pl} (exp.) = predicted concentration the plant (Bq/ g fresh weight)

C_{pl} = measured concentration in the plant (Bq/g fresh weight)

T.C. = $\dfrac{C_{pl}}{C_{wv}}$

This means that the concentration of TcO_4^- (in the nutrient solution) at the moment the plants are in the steep phase of their exponential growth, is much more important than concentrations at other moments. The exponential growth function can be used to weigh the concentration of TcO_4^- in the nutrient solution.

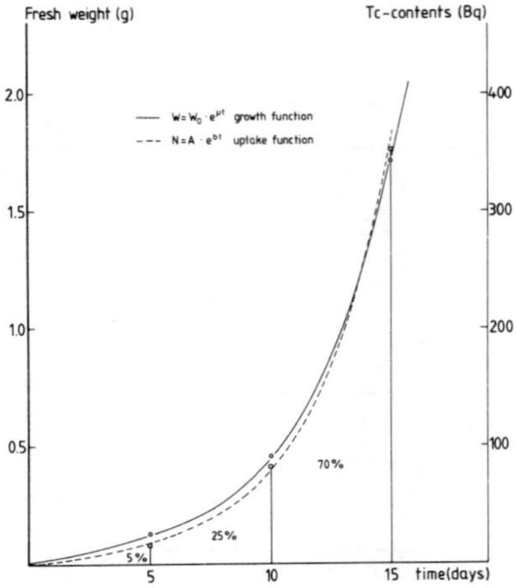

FIGURE 3 Growth and TcO_4^- uptake functions of spinach plants

3.3 Experiment 3

The total concentration of Tc in the nutrient solution is constant during the growth period. The concentration of TcO_4^-, on the other hand, continuously increases as a function of time due to reoxidation of the Tc-cysteine complex :

$$\text{Tc(IV) cysteine} \longrightarrow \text{Tc(VII)}O_4^-$$

Weighed mean concentrations of TcO_4^- were calculated. The weighed mean concentrations were related to the amounts of Tc in the plants by the same relation as in exp. 1 (table 1). This means that Tc is mainly taken up as TcO_4^- and the contribution of the Tc-cysteine complex can be neglected.

Another indication for the uptake of only TcO_4^- is the speciation of Tc in the plants which can tell us something about the chemical form in which Tc is taken up.[5] The metabolization of Tc is dependent on the chemical form of the administered Tc and results in a typical speciation of Tc in the plants. Plants grown on a mixture of TcO_4^- and Tc-cysteine, show the same relative amount of different Tc complexes as plants grown on TcO_4^- only.[5]

3.4 Experiment 4

In this experiment, the Tc-tetren is the only species of Tc present in the nutrient solution, due to the presence of an anion exchange resin that acts as a scavenger for TcO_4^-.

Plants grown in a nutrient solution containing only TcO_4^-, show 40 % of the Tc in the leaves being TcO_4^- and 60 % bio-organic Tc-compounds. Plants grown on Tc-tetren, show a pronounced shift in this distribution pattern in favour of the bio-organic compounds (83 %) (fig. 4 and fig. 5).

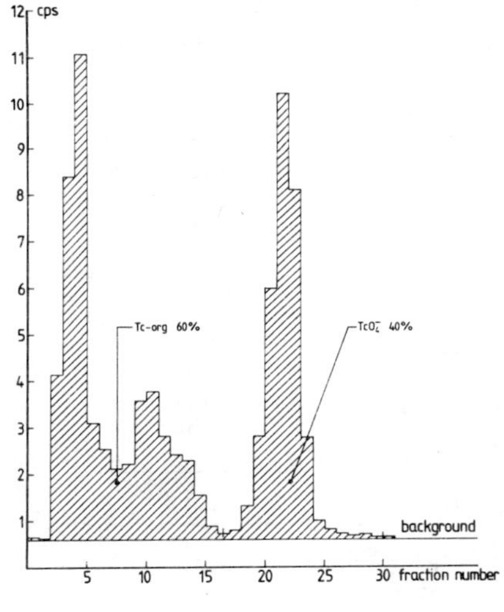

Fig. 4

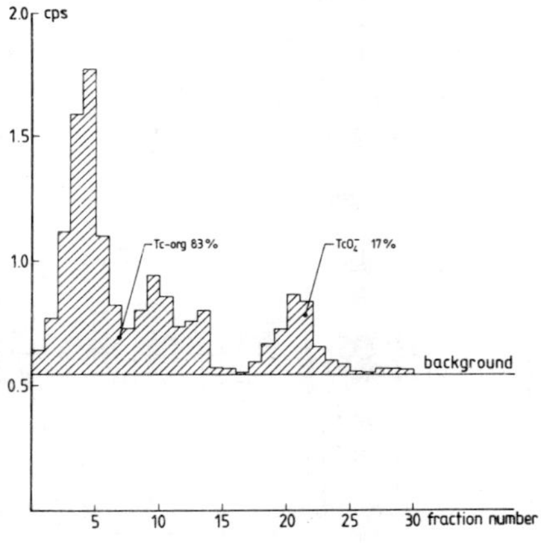

Fig. 5

FIGURE 4 Speciation of ^{99}Tc in spinach plants grown on TcO_4^-.

FIGURE 5 Speciation of ^{99}Tc in spinach plants grown on a Tc-tetren complex.

This means that the Tc-tetren complex is taken up by the spinach plants. The uptake rate, however, is lowered by two order of magnitude as compared to TcO_4^-. The transferfactor for the

Tc-tetren complex is also lowered by two order of magnitude.

4 CONCLUSIONS

- In case of a constant TcO_4^- concentration, the amount of Tc in the plants (Bq/g fresh weight) is related to the concentration of TcO_4^- in the nutrient solution by a linear relationship.
- The uptake of TcO_4^- by plants can be approximated by an exponential function and is related to plant growth.
- In case of varying TcO_4^- concentrations, a weighed mean concentration can be calculated and used in the linear relationship to predict the concentration of Tc in the plant and to calculate transferfactors.
- The uptake of Tc-complexes is possible, but the uptake rate is lowered by two order of magnitude. Its contribution to the Tc-content of plants can be neglected.

5 REFERENCES

1. G. Sposito, Soil Sci. Soc. Amer. J., 48, 531 (1984).
2. D.L. Sparks, Soil Sci. Soc. Amer. J., 48, 514 (1984).
3. F.T. Bingham, J.E. Strong and G. Sposito, Soil Science, 135, 160 (1983).
4. W.G. Sunda, D.W. Engel and R.M. Thuotte, Env. Sci. Techn., 12, 409 (1978).
5. L.R. Van Loon and J.F. Lembrechts, Symposium on the behaviour of technetium in the environment, Cadarache, France, 23 - 26 Octobre 1984.

ACCUMULATION OF ^{113}SN BY A MARINE DIATOM

N.S. FISHER, F. AZAM, and J.-L. TEYSSIE

International Laboratory of Marine Radioactivity
Musée Océanographique
MC 98000 Monaco

ABSTRACT

The bioconcentration of ^{113}Sn by a marine centric diatom, Thalassiosira pseudonana, was studied in laboratory culture experiments. The tin content of water and algal samples was determined by detecting gamma emissions of ^{113}In in radioactive equilibrium with the ^{113}Sn. Log-phase cells accumulated Sn linearly over time in the dark, so that volume/volume concentration factors (VCFs) of 5×10^4 were attained after 1 day. In the light, the cells were apparently able to regulate their cellular Sn content, and a steady-state VCF value of 2×10^4 was observed. The accumulation of Sn by Si-limited cells washed with acid was found to increase inversely with the ambient Si concentration to which the cells were exposed. The experimental results suggest that Sn may be incorporated along with Si in the diatom frustule, similar to the behavior of Ge. The VCF of Sn in Si-limited cells reached 1×10^5, indicating that Sn is one of the more reactive elements for diatoms. Cells at 0 C in the dark accumulated equivalent amounts of Sn to illuminated cells at 18 C (approximately), suggesting that Sn accumulation in the cells proceeds passively (ie., requiring no metabolic activity). Phytoplankton could influence the distribution of Sn in natural waters by introducing it into food webs and/or mediating its vertical transport in the water column.

INTRODUCTION

Interest in the behavior of tin in marine systems stems from three geochemical characteristics: (a) its mobilization by man exceeds by tenfold the natural erosion rate; (b) it is one of the three most highly enriched metals in atmospheric particulate matter relative to the earth's crust; and (c) it can be biomethylated to highly toxic species[1]. Its presence in the world's oceans, primarily attributable to atmospheric input, has been documented[1]. Few studies, however, have addressed the question of bioaccumulation of tin in marine biota. An early field study[2] suggests that some marine organisms, including phytoplankton, may concentrate Sn appreciably from English coastal waters, and Sn was also found to concentrate in various macroalgal species[3] and fish species[4]. In addition, a series of experimental laboratory studies on the biokinetics of Sn in select macroalgal species and marine invertebrates has recently been conducted; results to date suggest that Sn may be very reactive for certain species (S.W. Fowler, personal communication). Interest in phytoplankton interactions with pollutants, including Sn, stems from the fact that these organisms are at the base of most marine food webs, are known to concentrate certain toxic substances substantially from ambient water, and may serve to introduce such toxicants into food chains and/or significantly mediate their vertical transport in the water column. Diatoms may be of particular interest here since they are not only extremely important in many marine ecosystems but make siliceous cell walls. Given its position in the periodic table of the elements, it seems entirely possible that Sn may act as a Si substitute, much as Ge does in marine diatoms[5]. We have therefore examined the accumulation of Sn by a cultured marine diatom under various Si concentrations, using radiotracer (^{113}Sn) methodology.

MATERIALS AND METHODS

All experiments used the small centric marine diatom, *Thalassiosira pseudonana*, maintained in unialgal clonal (clone 3H) culture in f/2 medium[6] prepared with Mediterranean surface seawater, minus Cu, Zn, and EDTA. Experimental media, consisting of 40 ml sterile-filtered Mediterranean surface seawater enriched with f/2 levels of N, P, and vitamins, and varying levels of silicate, were inoculated with 3×10^4 log-phase cells ml^{-1} and received 1.5×10^{-10} M Sn by Eppendorf micropipet. The cultures were contained in 50 ml polystyrene vessels sealed with polypropylene caps. The Sn, in dilute HCl, was supplied by Amersham, U.K. and, after dilution, contained 2.9×10^4 kBq ^{113}Sn l^{-1} ($t_{1/2}$ = 115 d), so that the cultures contained 6.6 kBq ^{113}Sn l^{-1}. The cultures were incubated at 18 C in constant darkness or constant light (5.2×10^{-2} ly min^{-1}) provided by "cool-white" fluorescent lamps. Some cultures, pre-equilibrated at 0 C, were also incubated at 0 C in the dark to assess the significance of metabolic activity in mediating the accumulation of Sn. All treatments were run in triplicate. Periodically, the cultures were sampled for cell growth and for cellular accumulation of Sn using filtration methods described elsewhere[7]. In another experiment using 2.8×10^{-9} M Sn and inocula of 1.1×10^5 late log-phase cells ml^{-1}, aliquots of filtered cells were also washed with 3 N HCl to remove all cellular Sn except frustular Sn. ^{113}Sn decays by electron capture to ^{113}In ($t_{1/2}$ = 99 min); filter and water samples were therefore counted several days after sampling by detecting photon emissions of ^{113}In from 350 to 440 keV with a Packard 5650 gamma counter equipped with a NaI(Tl) well-type crystal. Counting times were such that 1σ propagated counting errors were <5%. The ^{113}Sn content of filters from the uninoculated control cultures (treated identically) were subtracted to give net values[7]. The control filter values were generally negligibly small relative to the cell culture values. Volume/volume concentration factors (VCFs) were determined as moles Sn/μm^3 ÷ moles dissolved Sn/μm^3 water at time of apparent equilibrium with respect to isotope partitioning between dissolved and particulate states.

RESULTS

All cells accumulated Sn, irrespective of culture conditions (Fig. 1). Cells did not grow at 0 C or at 18 C in the dark, but the cells which came from a log-phase stock culture did divide at 18 C in the light (Fig. 1), even in the absence of added Si. The mean coefficient of variation of cell division rates among replicate cultures was 5.7%. In the light, the cells were apparently able to regulate the amount of cellular Sn, irrespective of external Si concentration (Fig. 1). There was no apparent trend for cellular Sn uptake as a function of external Si concentration in either the light or the dark (Fig. 2). The late log-phase cells, by contrast, were apparently Si-limited, as there was no growth of cells in the culture which received no Si addition, while cells receiving Si enrichment grew (Fig. 3). ^{113}Sn uptake increased over the first 4.5 h in all cultures but subsequently declined in the Si-enriched media, while in the unenriched water the cells continued to accumulate ^{113}Sn (Fig. 3). Moreover, in the acid-washed cells, the Sn content increased linearly with time for the Si-limited cells but stabilized within 2.3 h for the Si-enriched cells (Fig. 3). Within the first 4.5 h incubation Si enrichment enhanced the ^{113}Sn accumulation in the cells, although this was only seen in the cells which were not acid-washed (Fig. 4). The Sn remaining after acid washing (ie., in the cell walls) was relatively unaffected

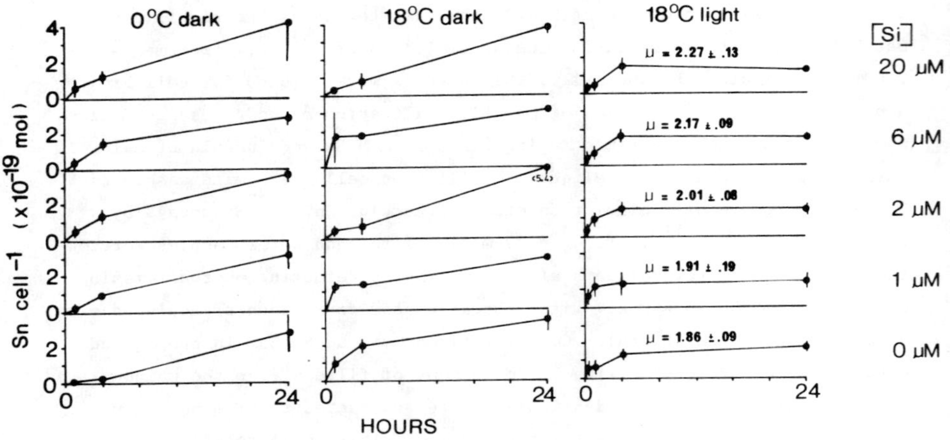

Fig. 1 Sn accumulation by log-phase T. pseudonana as a function of time and under different Si concentrations and culture conditions. Data points are means ± 1 SD of 3 replicate cultures. Cell growth rates (µ, or number of divisions per day) are given for 18 C cultures in the light.

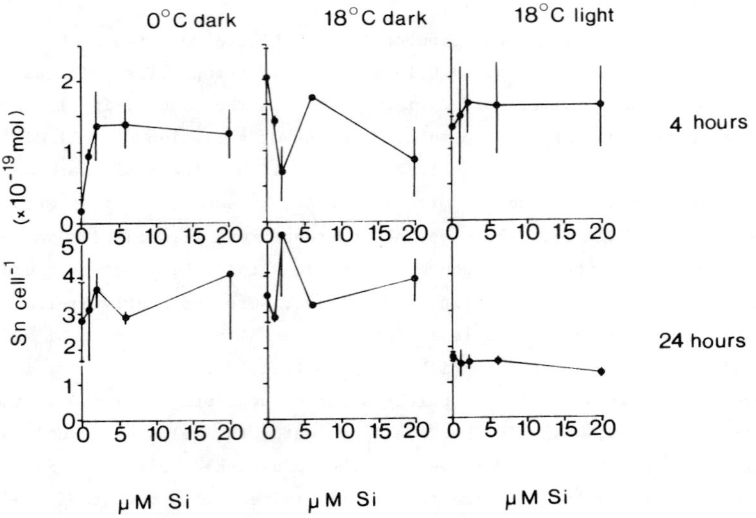

Fig. 2 Sn accumulation by log-phase T. pseudonana under different culture conditions at two sample times, as a function of Si concentration. Data points are means ± 1 SD of 3 replicate cultures.

by Si enrichment at the beginning of the incubation period. After 22.5 h incubation, it was clear that Si enrichment substantially depressed ^{113}Sn uptake by the cells, being particularly evident in the acid-washed cells (Fig. 4). The VCF of Sn in the log-phase cells stabilized at 18 C in the light at about 2×10^4, while in the dark the VCF continued to rise with time; at 24 h, the VCF was around 5×10^4. In the Si-limited cells, the VCF reached 1×10^5 at 22.5 h in the absence of Si enrichment.

DISCUSSION

The likeliest dominant species of Sn in seawater is $Sn(OH)_4$[1], analogous to $Si(OH)_4$, and Sn might therefore be expected to be highly reactive for marine particulate matter. The results of our experiments (VCFs ranging from 2×10^4 to 10^5, depending on physiological state of the cells and Si enrichment) suggest that Sn is a particle reactive element, comparable with such metals as Zn, Ag, and Hg in their reactivity for phytoplankton[8], but somewhat less reactive than such transuranic elements as Pu, Am, Cm, and Cf[7]. The bioaccumulation of Sn in diatoms appears to proceed passively by surface adsorption, although in the light the cells may be able to regulate their cellular Sn content. Sn is also concentrated passively (VCF $\approx 9 \times 10^4$) by the freshwater green alga *Ankistrodesmus falcatus*, in which Sn primarily associates with carbohydrate[9]. In comparison, a VCF in net phytoplankton of $\approx 9 \times 10^3$ can be estimated from the marine field data of Smith and Burton[2]. In the cultured picoplanktonic cyanobacterium, *Synechococcus* sp., Sn was also found to concentrate passively and in accordance with Freundlich adsorption isotherms, with a VCF of $\approx 8 \times 10^5$ [10]. This higher VCF simply reflects the greater surface:volume ratio of the *Synechococcus* cell ($\approx$ 1 μm diameter) than of the *T. pseudonana* cell (4 - 6 μm diameter)[10].

It appears that diatoms may incorporate Sn into their frustules much as they do Ge[5]. The uptake of Si in diatoms is an energy-dependent process, whereas diatoms generally accumulate Ge at similar rates in the light and the dark[11]. Similarly, *T. pseudonana* continued to accumulate Sn at 0 C and 18 C in the dark, indicating that Sn can accumulate in cells in the absence of photosynthesis or other metabolic processes. In the light, the Sn content per cell was maintained at a constant level, possibly reflecting the regulation of Si in the cells under these conditions. The elevated Sn uptake in Si-starved cells was probably a result of the lack of competition from Si, suggesting that Sn and Si are incorporated together in these diatoms. Along these same lines, it is noteworthy that in preliminary experiments we have observed that Sn showed

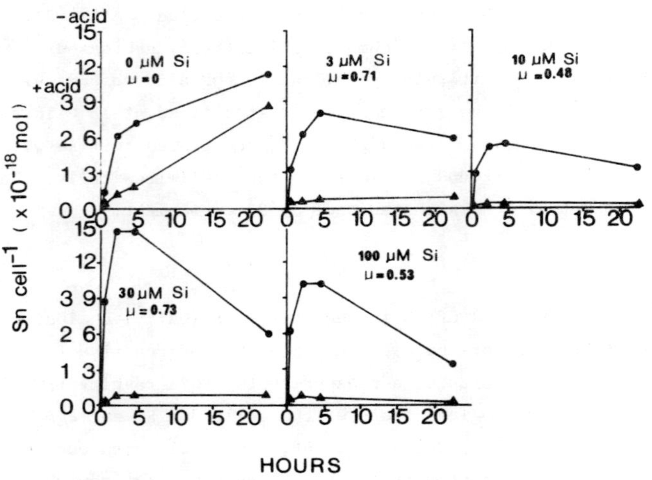

Fig. 3 Sn accumulation by Si-limited T. pseudonana, at 18 C in the light, as a function of time and Si concentration. Cell growth rates (µ) are also shown; note that cells receiving no Si enrichment failed to divide. (•) denotes filtered cells washed with seawater only; (▲) denotes filtered cells washed with seawater and acid (see text).

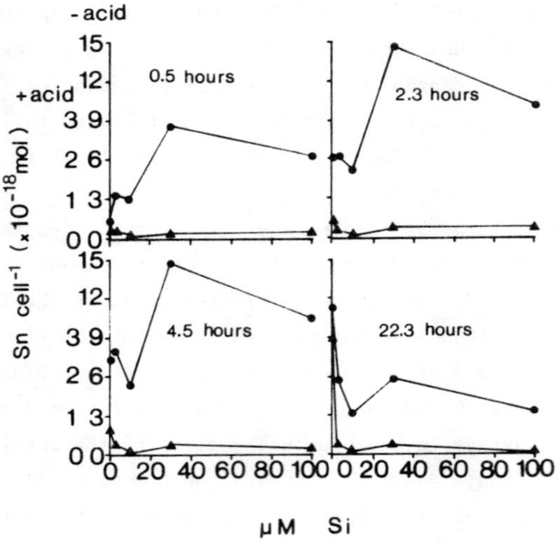

Fig. 4 Sn accumulation by Si-limited T. pseudonana, at 18 C in the light, at four sample times, as a function of Si concentration. (•) denotes filtered cells washed with seawater only; (▲) denotes filtered cells washed with seawater and acid (see text).

much greater reactivity for the walls of borosilicate glass vessels than did other metals, with 25 - 30% of the added Sn associating with walls within one day, in contrast with typically less than 1% for most metals.

These experiments were designed to use ^{113}Sn only as a tracer, and the low atom concentrations employed were probably well below toxic levels. By analogy with Ge, it may be expected that molar ratios of Sn:Si of approximately 0.1 would be sufficient to depress significantly cell division rates in diatoms[11]. Given a Si content of ≈ 1 pg $cell^{-1}$ of _T. pseudonana_[12], the highest cellular Sn:Si molar ratio in our experiments was ≈ 4×10^{-4}. Relatively few studies have thus far addressed the question of Sn toxicity to marine phytoplankton. Of particular interest to our study is the finding that the marine diatom, _Nitzschia liebethrutti_, exposed for 14 days to a comparatively high Sn concentration (1.5 μM), displayed frustular abnormalities[13], probably as a result of the deposition of the Sn in the cell walls as a Si substitute.

REFERENCES

1. J.T. Byrd and M.O. Andreae, _Science_, _218_, 565 (1982).
2. J.D. Smith and J.D. Burton, _Geochim. Cosmochim. Acta_, _36_, 621 (1972).
3. V.F. Hodge, S.L. Seidel, and E.D. Goldberg, _Anal.Chem._, _51_, 1256 (1979).
4. R. Eisler, _Trace Metal Concentrations in Marine Organisms_ (Pergamon, N.Y., 1981).
5. F. Azam, B.B. Hemmingsen, and B.E. Volcani, _Arch. Microbiol._, _92_, 11 (1973).
6. R.R.L. Guillard and J.H. Ryther, _Can. J. Microbiol._, _8_, 229 (1962).
7. N.S. Fisher, P. Bjerregaard, and S.W. Fowler, _Limnol. Oceanogr._, _28_, 432 (1983).
8. N.S. Fisher, M. Bohé, and J.-L. Teyssié, _Mar. Ecol. Prog. Ser._, _18_, 201 (1984).
9. P.T.S. Wong, R.J. Maguire, Y.K. Chau, and O. Kramar, _Can. J. Fish. Aq. Sci._, _41_, 1570 (1984).
10. N.S. Fisher, _Mar. Biol._, in press.
11. D. Werner, in _The Biology of Diatoms_, edited by D. Werner (Blackwell, Oxford, 1977), pp. 110-149.
12. E. Paasche, _Norw. J. Bot._, _20_, 197 (1973).
13. E.M. Saboski, _Water Air and Soil Poll._, _8_, 461 (1977).

CHEMICAL SPECIATION OF TECHNETIUM IN SOIL AND PLANTS : IMPACT ON SOIL-PLANT-ANIMAL TRANSFER.

VANDECASTEELE C.M. [a], C.T. GARTEN Jr. [b], R. VAN BRUWAENE †[c], J. JANSSENS [c], R. KIRCHMANN [c] and C. MYTTENAERE [d].

a. Laboratoire de Physiologie Végétale, Université Catholique de Louvain, B-1348 Louvain-La-Neuve, Belgium.

b. Environmental Sciences Division, Oak Ridge National Laboratory, Oak Ridge, TN 37831, USA.

c. Département de Radiobiologie, Centre d'Etude de l'Energie Nucléaire, B-2400 Mol, Belgium.

d. Programme "Biologie - Radioprotection", CCE-DG XII, Rue de la Loi, 200, B-1049 Bruxelles, Belgium.

INTRODUCTION.

Considerable uncertainties are associated with the environmental behaviour of technetium-99 and its transfer from soil to plants and then to animals and man. For this reason, most of the mathematical models built to simulate the environmental transport of Tc and to calculate the dose to man are associated with conservative simplifications and produce overestimates of the calculated dose {1}.

In order to follow the new ICRP recommendations, transfer models are needed that estimate as accurately as possible the dose to the population; this implies a better knowledge of the behaviour of Tc in the environment, especially concerning its long-term behaviour. At this time, most of the available data deal with the short-term and only scanty results have been obtained regarding the plant-animal transfer, especially in the case of polygastric mammals.

More information is thus required principally concerning the long-term, from which model parameters reflecting reality as closely as possible will be deduced and will take into account the overall effects of Tc chemical form modifications on the transfer of this nuclide from source to man.

MATERIAL AND METHODS.

The technetium long-term behaviour in soil is under investigation in four lysimeters installed in August 1983 in open field, submitted to natural climatic conditions. Each unit consists of a stainless steel cuve (area = 1 m^2) filled with 200 l of a disturbed soil laying above a granulometric stratification of sand, gravels and stones in order to obtain a good percolation of the rain water. The lixiviation water is collected from the lowest level of the cuve and drained by gravity to a 50 l collecting reservoir. Percolates are collected and sampled regularly.

The soil was sown in autumn 1983 with a pasture grass mixture and allowed to equilibrate during autumn and winter. Immediately after the first spring cut (04-02-1984), each soil was contaminated by surface spraying with 0.5 mCi ^{99}Tc as NH_4TcO_4 (Amersham). Three successive grass harvests and two soil core samples were collected before a second Tc deposit (07-13-1984) of an equal amount of activity (0.5 $mCi.m^{-2}$) and the vegetation was harvested three more times before winter.

Plant and soil samples were analysed for their Tc content by liquid scintillation counting (Tri-Carb 2450, Packard). Plant material was digested with HNO_3 and H_2O_2 {2}; the activity present in soil samples was extracted by four successive solvents {3} in order to distinguish the water soluble and exchangeable fractions (bioavailable fraction) from the forms that are less readily available for plant uptake.

Soil-plant transfer were calculated as the ratio of the activity per gram of plant dry weight to the soil Tc concentration at the harvesting time or at the start of the vegetative growth period, assuming a soil density of 1.55 $g.ml^{-1}$.

Plant material used for autoradiography, speciation and animal feeding were grown on diluted (2/5) Hoagland-Arnon nutrient solution {4} contaminated with various subtoxic Tc levels, using ^{99}Tc as NH_4TcO_4 (Amersham), eventually traced by ^{95m}Tc ($KTcO_4$, produced at the Louvain-La-Neuve cyclotron).

Autoradiographs of contaminated leaves, previously dried at 105°C, were produced on X-ray films (Definix Medical, Kodak).

Chemical speciation of Tc in plant material was determined on fresh and freeze-dried material following the method described by Bowen *et al.* {5}. The distribution of the activity at the subcellular level was investigated by filtration of ground fresh and freeze-dried material and successive centrifugations of the filtrate : 2 min x 500 g to pellet cell wall debris and nuclei, 5 min x 2000 g for plasts, 20 min x 10 000 g for mitochondria, and finally 70 min x 105 000 g to separate the cytoplasmic fraction from the microsomes. A fraction of the postmicrosomal supernatant was submitted to gel filtration chromatography on various sieves (Sephadex G-75, G-50 and G-25, Pharmacia) and to electrophoresis on Whatman 3MM paper bands (1 x 40 cm^2) in Tris HCl buffer 20 $mM.l^{-1}$ pH 8.0 at 300 V.

Plant-animal transfer was studied on rats (adult female Wistar rats, R/cnb imbred strain) and sheeps (Suffolk strain, 8 months old). Rats each received one of the following treatments : 1) a single intravenous injection of TcO_4^-, 2) a single overnight feeding of TcO_4^- added to pulverized standard laboratory ration (chow), 3) a single overnight feeding of bioincorporated Tc in freeze-dried corn leaves or algae (*Fucus vesiculosus*) mixed to the chow. Urine and feces were collected and analysed for their Tc content for seven days following Tc administration. On days 1, 3, 7 and 14 after administration, three rats from each experimental group (12 rats) were killed, dissected and analysed to determine the Tc concentration in 18 different body tissues. Tc excretion and concentration in various organs and tissues were investigated in sheep after intravenous injection of TcO_4^- and after injection directly into the rumen of TcO_4^- and of Tc biologically bound in freeze-dried *Fucus*. Excretion was followed for 93 days after administration and animals were sacrified at different times after application, dissected and analysed for the activity within different tissues.

SOIL.

The technetium released from the nuclear fuel cycle is generally considered to reach the soil in its most oxidized form (TcO_4^-) and to remain in this very soluble and very mobile chemical form in most agricultural soils characterized by an aerobic environment {6,7}.

Kd values reported by Wildung *et al.* {8} for 22 different soils incubated during 24 hours with TcO_4^- trace amounts ranged from 0.007 to 2.8 attesting to the high solubility of this element. Similar values are calculated by Balogh and Grigal {9}, Routson *et al.* {10} and Mousny and Myttenaere {11} from column elution experiments or from classic estimation of the distribution coefficient between soil and contaminated water. In contrast, Landa *et al.* {12} observed very high sorption onto the solid phase, up to 98 % after 2 to 5 weeks incubation time; these results could be explained by the creation of a reducing environment due to anaerobic conditions in the closed soil-water system used by these authors, providing reduced Tc species that could be precipitated or chelated by the soil organic matter {6,9}.

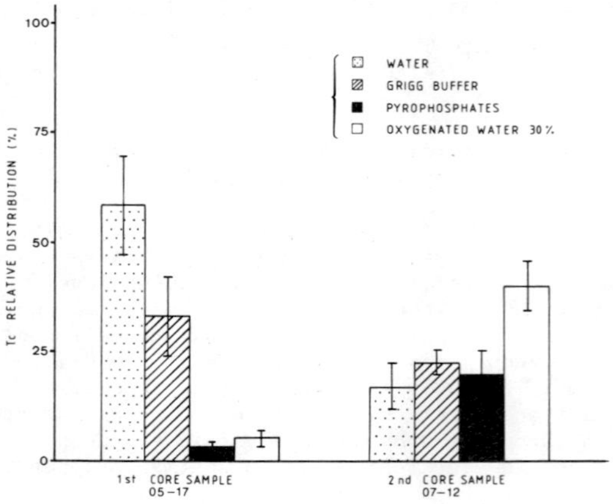

Fig. 1. *Technetium speciation in two soil core samples collected from open field lysimeters at two different times : relative distribution (%) of the activity following its extractability by four solvents used successively. Mean and standard deviation of four samples.*

From Kd measurements and soil column elution experiments, it appears that the Tc sorption onto the solid phase depends on various parameters including soil mineral composition, organic matter and iron content, moisture content, pH, redox potentiel, temperature and biological activity, itself governed by the former parameters {3,6,11,12,13,14,15,16,17}.

These biogeochemical factors promote chemical modifications of the Tc form in the soil (reduced and/or chelated forms) as the result of dynamic processes. The time dependence of these transformations has been confirmed in field lysimeter experiments investigating the distribution of the activity in soil core samples taken at two different times by successive extractions with four solvents (Fig.1.). Forty-five days after contamination with TcO_4^-, 89 to 94 % of the activity is recovered in the two plant available fractions while, one month later, only 30 to 47 % is still bioavailable, leaving more than one half of the Tc strongly bound to the soil matrix in very poorly to non available forms for plant uptake.

SOIL-PLANT TRANSFER.

Biogeochemical parameters in combination with the time, as they modify the technetium chemical form, affect its transfer from soil to plant.

Numerous short-term studies on plant grown in nutrient solution or in well aerated soils contaminated with TcO_4^- have indicated a very high Tc bioaccumulation {2,12,13,18,19,20,21,22}. For longer periods of time, however, Mousny and Myttenaere {3,11} and Routson and Cataldo {20} have demonstrated that a progressive evolution of technetium form from TcO_4^- to less available forms leads to a noticeable decrease of its transfer from soil to plant from one harvest to the next.

Under naturel climatic conditions, in field lysimeter experiments, the same behaviour is observed (Fig.2.) : the calculated transfer factors decrease markedly from the first harvest, 35 days after initial soil contamination, to the next two. A new deposit of TcO_4^- restores high transfer values in the next cut that rapidly decrease in following harvests. The transfer factor obtained for the last harvest is very low as compared to what is generally expected for aerated soils but could be explained by the high amount of precipitation during the preceding months that induces reducing conditions in a more hydrated soil and could also have leached a fraction of the activity taken up by the plant {23}.

During the first year, almost 70 % of the deposited Tc was removed from the soil, 40 % being exported by the vegetation harvesting and 30 % eliminated from the plough layer by lixiviation. This result is far from the concept of soil decontamination by the simple physical decay process, as generally accepted by most of mathematical models, and indicates the need for better knowledge of the nuclide exportation balance in model building.

PLANT.

The distribution of technetium between the different parts of the plant is also affected by the transformation of its chemical form within the soil {11} : with time and successive harvests, a higher proportion of the activity transfered to the aerial organs of pea plants grown on contaminated soil is found in shoots and seeds and the fraction recovered in leaves is

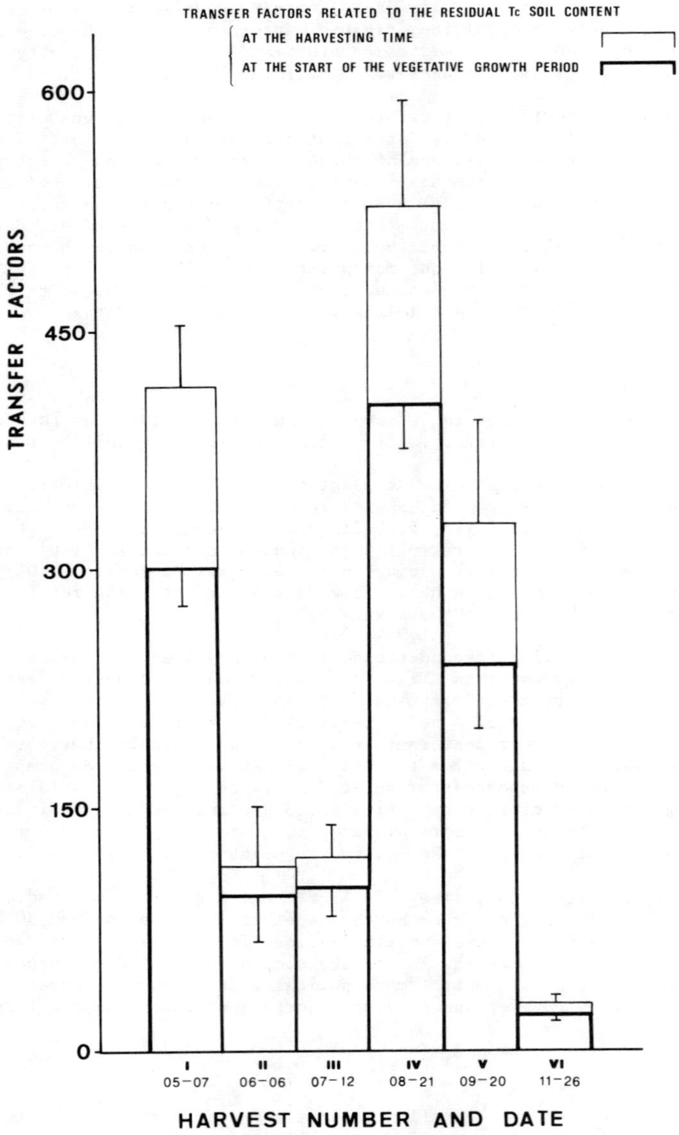

Fig. 2. *Time evolution of soil-plant transfer factors in open field lysimeters.*

slightly lowered, attesting to changes in the form of Tc in the soil. However, most of the plant activity is accumulated in the leaves {2,12,13, 18,19,21,22}. The reason for this accumulation in the green parts seems to be the reducing environment generated by the photosynthetic metabolism of the leaf. This hypothesis is favoured by a preferential localization of the activity in the chlorophyllian parenchyma of variegated leaves demonstrated by autoradiography of pathogenic variegation (photo 1) and genotypic variegation (photo 2). In addition, Myttenaere *et al.* {24} report a direct proportionality between accumulation and the length of the light period.

The cellular localization of technetium in leaves has been investigated in contaminated soybean, maize and *Fucus vesiculosus*, a brown alga, by chemical extraction : in order to compare the results presented in table I for the three different plant materials, it seems reasonable to consider together instead of separately the two first extracts obtained with quite soft extractants and representing the technetium easily available for gastrointestinal uptake by monogastric mammals; this fraction represents 51 to 62 % of the total activity of leaf. 31 to 41 % is poorly extractable, only by harsh NaOH 2N treatment or remain associated with the residue. In fact, the fraction corresponding to the first two treatments is already extractable by water : 60 % of the activity is removed by three successive overnight soakings of chopped plant material at 4°C, confirming the high availability of this part of the accumulated activity.

Filtration of ground material followed by differential centrifugations of the filtrate (Table II) shows that 45 to 76 % of the activity present in the leaves was recovered in the postmicrosomal supernatant as free Tc or Tc bound to cytosoluble compounds. The electrophoretic behaviour of this Tc is, however, quite different from TcO_4^- (Fig. 3.). It does not migrate very far from the deposition point suggesting that it is in an uncharged reduced form or bound to large molecules; by protein precipitation with $(NH_4)_2SO_4$, 74 % of the activity of the postmicrosomal supernatant is coprecipitated, suggesting binding of this nuclide with protein compounds. Gel filtration chromatography on Sephadex (Fig. 4.) associates the technetium with compounds of molecular weight ranging from 2 to 30 kD approximately for one half of the activity while the second half is eluted in a single peak at V_e/V_o ratios of 3 on G-50 and 4 on G-25 so that it could correspond to TcO_4^-; an appreciable UV absorbance at 280 nm is however associated with this peak indicating that Tc in this fraction could be bound to small organic molecules (peptides).

PLANT-ANIMAL TRANSFER.

The absorption of technetium by monogastric mammals (rat) depends on the administrated form of the nuclide : comparing intravenous injection of TcO_4^-, TcO_4^- spread on the chow or biologically incorporated Tc in maize, the cumulative excretion in urine and feces combined during the first seven days after administration was as follows : I.V. treatment 91 %; TcO^- + chow 82 % and Tc biologically incorporated in corn 74 %. Moreover, the distribution of the activity between urinary and fecal excretion varied depending upon the administration route : the urinary pathway was predominant for I.V. injection, equal to the fecal pathway in the case of TcO_4^- spread on chow and greatly decreased in importance in rats fed with bioincorporated Tc (Fig. 5.). Similar results have been obtained for Tc biologically bound to brown algae.

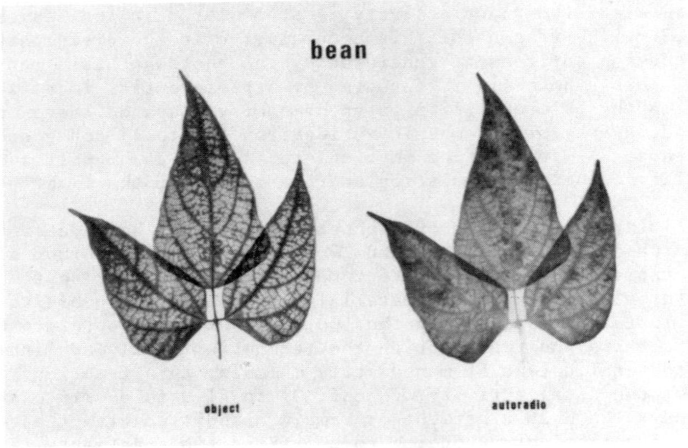

Photo 1. *Distribution of Tc in variegated bean leaf (pathogenic variegation) visualized by autoradiography.*

Photo 2. *Distribution of Tc in variegated coleus leaf (genotypic variegation) visualized by autoradiography.*

Table I. *Relative distribution (%) of Tc accumulated in plant material between the fractions extracted by different solvents following the method of Bowen et al.* {5}.

EXTRACTANT	PLANT MATERIAL				
	SOY BEAN	MAIZE		FUCUS	
	FRESH	FRESH	FREEZE DRIED	FRESH	FREEZE DRIED
- ETHANOL 95 %	7	40	24	25	10
- CHLORHYDRIC ACID 0.2 N	51	22	28	29	41
- PERCHLORIC ACID 0.5 N	2	6	7	7	9
- Na HYDROXIDE 2N	34	29	37	37	38
- RESIDUE	5	2	3	2	3

Table II. *Relative distribution (%) of the Tc accumulated in plant material between fractions obtained by filtration of ground fresh and freeze-dried material and differential centrifugation of the filtrate.*

	SOY BEAN	MAIZE		FUCUS	
	FRESH	FRESH	FREEZE DRIED	FRESH	FREEZE DRIED
FILTRATION					
RESIDUE	16	29	33	42	36
FILTRATE	84	71	67	58	64
CENTRIFUGATION OF FILTRATE (= 100 %)					
CELL DEBRIS AND ORGANELLES	9	33	20	23	19
POSTMICROSOMAL SUPERNATANT	91	67	80	77	81

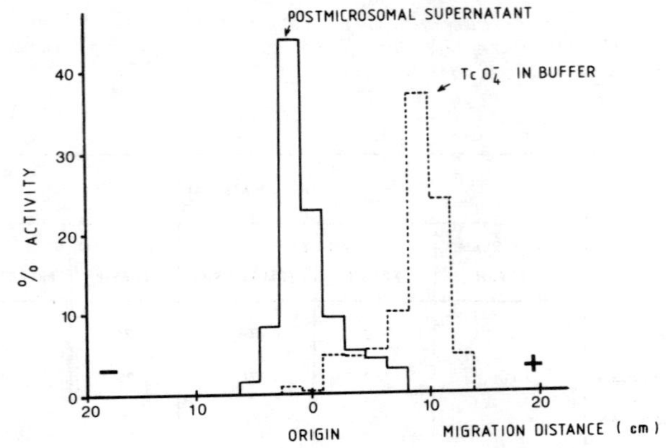

Fig. 3. *Electrophoretic behaviour of TcO_4^- and of Tc in the postmicrosomal supernatant from soybean leaves.*

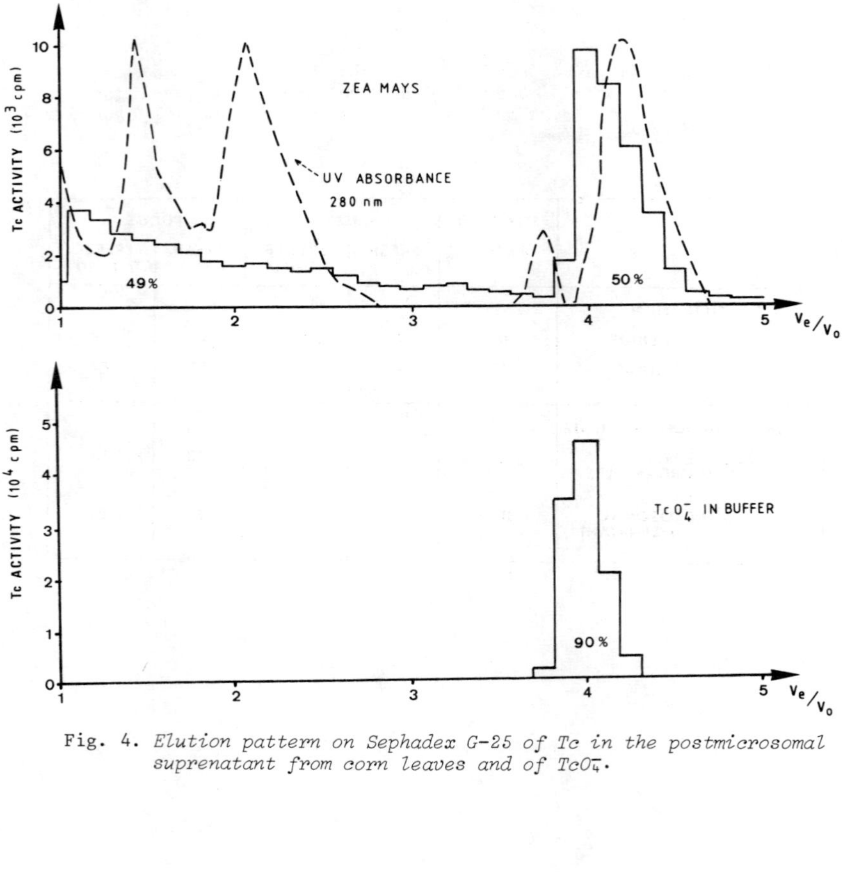

Fig. 4. *Elution pattern on Sephadex G-25 of Tc in the postmicrosomal suprenatant from corn leaves and of TcO_4^-.*

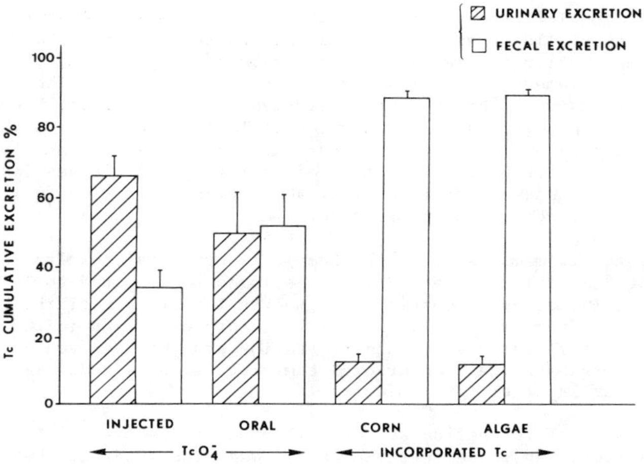

Fig. 5. *Relative distribution (%) of the Tc cumulative excretion in rat, between urinary and fecal pathway seven days after administration of the activity as TcO_4^- or as bioincorporated Tc in plant tissues.*

Tc FORM APPLIED / EXCRETION PATHWAY	TcO_4^-	Tc IN ALGAE	PERIOD FOLLOWING ADMINISTRATION
URINE	0.504	3.83	day 0 to day 4
	0.024	0.310	4 8
	0.018	0.090	8 30
	0.004	0.020	30 ∞
	0.610	**4.25**	**0 ∞**
FECES	88.7	58.4	day 0 to day 4
	0.703	9.75	4 8
	0.024	0.446	8 30
	0.003	0.004	30 ∞
	89.4	**68.6**	**0 ∞**

Table III. *Time evolution of the Tc excretion (as % of the administrated dose) by urinary and fecal pathway in sheep after direct injection into the rumen of TcO_4^- and of Tc biologically incorporated in a brown alga.*

Although the Tc concentration in different organs was lower for bioincorporated radionuclide than for I.V. injection, reflecting a lower gastrointestinal absorption in the case of Tc metabolised by plants, the distribution of this element was similar for the three treatments. The most contaminated tissues were thyroid, hair, kidneys, liver and skin. The high hair content could be used as a bioindicator of Tc environmental contamination. The present data are in good agreement with the results of Sullivan *et al.* {25}, wich showed that incorporation of Tc into plant tissues (soybeans) resulted in decreased gastrintestinal absorption by rats and guinea pigs relative to the absorption of inorganic Tc.

In polygastric mammals (Suffolk sheep), experimental results concerning the metabolism of TcO_4^- injected into the rumen as compared to TcO_4^- by intravenous injection demonstrate an important influence of the rumen microflora : after oral dose, the total urinary excretion is 100 times lower than after intravenous administration and the observed concentration in various organs 29 days after oral administration is generally 10 times lower than it is after I.V. injection.

Assimilation of technetium as TcO_4^- or bound to brown algae (*Fucus vesiculosus*), directly injected into the rumen, was also studied. Whereas in rat the biologically bound Tc appears to be less well absorbed, the data in sheep indicate a lower excretion into feces, a greater excretion into urine and higher residual activities in tissues 3 months after exposure to Tc bound to algae than after exposure to TcO_4^-. These differences are most likely due to different processing of food in monogastric compared to polygastric mammals, and more probably to the presence of a rumen microflora affecting the Tc chemical form.

CONCLUSIONS.

The technetium chemical form in soil is affected by the soil biogeochemical conditions and tends, with time, to evolve in less available forms for plant uptake; the observed transfer factors between soil and plant in field conditions are more or less reduced compared to the values first reported for short-term experiments in nutrient solution and well aerated soils.

Moreover, the Tc metabolism in plants also reduces the transfer of this element from vegetation to animal as compared to absorption after *per os* administration of TcO_4^-, at least for monogastric mammals.

These data suugest that the radiological impact of Tc releases to the environment may be substantially less than predicted by assessment models that consider pertechnetate as the transfered form and use data derived from laboratory experiments.

BIBLIOGRAPHY.

{1} SCHWARZ G. and F.O. HOFFMAN, "An examination of the effect in radiological assessment of high soil-plant concentration ratios for harvested vegetation", Health Physics, 39 : 983-986 (1980).

{2} VANDECASTEELE C.M., A. DELMOTTE, P. ROUCOUX and C. VAN HOVE, "Technétium et organismes diazotrophes : toxicité, localisation, facteurs de transfer", *in* "Environmental migration of long-lived radionuclides",

Proc. Int. Symp. on the "Migration in the terrestrial environment of long-lived radionuclides from the nuclear fuel cycle", Knoxville, Tenn. (USA), July 27-31, 1981, IAEA-SM-257/76 : 275-286, IAEA (Vienna), 1982.

{3} MOUSNY J.M. and C. MYTTENAERE, "Transfert du technétium du sol vers la plante en fonction du type de sol, du mode de contamination et de la couverture végétale", *in* "Environmental migration of long-lived radionuclides", Proc. Int. Symp. on the "Migration in the terrestrial environment of long-lived radionuclides from the nuclear fuel cycle", Knoxville, Tenn. (USA), July 27-31, 1981, IAEA-SM-257/75P : 353-358, IAEA (Vienna), 1982.

{4} HOAGLAND D.R. and I.R. ARNON, "The water-culture method for growing plants without soil", Univ. Calif. Ag. Experimental Station Circular, 347 (1950).

{5} BOWEN H.J.M., P.A. CAWSE and J. THICK, "The distribution of some inorganic elements in plant tissues extracts", J. Exp. Bot., 13 : 257-267 (1962).

{6} WILDUNG R.E., K.M. McFADDEN and T.R. GARLAND, "Technetium sources and behaviour in the environment", J. Environ. Qual., 8 : 156-161 (1979).

{7} TRABALKA J.R. and C.T. GARTEN Jr., "Behaviour of the long-lived synthetic elements and their natural analogs in food chain", Adv. Radiat. Biol., 10 : 39-104 (1983).

{8} WILDUNG R.E., R.C. ROUTSON, R.J. STERNE and T.R. GARLAND, "Pertechnetate, iodide and methyl iodide retention by surface soils", USERDA Rep., BNWL-1950 Part 2 : 37-41, Natl. Tech. Inf. Serv., Springfield, Va. (1974).

{9} BALOGH J.C. and D.F. GRIGAL, "Soil chromatographic movement of technetium-99 through selected Minnesota soils", Soil Science, 130 : 278-282 (1980).

{10} ROUTSON R.C., G. JANSEN and A.V. ROBINSON, "^{241}Am, ^{237}Np and ^{99}Tc sorption on two United States subsoils from differing weathering intensity areas", Health Physics, 33 : 311-317 (1977).

{11} MOUSNY J.M. and C. MYTTENAERE, "Absorption of technetium by plants in relation to soil type, contamination level and time", Plant Physiol., 61 : 403-412 (1981).

{12} LANDA E.R., L.J. THORVIG and R.G. GAST, "Effect of selective dissolution, electrolytes, aeration and sterilization on technetium-99 sorption by soils", J. Environ. Qual., 6 : 181-187 (1977).

{13} SAAS A., J.L. DENARDI, C. COLLE and J.M. QUINAULT, "Cycle du molybdène et du technétium dans l'environnement; évolution physico-chimique et mobilité dans les sols et les végétaux", Actes du II° Symp. Int. de Radioécologie, CEA-EDF, Cadarache (F), 19-22 Juin 1979, Cadarache CEN : 443-489 (1980).

{14} TANG VAN HAI and M. VAN DOORSLAER DE TEN RYEN, "Movement of pertechnetate through soils : hydrodynamic dispersion", J. Environ. Radioactivity, 1 : 41-50 (1984).

{15} STALMANS M., A. MAES and A. CREMERS, "Role of organic matter as a geochemical sink for technetium in soils and sediments", Proc. Scientific Seminar on the "Behaviour of technetium in the environment", Cadarache (F), October 23-26, 1984, CEA-CCE-DOE(USA), in press.

{16} VAN LOON L., M. STALMANS, A. MAES and A. CREMERS, "Soil humic acid complexes of technetium synthesis and characterization", Proc. Scientific Seminar on the "Behaviour of technetium in the environment", Cadarache (F), October 23-26, 1984, CEA-CCE-DOE(USA), in press.

{17} RANCON D. and J. ROCHON, "Rétention des radionucléides à vie longue par divers matériaux naturels", Réunion de travail CCE sur "la migration des radionucléides à vie longue dans la géosphère", Bruxelles (B), 29-31 Janvier 1979.

{18} CATALDO D.A., R.E. WILDUNG and T.R. GARLAND, "Technetium accumulation, fate and behaviour in plants", Chap.28, *in* Proc. of the Environ. Chemistry and Cycling Processes Symp., Augusta, Ga., Conf-760429 Tech. Inf. Center, Oak Ridge, Tenn. (USA), 1978.

{19} MOUSNY J.M., P. ROUCOUX and C. MYTTENAERE, "Absorption and translocation of technetium in pea plant", Environ. Exp. Bot., 19 : 263-268 (1979).

{20} ROUTSON R.C. and D.A. CATALDO, "Accumulation of ^{99}Tc by tumbleweed and cheatgrass grown on acid soils", Health Physics, 34 : 685-690 (1978).

{21} WILDUNG R.E., T.R. GARLAND and D.A. CATALDO, "Accumulation of technetium by plants", Health Physics, 32 : 314-317 (1977).

{22} VANDECASTEELE C.M., R. DE BECKER, TANG VAN HAI and C. MYTTENAERE, "Etude de l'absorption du technétium-99 par le riz irrigué (*Oryza sativa* L.)", Radioprotection, 18 : 19-30 (1983).

{23} MYTTENAERE C., C. DAOUST and P. ROUCOUX, "Leaching of technetium from foliage by simulated rain", Environ. Exp. Bot., 20 : 415-419 (1980).

{24} MYTTENAERE C., C.M. VANDECASTEELE, P. ROUCOUX, E.A. LIETART, A. ITSCHERT and J.M. MOUSNY, "Processus biologiques responsables de l'accumulation du Tc-99 par les végétaux", Proc. Scientific Seminar on the "Behaviour of technetium in the environment", Cadarache (F), October 23-26, 1984, CEA-CCE-DOE(USA), in press.

{25} SULLIVAN M.F., T.R. GARLAND, D.A. CATALDO and R.G. SCHRECKHISE, "Absorption of plant-incorporated nuclear fuel cycle elements from the gastro-intestinal tract", *in* "Biological implications of radionuclides released from nuclear industries", Vol. I, IAEA-SM-237/58 : 447-457, IAEA (Vienna), 1979.

DISCUSSION

BULMAN:

Do you think that molydate and pertechnate could compete for transport mechanisms?

VANDECASTEELE:

In *Azotobacter*, a nitrogen fixing free bacteria, Tc reduces the uptake of Mo by the cells but in these micro-organisms the molybdenum absorptions does not seem to be regulated. In plants we have not observed any competition for absorption between these two elements.

BONOTTO:

You showed clearly, by autoradiography, that Tc is present in the leaf regions, where photosynthesis occurs. Do you know if Tc is present inside and/or outside the chloroplasts?

VANDECASTEELE:

Tc is associated with the chloroplast fractions isolated by differential centrifugation but it does not represent an important fraction of the cellular Tc.

BEHAVIOUR OF TECHNETIUM IN MARINE ALGAE

S. BONOTTO(1), R. KIRCHMANN(1), J. VAN BAELEN(1) C. HURTGEN(2), M. COGNEAU(3), D. VAN DER BEN(4), C. VERTHE(5) and J.M. BOUQUEGNEAU(5)

(1) Département de Radiobiologie, C.E.N.-S.C.K., B-2400 Mol ;
(2) Département de Métrologie Nucléaire, C.E.N.-S.C.K., B-2400 Mol ;
(3) Laboratoire de Chimie Inorganique et Nucléaire, Université Catholique de Louvain, B-1348 Louvain-la-Neuve ;
(4) Institut Royal des Sciences Naturelles de Belgique, B-1040 Bruxelles
(5) Laboratoire d'Océanologie, Université de Liège, Sart Tilman, B-4000 Liège

ABSTRACT

Uptake and distribution of technetium were studied in several green (Acetabularia acetabulum, Boergesenia forbesii, Ulva lactuca) and brown (Ascophyllum nodosum, Fucus serratus, Fucus spiralis and Fucus vesiculosus) marine algae. Technetium was supplied to the algae as Tc-95m-pertechnetate. Under laboratory conditions, the algae were capable of accumulating technetium, with the exception, however, of Boergesenia, which showed concentration factors (C.F.) comprised between 0.28 and 0.71. The concentration of technetium-99 in Fucus spiralis, collected along the Belgian coast, was measured by a radiochemical procedure. The intracellular distribution of technetium was studied by differential centrifugation in Acetabularia and by the puncturing technique in Boergesenia. The chemical forms of technetium penetrated into the cells were investigated by selective chemical extractions, molecular sieving and thin layer chromatography.

1. INTRODUCTION

In the marine environment, unicellular and pluricellular algae are important intermediaries in the food chain. It is, thus, of interest to know to which extent they incorporate technetium, a long-lived radionuclide (T1/2 = 2.1 x 10^5 years), mainly produced during the fission of uranium-235 (6% yield). Recent work, done in our and other laboratories[1-12], has demonstrated that several species of marine brown algae (Alaria esculenta, Ascophyllum nodosum, Cystoseira compressa, Ectocarpus confervoides, Fucus serratus, Fucus spiralis, Fucus vesiculosus, Laminaria digitata, Laminaria flexicaulis and Pelvetia canaliculata) have a relatively high affinity for technetium. On the contrary, green (Boergesenia forbesii, Caulerpa prolifera, Cladophora rupestris, Codium tomentosum, Dunaliella bioculata, Enteromorpha sp. and Ulva lactuca) and red species (Chondrus crispus, Corallina officinalis, Porphyra umbilicalis and Rhodymenia palmata (= Palmaria palmata)) incorporate low amounts of this element. Between the green marine algae, Acetabularia acetabulum (= mediterranea) is an exception, showing concentration factors (C.F.) up to about 400, at low technetium mass concentrations in the medium[10]. The reason why some marine algae concentrate technetium and others do not, remains to be elucidated.

The aim of this work was to obtain more information on the distribution of technetium at the cellular and molecular level and on its chemical forms.

2. MATERIAL AND METHODS

The giant unicellular algae Acetabularia acetabulum (= mediterranea) and Boergesenia forbesii were cultivated, under sterile conditions, as previously reported[13-14]. The other algae (Ulva lactuca, Ascophyllum nodosum, Fucus serratus, Fucus spiralis and Fucus vesiculosus were collected along the Belgian and/or the Dutch coast. Technetium distribution was studied by differential centrifugation (Acetabularia)[10], by the puncturing technique (Boergesenia)[15] and by the extraction procedure of Bowen et al.[16], which permits to separate seven main fractions[10-11]. Molecular sieving chromatography of ^{95m}Tc-labeled algal extracts was done on Sephacryl S-300, as previously reported[10]. Tc-pertechnetate ($^{95m}TcO_4^-$)

was separated by thin layer chromatography (methanol-water, 85-15%, DC-Alufolien Kiezelgel 60F/254 Merck, 20 x 20 cm). ^{99}Tc (which transforms into stable ^{99}Ru) was measured in Fucus spiralis by a radiochemical analysis combining two published procedures[17-18]. ^{95m}Tc (which transforms into stable ^{95}Mo) was supplied to the algae as potassium-pertechnetate. ^{3}H-Mannose and ^{14}C-Leucine (NEN) were used as internal markers in column chromatography.

3. RESULTS AND DISCUSSION

Several experiments have shown that technetium enters Acetabularia and Boergesenia cells. In Acetabularia, ^{95m}Tc was found mostly in the cytoplasmic fraction (68.8%), the remaining part being distributed in the chloroplasts (26.6%), mitochondria (0.6%) and cell wall (3.6%)[10]. In Boergesenia, ^{95m}Tc crosses the cell wall and the cytoplasmic layer and enters the vacuole, without being accumulated (C.F. < 1). Thin layer chromatographic analysis has revealed that ^{95m}Tc present in the vacuolar sap of Boergesenia behaves as TcO_4^- (+ VII) (results to be published elsewhere).

The results obtained by successive chemical extractions[16] showed some differences, between the 5 investigated species of marine algae, for ^{95m}Tc distribution in the seven fractions (fig.1 and 2). However, with the exception of Acetabularia (fig.1B), fraction 6, constituted mainly by proteins, contained the highest percentage (about 30-50%) of radioactivity. Results obtained by molecular sieving chromatography of labeled extracts of Acetabularia acetabulum (fig.3), Fucus vesiculosus and Ascophyllum nodosum suggested also that part of ^{95m}Tc taken up by the algae may be bound to proteins. Figure 3A shows the optical density (280 nm) and the ^{95m}Tc-radioactivity profiles of compounds eluted from a Sephacryl S-300 column for a sample of postmitochrondrial supernatant from labeled vegetative cells of Acetabularia. Six radioactive peaks are recognizable : the first three of them probably correspond to ^{95m}Tc bound to macromolecules (mainly proteins), whereas the fourth might represent ^{95m}Tc bound to small molecules, as suggested by the fact that ^{3}H-Mannose (fig.3B) and ^{14}C-Leucine (fig.3C) are eluted at about the same position. The fifth small peak corresponds to material not yet identified and the sixth one should contain $^{95m}TcO_4^-$, as inferred from the fact that marker pertechnetate is eluted at the same position.

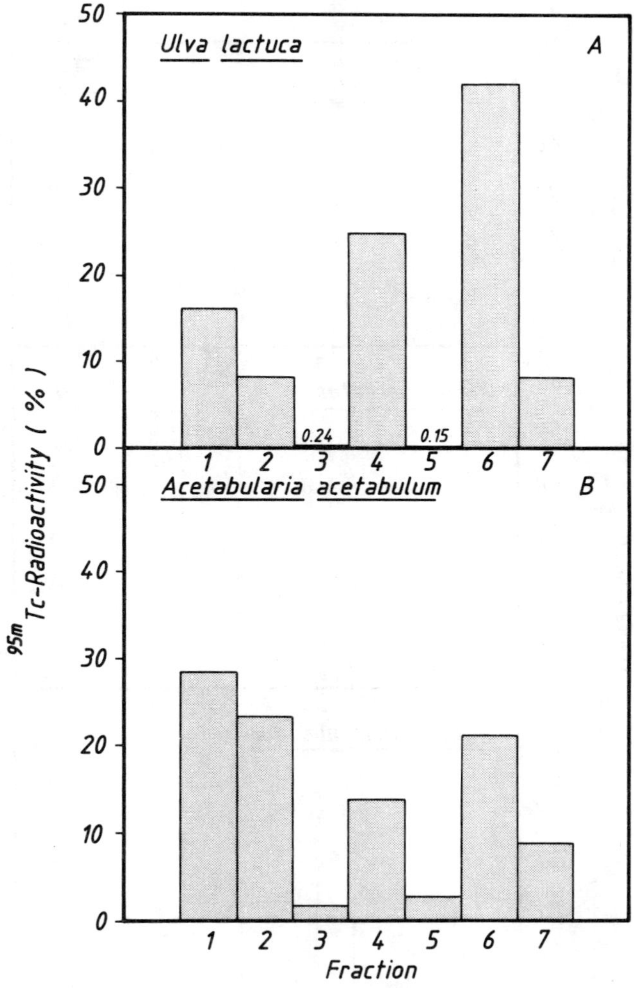

Fig.1. Distribution of ^{95m}Tc in the green marine algae Ulva lactuca (A) and Acetabularia acetabulum (B). The algae were labeled with $K^{95m}TcO_4$ and then extracted according to the procedure of Bowen et al.[16]. The seven fractions would contain : (1) lipids, lipophyllic pigments, free amino acids and possibly other small molecules ; (2) ionic forms including salts of organic acids, phosphates and carbonates ; (3) pectates and some proteins ; (4) some remaining ionic forms ; (5) nucleic acids and some proteins ; (6) most proteins and some polysaccarides ; (7) residual cell wall material.

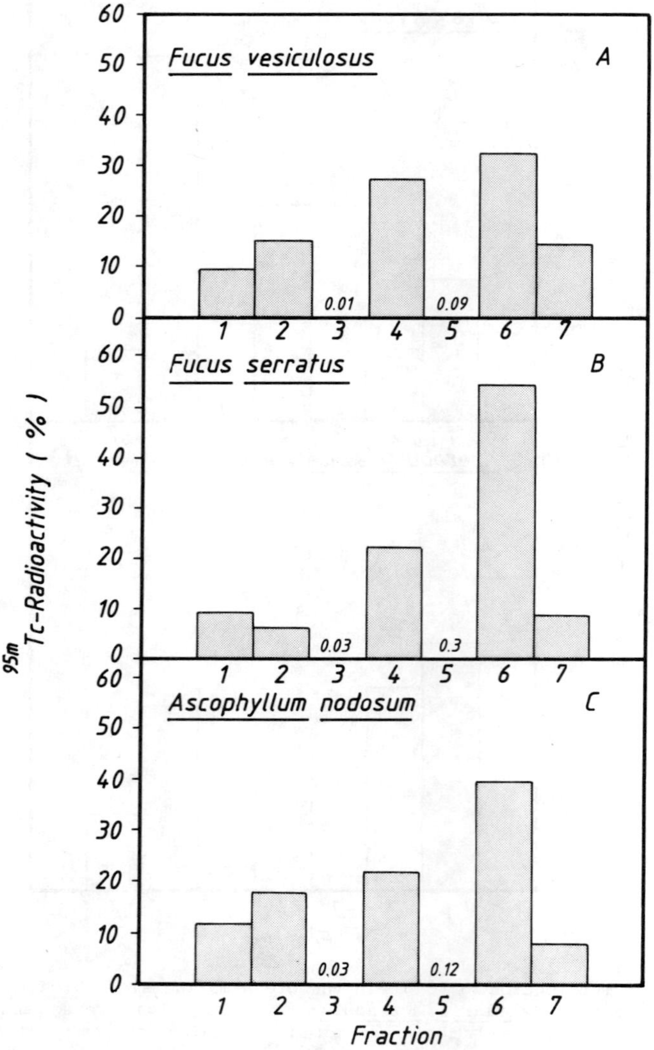

Fig.2. Distribution of ^{95m}Tc in the brown marine algae Fucus vesiculosus (A), Fucus serratus (B) and Ascophyllum nodosum (C). The algae were labeled with $K^{95m}TcO_4$ and then extracted according to the procedure of Bowen et al.[16]. See legend of fig.1 for the content of fractions 1-7.

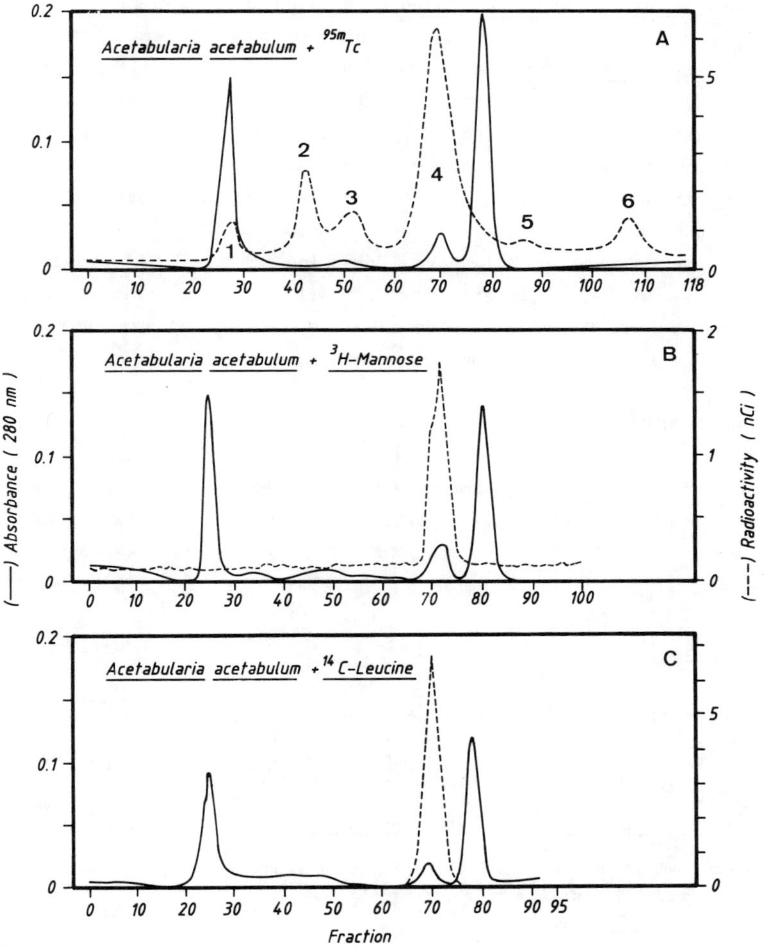

Fig.3. Molecular sieving chromatography of extracts of Acetabularia acetabulum. The cells were labeled with K^{95m}TcO$_4$ and their extract (postmitochondrial supernatant) was chromatographed on a Sephacryl S-300 column, with automatic recording of the optical density (280 nm). The radioactivity was measured in a Packard Auto-Gamma spectrometer. ^{3}H-Mannose and ^{14}C-Leucine were used as molecular markers. A : ^{95m}Tc-labeled extract ; B : unlabeled extract added with ^{3}H-Mannose ; C : unlabeled extract added with ^{14}C-Leucine.

In nature several marine brown algae were found to accumulate technetium (^{99}Tc)[1-11], with concentration ratios of dry material/water attaining 30,000 in Fucus vesiculosus[9]. A related species, Fucus spiralis, from the Belgian coast, was found to contain measurable amounts of ^{99}Tc (Table 1). The concentrations of ^{99}Tc found in Fucus spiralis are in good agreement with those reported by other authors[4,5,9]. However, it is not yet clear why some algae (for instance sample 3 in table 1) contain more ^{99}Tc than others. Since the accumulation of technetium in Fucus is probably integrated over a time of 2-4 years, depending on the life span of the plant[9], it would be of interest to investigate how physical (light, temperature, salinity) and physiological (age, biological stage) factors affect its capability of incorporating this artificial element.

4. ACKNOWLEDGEMENTS

Work supported in part by contract CEC nr BIO-B-485-82-B. Miss Carole Verthé and Mr J. Van Baelen are supported by the Belgian Ministry of Labour (BTK Project nr 17.003). We thank Mr A. Bossus, Mr G. Nuyts and Mr K. Hofkens for their excellent help.

5. REFERENCES

1. R.J. Pentreath, D.F. Jefferies, M.B. Lovett and D.M. Nelson, in Marine Radioecology, NEA-OECD Paris, pp. 203-221, (1980).
2. R.J. Pentreath, in Impacts of Radionuclide Releases into the Marine Environment, IAEA, Vienna, pp. 241-272, (1981).
3. M. Masson, G. Aprosi, A. Lanièce, P. Gueguenіat and Y. Belot, in Impacts of Radionuclide Releases into the Marine Environment, IAEA, Vienna, pp. 341-359, (1981).
4. L. Jeanmaire, M. Masson, F. Patti, P. Germain and L. Cappellini, Marine Pollution Bulletin, 12, 29-32, (1981).
5. M. Masson, F. Patti, L. Cappellini, P. Germain and L. Jeanmaire, J. Radioanal. Chem., 77, 247-255, (1983).
6. S. Bonotto, G.B. Gerber, M. Cogneau and R. Kirchmann, Ann. Ass. Belge de Radioprotection, 8, 281-292, (1983).
7. G. Aprosi and M. Masson, Radioprotection, GEDIM, 19, 89-103, (1984).
8. G. Aprosi and M. Masson, Rapport EDF-HE31/84-31, pp. 1-29.
9. E. Holm, J. Rioseco and G.C. Christensen, in Report EUR 9214 EN, A. Cigna and C. Myttenaere, Eds., pp. 357-367, (1984).
10. S. Bonotto, G.B. Gerber, C.T. Garten,jr., C.M. Vandecasteele, C. Myttenaere, J. Van Baelen, M. Cogneau and D. van der Ben, in Report EUR 9214 EN, A. Cigna and C. Myttenaere, Eds., pp. 381-396, (1984).
11. S. Bonotto, J. Van Baelen, G. Arapis, C.T. Garten, C.M. Vandecasteele, C. Myttenaere and M. Cogneau, Arch. Intern. Physiol. Biochim., 92, V14-V15, (1984).

12. B. Mania, Z. Moureau, D. van der Ben, J. Van Baelen, C. Verthé, J.M. Bouquegneau, M. Cogneau, S. Bonotto, C.M. Vandecasteele, L. Pignolet and C. Myttenaere, Paper presented at the Scientific Seminar on the Behaviour of Technetium in the Environment, Cadarache, France, October 1984 (in press).
13. S. Enomoto and H. Hirose, Phycologia, 11, 119-122, (1972).
14. L. Lateur and S. Bonotto, Bull. Soc. Roy. Bot. Belgique, 106, 17-38 (1973).
15. S. Bonotto, A. Bossus, G. Nuyts, R. Kirchmann, P. Mathot, J. Colard and F. Cinelli, in Wastes in the Ocean, Vol.3. : Radioactive Wastes in the Ocean, P.K. Park, D.R. Kester, I.W. Duedall and B.H. Ketchum, Eds., John Wiley & Sons, New York, pp. 287-300, (1983).
16. H.J.M. Bowen, P.A. Cawse and J. Thick, J. Exp. Bot., 13, 257-267, (1962).
17. F. Patti, L. Cappellini and L. Jeanmaire, Note CEA nr 2140, pp. 1-10, (1980).
18. J.P. Riley and S.A. Siddiqui, Anal. Chim. Acta, 139, 167-176, (1982).

TABLE I ^{99}Tc content in samples of Fucus spiralis collected at different times of the year along the Belgian coast.

Sample	Place	Date	Dry weight(g)	Radioactivity (Bq kg^{-1})*
1	Oostende	01.03.1982	49.7	60
2	Niewpoort	01.03.1982	50.8	90
3	Oostende	10.06.1982	33.8	380
4	Oostende	10.06.1982	46.5	110
5	Blankenberge	10.06.1982	43.2	70
6	Oostende	16.05.1983	42.4	50
7	Oostende	11.10.1983	38.8	70
8	Blankenberge	11.10.1983	48.5	50
9	Niewpoort	12.10.1983	38.3	110

* Standard error (2 σ) : 10%

DISCUSSION

HARVEY:

You noted significant differences between the concentration factors for ^{95m}Tc and ^{99}Tc in marine algae presumably due to mass differences. Have you been able to identify differences in the sub-cellular distributions of these nuclides?

BONOTTO:

We have figures only for ^{95m}Tc, for which the subcellular distribution is respectively: 68.8% in cell cytosol, 26.6% in chloroplasts, 0.6% in mitochondria and 3.6% in cell wall material. It seems, thus, that ^{95m}Tc is not absorbed very much by cell wall. Figures for ^{99}Tc shall, hopefully, be available in the future.

PRODUCTION OF CHELATING AGENTS BY *PSEUDOMONAS AERUGINOSA* GROWN IN THE PRESENCE OF THORIUM AND URANIUM

E.T. PREMUZIC,[1] M. LIN,[1] A.J. FRANCIS,[1] and J. SCHUBERT[2]

[1]Department of Applied Science, Brookhaven National Laboratory, Upton, NY 11973

[2]Department of Chemistry, University of Maryland, Baltimore County, Catonsville, MD 21228

ABSTRACT

Chelating agents produced by microorganisms enhance the dissolution of iron increasing the mobility and bioavailability of the metal. Since some similarities exist in the biological behavior of ferric, thorium and uranyl ions, microorganisms resistant to these metals and which grow in their presence may produce sequestering agents of Th and U, and other metals in a manner similar to the complexation of iron by siderophores. The ability of *P. aeruginosa* to elaborate sequestering agents in medium containing thorium or uranium salts was tested. Addition of 10, 100, and 1000 ppm of uranium or thorium to culture medium increased the lag period of the organism as the concentration of the metal increased. At concentrations of 1000 ppm and higher, there was an extended lag period followed by reduction in growth. Uranium has a stronger inhibitory effect on growth of the organism than thorium at similar concentrations. Analyses of the culture media have shown, that relative to the control, and under the experimental conditions used, the microorganisms have produced several new chelating agents for thorium and uranium. Extracts containing these chelating agents have been tested for their decorporation potential. *In vitro* mouse liver bioassay and *in vivo* mouse toxicity tests indicate that their efficiency is comparable to DTPA (Diethylenetriaminepentaacetic acid) and DFOA (Desferrioxamine) and that they are virtually non-toxic to mice. The bacterially produced compounds resemble, but are not identical to the known iron chelating siderophores isolated from microorganisms. Some of their chemical properties are also discussed.

INTRODUCTION

Certain microorganisms are resistant to high levels of toxic metals.[1,2] Several mechanisms exist by which microorganisms interact with metals, one of which is the production of chelating agents.[3] For example, microorganisms requiring iron for growth produce chelating agents which enhance the distribution, mobility and bioavailability of iron.[4-7] A number of such chelating agents have been isolated and characterized. They are cyclic and acyclic compounds containing hydroxamate, amino, catechol and phenolic components, commonly known as siderophores. Because of known biochemical similarities between ferric iron, thorium and uranium,[8] it appeared reasonable to expect that microorganisms resistant to toxic metals may also produce sequestering agents in response to the presence of toxic metals in their environment analogous to the production of iron siderophores. If this were the case, then such chelators may also bind given toxic metals selectively. Such compounds could serve to develop highly efficient and selective or specific therapeutic agents for the decorporation of toxic metals.

To test these possibilities, two organisms have been chosen, *Pseudomonas aeruginosa-CSU* and *P. aeruginosa PAO-1*. These organisms have been chosen because *P. aeruginosa* sp. are known to elaborate strains resistant to toxic metals, and one of which (*P. CSU*) has been isolated from a soil contaminated with plutonium.[9,9a]

MATERIALS AND METHODS

Details of the procedures used in this work have been reported elsewhere,[10,11] and will be mentioned only briefly here with additional cross-references when necessary. *P. aeruginosa-CSU* and *P. aeruginosa-PAO-1* (ATCC 15692) were grown in a defined medium[12,13] in the presence of increasing concentrations of thorium or uranium. Growth of bacteria and changes in the medium were monitored by absorption spectroscopy and by direct counts.[14] Thorium or uranium concentrations were determined spectroscopically.[15] All supernatants from the culture media were ultrafiltered to remove large molecular weight (>5000 daltons) species and then freeze-dried. The concentrated samples were analyzed by gel-permeation, thin-layer and high pressure liquid chromatography[10,11] followed by mass-spectroscopy and others (e.g., ultra violet, infra-red, and HNMR) as needed. *In vitro* mouse liver assay to evaluate chelating effectiveness was carried out according to published procedures.[15] *In vivo* mouse toxicity tests were conducted with Hale-Stoner Swiss Albino mice and the test solutions injected interperitoneally.[16]

RESULTS AND DISCUSSION

Bioavailability of Thorium and Uranium

Bioavailability of thorium and uranium depends on the speciation of the metal, which in turn depends on the concentration and pH. The effects of thorium and uranium concentrations are shown in Figure 1. As the concentration of thorium increases, the lag period in the growth of organisms also increases. With time, however, the delayed rate of growth recovers and nearly equals that of the control. In cultures grown in presence of uranium the concentration of the metal has a significant inhibitory effect. In both cases, the pH during the growth of the organism increases from near neutral to alkaline (~pH 8).

Production of Fluorescent Products

During the growth of *P. aeruginosa* (CSU and PAO-1) in the presence of thorium or uranium, a formation of fluorescent highly pigmented substances is induced by the presence of the metals. The onset and the attainment of

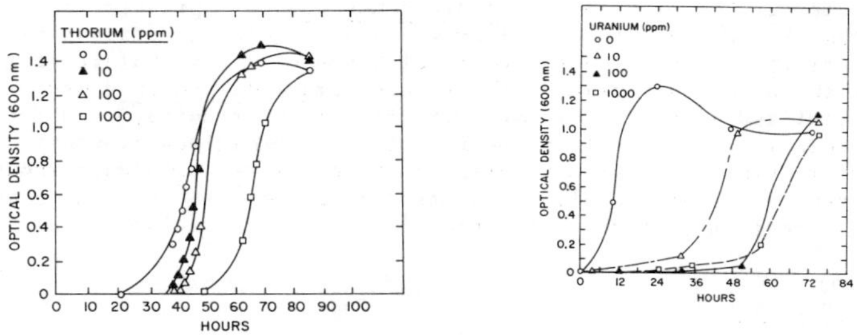

FIGURE 1 Growth curves of P. aeruginosa in the presence of different concentrations of thorium and uranium.

the maximum production of the pigments for thorium, λmax 368 nm, is much faster than the production of the corresponding uranium pigments, λmax 365 nm. Excitation at 360 nm yields a fluorescence spectrum with a single peak at λ = 433 nm for thorium and at λ = 440 nm for uranium.[11]

Effect of Supplemental Iron

Cultures prepared with reagent grade quality mineral salts used as nutrients in the defined media contain only small amounts of iron (<0.05 ppm). Addition of supplemental iron (e.g., 18 ppm) has little or no effect on cultures grown in the presence of thorium, but it has a pronounced effect when added to the uranium culture, as exemplified by the difference spectrum in Figure 2.

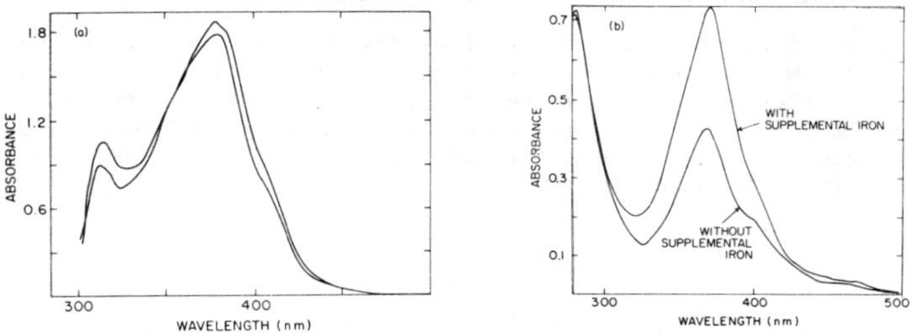

FIGURE 2 Effect of supplemental iron on thorium (a) and uranium (b) cultures.

Significance of Spectral Changes and the Production of Pigments

Several strains of P. aeruginosa when grown in iron deficient media exhibited spectral and compositional changes in the media during growth. These changes were monitored spectroscopically, by chromatographic separation and by tests for the presence of hydroxamates, catechols and related, known chelators. For example, *P. fluorescence putida*[17] produced a yellow-green fluorescent pigment, λmax 400 nm, containing hydroxamate and quinoline moieties, confirmed by the elucidation of the structure of the compound known as Pseudobactin.[18,19] Several other *Pseudomonas* have been characterized in a similar manner.[20-23] In our studies, chromatography and spectroscopy of culture media extracts at maximum growth in the presence of thorium or uranium identified various metal complexes induced by the presence of these metals during growth of the organism, which were not found in controls, grown in absence of thorium or uranium.

BIOLOGICAL TESTING

The complexes induced in presence of thorium or uranium are assumed to exhibit complexing properties in mammalian tissues. To test this, the ability of crude extracts to remove thorium in the liver assay[15] *in vitro* was compared to that of DTPA and DFOA. The results for thorium are shown in Table 1. The efficiency of thorium decorporation from liver cells by the crude extracts containing the chelating agents produced by *P. aeruginosa* grown in the presence of thorium is comparable to that of DTPA and DFOA. An *in vivo* toxicity study[16] with mice injected with 10%, 5%, and 1% solutions of freeze-dried extracts, as well as with methanol extracts, proved very promising. Thus, the methanolic extract of the culture medium of PAO-1 grown in the presence of 100 ppm thorium was non-lethal at the dose of about 100 mg of lyophilized culture medium at maximum growth. This dose is equivalent to about 0.24 liter of the original culture medium. An identical sample, however, from the batch in which the organism was grown in the absence of thorium was lethal. At doses less than 10%, i.e., 5% and 1%, aqueous extracts of thorium media were non-lethal.

PRELIMINARY CHARACTERIZATION OF THORIUM INDUCED CHELATING AGENTS

Chromatography of an ethanol extract of acidified, succinate free PAO-1-thorium culture media yielded ten fractions. Mass spectroscopy of these fractions indicates a presence of several compounds with molecular weights ranging from 208 to >600 daltons. Fraction containing $M+H^+$ 209 closely resembles pyrimine[24], $C_{10}N_2O_3H_{12}$, Fig. 3, with the following fragmentation pattern: M/e 209, 163, 130, 79, 75.

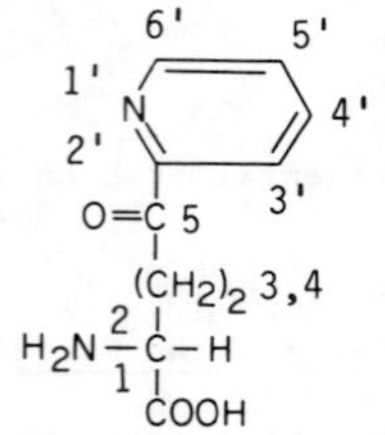

FIGURE 3 Pyrimine.

TABLE I Liver assay of P. aeruginosa extracts.

Sample type	% Th found Supernatant	Liver
Liver assay of PA-CSU extracts.		
1. $Th(NO_3)$, 100 ppm in Th	7	93
2. Th-citrate, 100 ppm in Th	64	32
3. Culture medium* zero growth with 100 ppm of Th added immediately prior to the analysis	20.5	80
4. Culture medium, maximum growth with 100 ppm of Th added immediately prior to the analysis	17.6	82
5. Culture medium, maximum growth in the presence of 100 ppm Th	97.3	3
6. Culture medium, maximum growth, preformed Th-DTPA added prior to analysis (equimolar; 100 ppm with respect to Th)	97	3
7. Culture medium, maximum growth to which Th DFOA was added prior to analysis (equimolar; as under 6)	90	10
Liver assay of PAO-1 extracts.		
1. $Th(NO_3)_4$, 27 ppm in Th	1	99
2. Th-succinate, 27 ppm in Th	3	97
3. Culture medium*, maximum growth with 27 ppm of Th added immediately prior to the analysis	9	91
4. Culture medium, maximum growth in the presence of 100 ppm Th	98	2
5. Culture medium, maximum growth preformed Th-DTPA added prior to analysis (equimolar; 27 ppm with respect to Th)	84	6
6. Culture medium, maximum growth, preformed Th-DFOA added prior to analysis (equimolar, as 5)	27	75

*All culture media were centrifuged and ultrafiltered.

The ^{1}HNMR spectrum with signals (in ppm) at 3.8 (2-H), 2.12 (3-H), 2.33 (4-H), 8.2 (3'-H), 8.4 (4'-H), 7.5 (5'-H), and 9.04 ppm (6'-H) is also consistent with the structure given in Fig. 3. Fraction containing $M+H^+$ 325, with a fragmentation pattern of M/e 325, 223, 220, 219, 191, 178, 146, 137, 120, 102, and 100 is consistent with that reported for a pyochelin, $C_{14}H_{16}N_2O_3S_2$[20], Fig. 4. The corresponding ^{1}HNMR spectrum with signals at 9.18 (2-OH), 6.85-7.42 (3-H, 4-H, 5-H, 6-H), 4.93 (4'-H), 4.42 (2"-H), 3.78 (4"-H), 3.29 (5"-H), and 2.65 ppm (3"-CH_3) is consistent with the structure given for pyochelin. The remaining fractions contain analogues of pyochelin and schizokinen whose structures are currently being identified. Preliminary data indicate that at least three fractions contain compounds which do not resemble those reported to be present in *Pseudomonas* sp.

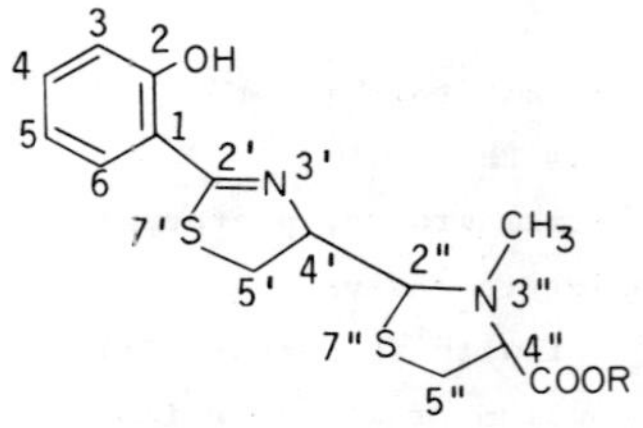

FIGURE 4 Pyochelin.

REFERENCES

1. M.M. Varma, W.A. Thomas and C. Prasad, *J. of Applied Bacteriology*, *41*, 347-349 (1976).
2. J.M. Wood and H.K. Wang, *Environ. Sci. Technol*., *17*(12), 582A-590A (1983).
3. D.P. Kelly, P.R. Norris and C.L. Brierley, *J. Soc. General Microbiology*, *29*, 263-308 (1979).
4. J.B. Neilands, in *Trace Metals in Health and Disease*, edited by N. Kharasch (Raven Press, New York, 1979), pp. 27-41.
5. A.L. Newsome and W.E. Wilhelm, *Appl. Environ. Microbiol*., 665-668 (1983).
6. H.A. Akers, *Appl. Environ. Microbiol*., *45*, 1706 (1983).
7. P.A. Vandenbergh, C.F. Gonzalez, A.M. Wright, and B.S Kunka, *Appl. Environ. Microbiol*., *46*, 128-132 (1983).
8. H.C. Hodge, J.N. Stannard and J.B. Hursh, Editors, *Handbook of Experimental Pharmacology* XXXVI (Springer-Verlag, New York, 1973).

9. G.W. Strandberg, S.E. Shumate, II and J.R. Parrott, Jr., Appl. Environ. Microbiol., 41, 237-245 (1981).
9a. J.E. Johnson, S. Svalberg and D. Paine, Department of Animal Sciences and the Department of Radiology and Radiation Biology, Colorado State University, Fort Collins, Colorado 80521, to Dow Chemical Co., Contract No. 41493-F (1974).
10. E.T. Premuzic, A.J. Francis, M. Lin, J. Schubert and H. Quinby, BNL Report 32972 (Brookhaven National Laboratory, Upton, NY, 1983).
11. E.T. Premuzic, M. Lin, J. Schubert and A.J. Francis, Archives of Environmental Contamination and Toxicology (In press, 1985).
12. R.M. Aickin and A.C.R. Dean, Microbios Letters, 9, 55-66 (1979).
13. C.D. Cox, J. of Bacteriology, 142(2), 581-587 (1980).
14. J.E. Hobbie, R.J. Daley and S. Jasper, Appl. Environ. Microbiol., 33, 1225-1228 (1977).
15. A. Lindenbaum and J. Schubert, Nature, 187(4737), 575-576 (1960).
16. D.N. Slatkin, M.S. Lin and E.T. Premuzic, BNL Report 35560 (Brookhaven National Laboratory, Upton, NY, 1984).
17. B. Maurer, A. Müller, W. Keller-Schierlein and H, Zähner, Archiv für Mikrobiologie, 60, 326-339 (1968).
18. M. Teintze, M.B. Hossain, C.L. Barnes, J. Leong and D. van der Helm, Biochemistry, 20, 6446-6457 (1981).
19. M. Teintze and J. Leong, Biochemistry, 20, 6457-6462 (1981).
20. C.D. Cox, K.L. Rinehart, Jr., M.L. Moore and J.C Cook, Jr., Proc. Natl. Acad. Sci., 78(7), 4256-4260 (1981).
21. C.D. Cox and R. Graham, J. of Bacteriology, 137(1), 357-364 (1979).
22. A.R. McCracken and T.R. Swinburne, Physiological Plant Pathology, 15, 331-340 (1979).
23. K.B. Mullis, J.R. Pollack and J.B. Neilands, Biochemistry, 10(26), 4894-4898 (1971).
24. R. Shiman and J.B. Neilands, Biochemistry, 4(10), 2233,-2236 (1965).

ACKNOWLEDGMENTS

This work was sponsored in part by the Defense Nuclear Agency under Task Code U99QAXMK and Work Unit 00029, Brookhaven National Laboratory under Contract No. DE-AC02-76CH00016 with the United States Department of Energy, and in part by the Office of Naval Research, Contract N00014-84-F-0106 and Work Unit No. NR 685-005.

RADIONUCLIDE COMPLEXATION IN XYLEM EXUDATES OF PLANTS

D. A. CATALDO, K. M. McFADDEN,
T. R. GARLAND, AND R. E. WILDUNG

Pacific Northwest Laboratory
P. O. Box 999
Richland, Washington 99352, U.S.A.

ABSTRACT

The plant xylem is the primary avenue for transport of nutrient and pollutant elements from the roots to aerial portions of the plant. It is proposed that the transport of reactive or hydrolyzable ions is facilitated by the formation of stable/soluble complexes with organic metabolites. The xylem exudates of soybean (Glycine max cv. Williams) were characterized as to their inorganic and organic components, complexation patterns for radionuclides, both in vivo and in vitro, and for class fractions of exudates using thin-layer electrophoresis.

The radionuclides Pu-238 and Fe-59 were found primarily as organic acid complexes, while Ni-63 and Cd-109 were associated primarily with components of the amino acid fraction. Technetium-99 was found to be uncomplexed and transported as the pertechnetate ion. It was not possible to duplicate fully complexes formed in vivo by back reaction with whole exudates or class fractions, indicating the possible importance of plant induction processes, reaction kinetics and/or the formation of mixed ligand complexes.

1 INTRODUCTION

Terrestrial plants, as components of food webs, represent an important link between their environment and humans. As such, plants can represent a significant source of radionuclides to humans, both directly and indirectly. While a significant effort has been expended over the past 40 years to evaluate and quantify the transfer of radioisotopes from soils to plants, relatively little attention has been given to the physiological controls that influence uptake, transport and, most importantly, the chemical forms of elements in plants. The chemical form of an element will not only affect its behavior in the plant, but also its solubility and ability to be bioconcentrated in food chains.

If one considers the functional and chemical constraints imposed on the soil/plant system, it becomes clear that chemical and metabolic mechanisms or processes must be present for the plant to absorb, store, and transport both nutrient and nonnutrient ions. A simple compartment model has been proposed to describe these processes (Figure 1). This model assumes 1) that components of the root apoplast and symplast are responsible for ion modification, ion uptake and compartmentalization, and that the root can act as a buffering organ to regulate ion fluxes to shoots; 2) that the bidirectional transport pathway, needed for transfer of ions between root and shoot, is adapted to transport complex mixtures of ions; and 3) that a means exists for the eventual partitioning of ions in leaves and/or edible fruit, which leads to changed chemical form as a result of either metabolic incorporation or chemical stabilization. It is assumed that any change in the chemical form of inorganic ions resulting from these processes, such as organic complexation, will affect their relative availability to animals. It is further assumed that the processes influencing the uptake and chemical form of nutrient ions in plants can also function for nonnutrient elements such as radionuclides.

While data can be found in the literature to support the presence of many of these processes in plants, it is relatively limited. For example, the results of Tiffin[2] and others[3-5] have shown that there is a requirement for the reduction of Fe^{+3} prior to membrane transport. A similar reduction process has been observed to occur[6,7] for Pu^{+6} valence. This

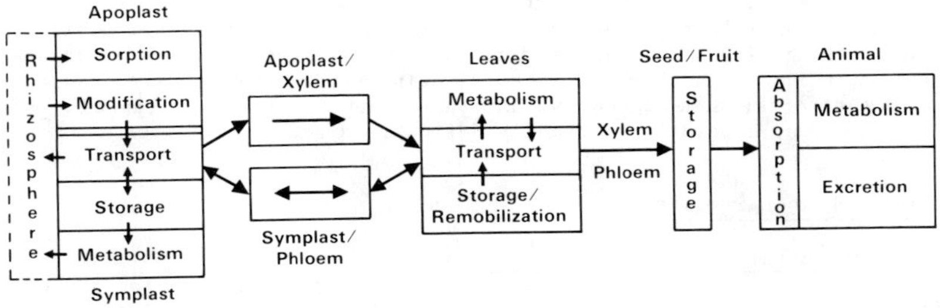

FIGURE 1 Process aspects of trace element transfer from soils through plants to animals.

reduction process apparently occurs either in the apoplast of the root or at the outer membrane surface, but prior to the actual membrane transport step. It is generally accepted that membrane transport of nutrient ions in higher plants exhibits multiphasic absorption isotherms[8,9] at physiological concentrations, and is considered to be an active process allowing some control over the rates of ion absorption in response to demand. These same absorption isotherms are seen for a range of radionuclides including Ni,[10] Tc,[11] Cd,[12] and Tl,[13] which would indicate that absorption mechanisms functioning for nutrient ions are also able to regulate nonnutrient ion absorption. It is believed that chemical similarities, at least with respect to the transport step, between specific nutrient ions and their nonnutrient analogs (Ni/Cu, Tc/S, Cd/Zn, Tl/K) account for their plant availabilities and subsequent chemical fates.

Once an essential element is absorbed into the root, some method must be present to prevent sorption, hydrolysis or nonspecific chemical reaction between the myriad of trace and macroions being stored, transported or metabolized. In plants, this problem is most likely avoided by organic complexation of cations. A substantial body of qualitative data indicate that the nutrient ions Ca, Co, Fe, Mn and Zn exist in organically complexed form in xylem exudates,[14,15] as do Mn and Zn in phloem exudates.[16] Recently, xylem exudates have been receiving increased attention in studies examining the role of chemical complexation in the chemical stabilization and subsequent availability to animals of potentially toxic elements. While these efforts were initiated with the work of Tiffin[17] on Fe complexes in exudates, more recent studies have addressed the behavior of pollutant elements such as Pu, Ni and Cd in plant exudates,[6,12,18,19] and the composition and role of xylem exudate constituents on cation complexation and chemical solubilization.[20,21] In general, these studies have shown that a wide range of cations, even the most hydrolytic polyvalent species such as Pu, are mobile and soluble in transport fluids such as xylem exudates, and exist primarily as organically complexed species. This, coupled with the fact that many nutrient and nonnutrient cations extracted from plant tissues (leaves, roots and seeds) are soluble and complexed by organic ligands of varying molecular weight,[2,6,11,19,22-27] would suggest that plant-produced organometal complexes may affect the transfer of potentially toxic elements from plants to animals.

Our ongoing research efforts are designed to identify and understand those soil/plant processes which influence the solubility, mobility and bioavailability of inorganic ions, and the impact of these processes on the environment and humans. The following study represents one aspect of these efforts; namely, an evaluation of the complexation capacity of xylem exudates, a relatively readily definable system.

2 METHODS AND MATERIALS

2.1 Plant Culture and Exudate Collection

Soybean plants were grown from seed and maintained on hydroponics as described previously.[22] Xylem exudates were collected from plants of various ages from 17 to 100 days; flowering occurred at 60 to 70 days post-germination. Exudates were collected by carefully severing the plant stem below the cotyledonary node, washing the cut stem to remove cellular debris, and fitting the stem with a tight-fitting gum rubber bushing connected to a length of Teflon tube. Exudates were then collected in cooled (4C) vials for various periods of time as noted. *In vivo* exudates were collected following root absorption of radiotracers, decapitation, followed by collection. *In vitro* exudates were collected, untreated, then amended with radionuclides, or were class-fractionated prior to amendment.

2.2 Characterization of Xylem Exudates

Inorganic analyses of xylem exudates were performed by inducive coupled argon plasma (Jarrell Ash Model 975) and corrected, when necessary, for interference. Anions were analyzed using a Dionex (Model 16 Ion Chromatograph).

Whole exudates were fractionated into several compound classes (organic acids, amino acids, neutrals, polyphosphates) to evaluate both their composition and complexation potential. Exudate fractionation involved passing an aliquot (1-2 mL) of exudate through two small ion exchange columns containing AG 50W-x12 and AG 1-x8 in the hydrogen and formate forms, respectively.[28] The neutral fraction, which was washed through both columns with 20 mL of water, contains mainly neutral carbohydrates. The basic fraction, consisting mainly of amino acids, was eluted from the cation resin with 20-30 mL 2N NH_4OH. The weakly acidic fraction, consisting of organic acids and sugar phosphates, was eluted from the anion resin with 20-30 mL 2N formic acid; this was followed by elution with 2N HCl to recover the strongly acidic fraction containing sugar diphosphates and other polyphosphates. Each of these fractions was freeze-dried, and reconstituted with water (1-2 mL) for evaluation of complexation or for organic analysis.

Only the organic acid and amino acid fractions were further characterized. Organic acids were separated by HPLC using a Bio-Rad (HPX-87H column), run isocratic at 45C, with 0.006 N H_2SO_4 as the mobile phase, and a flow rate of 0.5 mL/min. Detection, identification and quantification of specific acids was based on UV absorbance at 210 nm and refractive index, against standards. Amino acids were analyzed using a Glenco (Amino Acid Analyzer) operated in the physiological fluid mode. Identification and quantitation were based on retention time, absorbance and 440/570 nm ratios.

Electrophoresis was performed using Brinkmann (MN 300, 20x20cm - 0.1 mm, cellulose plates). Separations were performed using a 0.1 M HEPES buffer, pH 7.5; potential was held constant at 400v for 30 min. Components were visualized by autoradiography.

Amendments of radionuclides to *in vitro* systems, including whole exudates and class fractions, involved addition of these aqueous components to vials containing the radionuclide in dry form to alleviate pH effects. The latter involved placing aliquots of soluble radionuclides, in a matrix appropriate for the required valence, into vials. These were brought to dryness, and either whole exudates or class fractions were added to give the desired concentration. The concentrations of Pu-238(IV), Fe-59(III), Cd-109(II) and Ni-63(II) in the *in vitro* amendments were 0.2, 0.25, and 0.05 μM, respectively.

3 RESULTS AND DISCUSSION

3.1 Inorganic Composition of Xylem Exudates

The xylem in plants represents the major transport pathway for movement of ions from root to shoot. In addition, substantial amounts of organic carbon, consisting of sugars, amino acids, organic acids and hormones, are cycled back to shoots via the xylem. Through the use of readily collectable and manipulated xylem exudates as a model system, it may be possible to describe those processes employed in plants to maintain the solubility and mobility of complex mixtures of trace and macro-nutrients.

The typical xylem exudate, derived from soybean plants grown on nutrient solutions, contains a number of inorganic elements (Ba, Cd and Si) in addition to the nutrient species provided (Table I). These represent trace contaminants associated with the nutrients' salts that are absorbed and transported to the shoots. The major cations are found in mM quantities, while the micro elements are at µM levels. It is interesting to note that these exudates are chemically stable. They can be freeze-dried and reconstituted with water, but the ions remain soluble, despite the fact that a number of these ions, specifically Fe and Si, are at concentrations exceeding their solubility. Since the pH of these exudates range from 6.2 to 6.8, and the organic carbon concentration can range from 200 to 800 µg C/mL, there is a possibility that the solubility of this complex mixture of ions is the result of their complexation with plant-produced organic ligands.

3.2 Organic Composition of Xylem Exudates

The relative distribution of organic carbon in exudates was determined by anion/cation fractionation into several fractions; namely, amino acids, organic acids, sugars and polyphosphates. This permitted *in vitro* evaluation of the complexation capacity of these individual fractions to complex a wide range of nutrient and nonnutrient elements, and allowed detailed characterization of each fraction for individual ligands. Figure 2 shows the distribution of organic carbon between the various class fractions, and their changes with plant age. The majority of the organic carbon is associated with the amino acid fraction until approximately 15 days post-flowering (75 days), at which point the organic acids dominate. The fraction of carbon associated with the sugars and polyphosphates remains relatively constant at 15 and <1%, respectively.

TABLE I Concentration of inorganic elements in xylem exudates collected at flowering from soybean plants grown on nutrient solution.

Major Elements	Concentration (µM)	Trace Elements[1]	Concentration (µM)
Ca	4900	B	73
K	9100	Ba	0.85
Mg	1700	Cd	0.03
Na	470	Cu	0.09
Cl^-	590	Fe	6.0
PO_4^{-3}	2100	Mn	2.4
NO_3^-	13000	Mo	0.22
SO_4^{-2}	720	Si	17.0
Organic C	61000	Zn	34.0

[1]Al, As, Co, Cr, Ni and Pb were at or below detection limits.

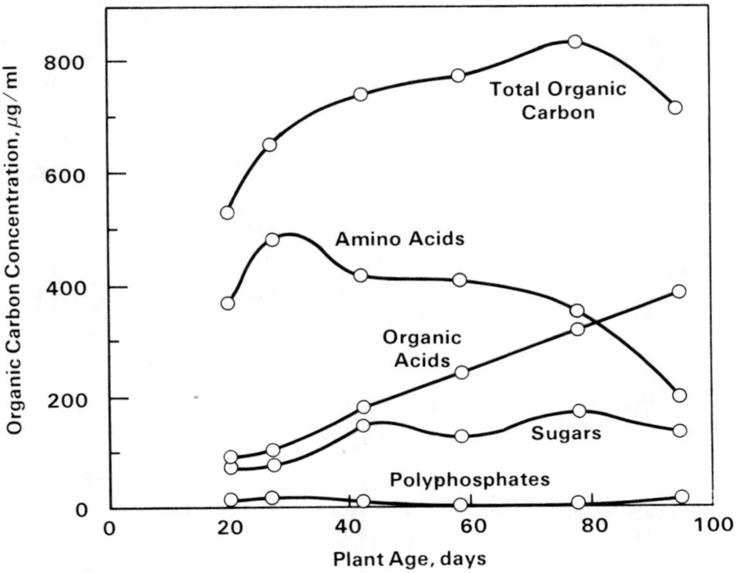

FIGURE 2 Organic fraction composition of soybean xylem exudates as a function of plant age.

Further organic characterization was limited to the amino acid and organic acid fractions of exudates collected at flowering (60-70 days). The organic acid fraction contained eight identified components which accounted for only 38% of the total carbon in this fraction (Table II). An additional 16 components were resolved but not identified. The two major identified acids were citric and malic acids, which can form stable complexes with multivalent cations. Characterization of the amino acid fraction resulted in identification of 12 major components and 15 minor components, which accounted for 84% of the carbon in this fraction. In addition, five major unidentified components were resolved. Hydrolysis of this fraction prior to reanalysis showed these to be small peptides.

3.3 Complexation of Radionuclides by Exudates and Their Class Fractions

Evaluation of the capacity of soybean xylem exudates (75-day-old plants) to complex radionuclides involved *in vitro* amendment procedures. Soluble species of Tc, Pu, Fe, Ni and Cd were amended with either whole exudates or class fractions, incubated for 24 hr, and complexation patterns determined. Of the five elements studied, TcO_4^- was the only one not exhibiting an affinity for complexation; its electrophoretic mobility on amendment to either whole exudates or their fractions indicated the presence of only TcO_4^-, as shown for *in vivo* exudates.[28] While this behavior

TABLE II Amino acid and organic acid composition of soybean xylem exudates at flowering.

Amino Acids	Organic Carbon Concentration (μgC/mL)	Organic Acids	Organic Carbon Concentration (μgC/mL)
ASP	3.6	Maleic	1.3
THR	4.5	Citric	20
ASPN	230	Malonic	2.4
GLN	55	Gluconic	9.4
ALA	1.0	Malic	28
VAL	6.6	Quinic	3.4
MET	1.2	Fumaric	0.8
GABA	2.2	Me-succinic	1.9
OH-LYS	0.24	Total Identified	67
LYS	3.7	Total Present	177
HIS	8.6		
ARG	6.5		
15 other AA	22		
Total Identified[1]	345		
Total Present	412		

[1]Total organic carbon = 736 μg/mL; in addition, the polyphosphate and neutral fractions contained 115 μgC/mL, accounting for 95% of the organic.

would not be unexpected for many oxyanions, it is somewhat surprising for this particular species, since TcO_4^- is felt to be readily reduced in the presence of organic matter. Yet it survived transport through the root and xylem. The behavior of multivalent cations is substantially different (Figure 3). Polyvalent Pu transported from root via the xylem (in vivo forms) is present as three anionic and a single cationic species. Two of the anionic species (b, c) are formed in vitro on amendment of Pu to whole exudates. Amendment of Pu to the organic acid fraction of exudates results in the formation of three anionic components with electrophoretic mobilities similar to those of the in vivo system. The formation of the cationic component d is not observed in any of the in vitro amendments, and may represent a mixed ligand, possibly formed prior to loading into the xylem. The amino acid, neutral (sugar) and polyphosphate fractions appear to have no complexation capacity, with only hydrolyzed or neutral species of Pu appearing at the origin.

By means of paper electrophoresis, iron has been shown to be transported in a number of plant species as a citric acid complex.[2,15,17] In our system, Fe^{+3} amended to whole exudates is associated with five anionic and two cationic components. Four of the anionic components (a, b, c, d) consist of ligands containing organic-acids. The least anionic component d in the whole exudate also contains a ligand with amino acid. As with Pu, the cationic species are not observed in the exudate class fractions and, again, may represent mixed ligands. No complexation of Fe is seen in either the polyphosphate or neutral sugar fractions.

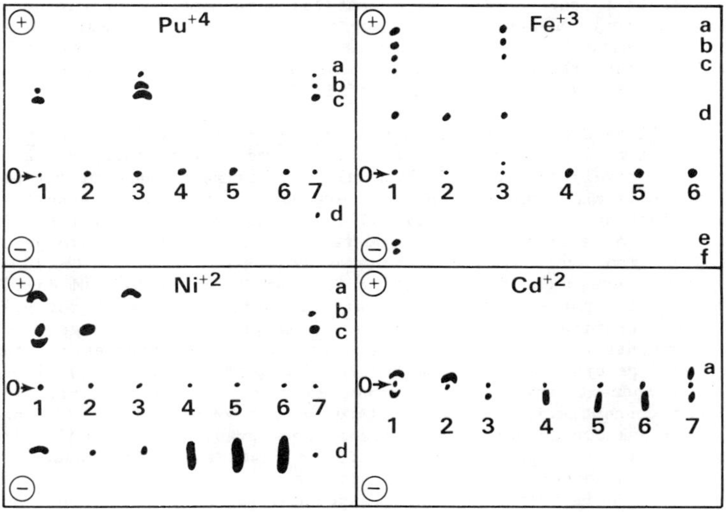

FIGURE 3 Comparative electrophoretic behavior of Pu, Fe, Ni and Cd in xylem exudates. Exudates from 75-day-old plants, incubated 24 hr prior to analysis (o, origin; 1, whole exudate; 2, amino acid fraction; 3, organic acid fraction; 4, neutral fraction; 5, polyphosphate fraction; 6, inorganic metal; 7, *in vivo* forms).

In whole exudates, Ni is associated with three anionic and one cationic component. One of these anionic components contains an amino acid ligand, and one contains an organic acid ligand. The cationic component *d* observed in whole exudates may, in fact, contain more than one form of Ni, since cationic components having similar mobility are seen in both the amino acid and organic acid fractions. The least anionic component found in whole exudates is not found in any of the class fractions, and occurs only on occasion in *in vivo* exudates. The *in vitro* anionic component *c* associated with whole exudates generally appears as the major Ni-containing component in *in vivo* exudates. This component has since been shown to be a tri-peptide, containing either glutamine/aspartic acid or glutamic acid/asparagine and an unknown 440 μm absorbing acid. Also, the presence of the organic acid-containing component *a*, observed in the *in vitro* whole exudates, is dependent on plant age and appears as the concentration of organic acids in exudates increases and the amino acids decrease (Figure 2).

The behavior of Cd differs from that of the other three cations. When Cd is absorbed through the roots, it is found in exudates (*in vivo*) as a single, slightly anionic species; the cationic species appears to be inorganic Cd. On amendment of Cd to whole exudates, one slightly anionic and one slightly cationic species is formed. However, the chromatographic shape of the anionic component suggests that it is nearly neutral in charge, is physically displaced towards the anodic pole by other non-Cd-containing components in the sample, and is not the same component as formed in vivo. The components formed in vitro appear to be stable complexes of amino acid and organic acid constituents of the exudate. The evidence would indicate that the in vivo anionic complex is formed in the root prior to xylem loading and is most likely a mixed ligand.

3.4 Influence of Plant Age On the Form of Ni and Pu in Xylem Exudates

It is quite clear from the data on whole and in vivo exudates that plants can effectively complex and chemically stabilize a range of cations. However, it is also apparent that the organic composition of exudates changes with plant age (Figure 2, Table II), and this alteration could affect the chemical forms of cations in exudates. To investigate this aspect of the complexation process, whole exudates collected from 17- to 100-day-old soybeans were amended with either 0.05 μM Ni-63 or 0.2 μM Pu-238. The complexation patterns for Ni (Figure 4), demonstrate that the previously observed peptide complex (component c) is present at all ages. However, the concentration of this component decreases with increasing plant age, as would be expected from the trends in organic composition. The organic acid component exhibits a maximum concentration at flowering, with a lesser concentration early and late in the growth cycle. In addition, several unknown anionic components are observed. The cationic component, thought to be either an amino acid or peptide complex, shows a slight increase in concentration with plant age.

When Pu is amended to whole exudates, the anionic, organic-acid-containing component appears in exudates from plants that are 33 to 79 days old. It appears to be absent in the 21-, 27- and 96-day-old samples. This may result from the changing organic composition of exudates with plant age, and also the possibility that these complexes are formed in the root prior to loading into the xylem. Under these conditions, one

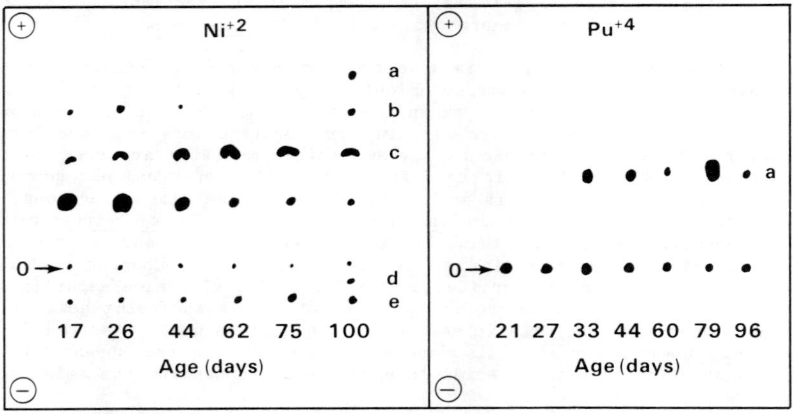

FIGURE 4 Influence of plant age on the complexation of Pu and Ni in amended whole exudates of soybean.

would expect that the Pu complexes formed in the in vitro system would vary somewhat from their in vivo behavior. As noted previously, a large fraction of the Pu amended to these exudates remains at the origin as either hydrolyzed or neutral species.

4 CONCLUSIONS

In the present studies, we have described the inorganic and organic composition of soybean xylem exudates, and postulated the need for organic complexation of reactive and/or hydrolytic elements to permit the transport of a complex mixture of inorganic elements from root to shoot. Each of the cations (Pu, Fe, Cd and Ni) studied is found to exist at least partially or totally as organic complexes in in vivo exudates. Many of these complexed forms can be formed in vitro using whole exudates amended with individual inorganic species. Based on class fractionation of exudates, Pu and Fe appear to be present primarily as organic acid complexes, while Ni exists in several complexed forms. The latter have been shown to consist of a tri-peptide, organic acid ligands and amino acid ligands. The behavior of Cd differed from that of the other cations studied, in that the in vivo complexation pattern could not be formed in vitro.

The use of plant xylem exudates provides a useful, if not unique, system for evaluating the complexation potential of plants. Since the xylem and its contents represent the major transport conduit for movement of both nutrient and pollutant elements to shoots in plants, it provides a readily accessible source of material for in vivo and in vitro evaluation of the potential and capacity of plants to chemically stabilize both essential and pollutant elements. This, in turn, can provide a point of reference for understanding solubility constraints, the influence of complexation on subsequent chemical incorporation processes in leaves and seeds, and the effect of these chemical transformations in plants on availability to animals ingesting chemically stabilized pollutant elements.

5 ACKNOWLEDGMENTS

This work is supported by the U.S. Department of Energy under contract DE-AC06-76RLO 1830.

6 REFERENCES

1. D.A. Cataldo and R.E. Wildung, Sci. Total Environ., 28, 159-156 (1983).
2. L.O. Tiffin, The form and distribution of metals in plants: An overview, in Biological Implications of Metals in the Environment, edited by H. Drucker and R.E. Wildung, TIC, Oak Ridge, Tennessee, (1977), pp. 315-334.
3. J.E. Ambler, J.C. Brown, and H.G. Gauch, Agron. J., 63, 95-97 (1971).
4. J.C. Brown and J.E. Ambler, Physiol. Plant., 31, 221-224 (1974).
5. V. Romheld and H. Marschner, Fine regulation of iron uptake by the iron-efficient plant Helianthus annuus, in The Root Soil Interface, edited by J.L. Harley and R.S. Russel, Academic Press, London (1979), pp. 407-417.
6. T.R. Garland, D.A. Cataldo and R.E. Wildung, J. Agric. Food Chem., 29, 915-920 (1981).
7. M.S. Delaney and C.W. Francis, Health Phys., 34, 492-494 (1978).
8. M.S. Vange, K. Holmern and P. Nissen, Physiol. Plant., 31, 292-301 (1974).
9. P. Nissen, Multiphasic ion uptake in roots, in Ion Transport in Plants, edited by W.P. Anderson, Academic Press, (1973), pp. 539-554.
10. D.A. Cataldo, T.R. Garland and R.E. Wildung, Plant Physiol., 62, 563-565 (1978).
11. D.A. Cataldo, R.E. Wildung and T.R. Garland, Technetium accumulation, fate and behavior in plants, in Environmental Chemistry and Cycling Processes, edited by D.C. Adriano and T.L. Brisbin, DOE Symposium Series 45, NTIS, (1978), pp. 537-549.
12. D.A. Cataldo, T.R. Garland and R.E. Wildung, Plant Physiol., 73, 844-848 (1983).
13. D.A. Cataldo and R.E. Wildung, Environ. Health Persp., 27, 149-159 (1978).
14. E.G. Bradfield, Plant Soil, 44, 495-499 (1976).
15. L.O. Tiffin, Plant Physiol., 42, 1427-1432 (1967).
16. B.J. van Goor and D. Wiersma, Physiol. Plant., 36, 213-216 (1976).
17. L.O. Tiffin, Plant Physiol., 45, 280-283 (1970).
18. M.C. White, R.L. Chaney and A.M. Decker, Plant Physiol., 67, 311-315 (1981).
19. D.A. Cataldo, T.R. Garland, R.E. Wildung and H. Drucker, Plant Physiol., 62, 566-570 (1978).
20. M.C. White, A.M. Decker and R.L. Chaney, Plant Physiol., 67, 292-300 (1981).
21. M.C. White, F.D. Baker, R.L. Chaney and A.M. Decker, Plant Physiol., 67, 301-310 (1981).
22. D.A. Cataldo, T.R. Garland and R.E. Wildung, Plant Physiol., 68, 835-839 (1981).
23. I. Bremner and A.H. Knight, Br. J. Nut., 24, 279-289 (1970).
24. R.L. Halstead, B.J. Finn and A.J. Mclean, Can. J. Soil Sci., 49, 335-342 (1969).
25. M.H. Timperley, R.R. Brooks and P.J. Peterson, J. Exper. Bot., 24, 889-895 (1973).
26. P.E. Dabin, J.M. Marafante, J.M. Mousny and C. Mytenaere, Plant Soil, 50, 329-341 (1978).
27. H.J. Weigel and H.J. Jager, Plant Physiol., 65, 480-482 (1980).
28. D.A. Cataldo and G.P. Berlyn, Amer. J. Bot., 61, 957-963 (1974).

REPORTS FROM CHAIRMEN OF SESSIONS

(chaired by Dr. Desmet)

BULMAN:

The invited paper in this session was specially selected so that the participants could be provided with a view of analytical procedures which might be outside their main area of research experience but may represent procedures they could adopt. Roger Brown's paper provided the stimulus we sought. Those of us who work with the actinides might be missing out on some of the procedures which are used to determine the chemical forms of elements with a lower atomic number. Instrumental methods continue to be developed and we must remain aware of new developments. Of course, the limitations of some of these methods to the problems of determining chemical forms in the environment is obvious. The experimental levels are generally well below the limits of detection of many of these instrumental methods. Mass spectrometry is perhaps one of the instruments which would be most useful to us. Unfortunately these mass spectrometers are exceedingly expensive. One to meet some of the requirements that we have would possibly cost £250,000. Mass spectrometry can detect elements at 10^{-9} g and in some cases even lower. Those laboratories that do have them are unwilling to have even low levels of plutonium in their analytical suite.

The joint paper from Southampton and AEE Winfrith was an excellent example of the excellent collaboration between a well established University Department which has long been noted for the quality of its studies in infra red spectroscopy and a nuclear research establishment.

Sequential extraction procedures continue to be criticised and I think that these criticisms will continue for a long time yet. The approach adopted in the paper presented by Dr. Nirel must go some way I think to answering some of these criticisms. The problems or drawbacks are obvious, how does the speciation of one element change as you pull out another element from a mineral. As Martin Frissel has pointed out the uptake of Pu, Am, Np and Cs by plants from soils can be correlated with their extractability.

It is reassuring for me that the people at Battelle share my views that humates are complexing the transuranic actinides. I recall 3 years ago being told by the Chairman of my session that humates did not complex plutonium. I had just spent 5 minutes of my talk referring to the various papers on this subject.

Computer modelling of speciation is an invaluable aid to the problems of determining the chemical forms of fission and activation products. Those of us doing the experimental work are, I hope, going to be less intimidated by computer modelling now that highly specialised microprocessors are being installed in our offices. More information needs to be sought out in the field and laboratory. Used with caution modelling can be a great aid. No doubt we all know of examples where the modellers have supposed to have got it wrong but I know of cases where modelling has certainly got it right.

Ravenglass sediment is evidently still much in demand as shown by this work from Manchester and by the work conducted by Martin Frissel and his colleagues. The removal of metals from these sediments poses problems. All the information we can gather on this material is essential. I don't know if we will ever see the Ravenglass Estuary turned into agricultural land. If land is going to be contaminated by an accident then we could possibly start getting some answers from a few studies of the Ravenglass Estuary.

The paper given by Brit Salbu is an interesting development. It is a rapid separation procedure and could well be be superior to gel permeation chromatography in many respects. Those of us who use gel permeation chromatography to separate humates should be aware of this work being conducted at Oslo.

Over the last few years membrane filtration devices and hollow fibre cartridges have developed and improved considerably. I am not fully conversant with the developments in polymer chemistry but it is possible that new materials could come along which would lead to new hollow fibre cartridges which could extend this work.

I mentioned earlier the humate studies from Battelle. Elsewhere in this meeting people have discussed humates as metal binding agents for actinides. Finally, I would like to take this opportunity of advertising the International Humic Substances Society (c/o W.L. Campbell, U.S. Dept. of Interior, Geological Survey, Box 25046, Federal Center, Denver, Colorado 80225, USA).

FRISSEL:

Three papers were involved with sequential extraction techniques. You see them here in shorthand.

Livens	Lowson	Wilkins
0.01 M $CaCl_2$ (exch)	0.1M NH_4Cl (exch)	$MgCl_2$ or Na acetate (exch)
---	---	---
0.5 M Acetic acid for (inorganic adsorption)	Ammonium oxalate for amorphous + ferrihydrite	Na acetate (pH 5) for carbonates
---	---	---
0.1 M Na pyrophosphate for organically bound	Citrate/dithionite for iron minerals	Citrate/dithionite for iron + manganese
---	---	---
Amm. oxalate (pH 5) for sesquioxides	Na_2CO_3 for Al + Si compounds	HNO_3 + H_2O_2 for organic matter
---	---	---
Residue	Residue (HF)	Residue (HF)

It is clear that if you have three methods that it is difficult to evaluate them. Of course we could now start asking which is the best but in a chemical sense I don't think it is very appropriate. In speciation studies we have to ask what is the impact on radiation protection in areas as diverse as migration through soils and strata and foodstuffs. Of course, we have to wonder if the sequential extraction technique is adequate. It should be realized that soil scientists worked for 50 years on chemical characterization of nutrient availability and they failed. Instead soil scientists are now determining by statistical procedures the correlation between chemical extraction technique and fertilizer experiments, a procedure which works. Of course in our speciation studies we can not do that if we lack sufficient data for our speciation problems. Have we gone about our speciation studies in the correct manner? I think not. Now Cremers made a fairly strong remark that if you are going to determine something by extraction techniques, you should take care that the extraction is as complete as possible. Only one extraction is not the proper way to do it. It can be improved.

We had a very nice presentation from Eriksson who showed that the availability of plutonium went down during the period of 8-10 years. It was really a very sharp decrease. To try to correlate our extraction with the data of Eriksson it would be necessary to do extraction procedures over 10 years. In a similar way we have also to correlate our extraction technique with migration experiments. This work would call for a coordinated effort.

Consider the extractants for studying Tc in soils.

0.01 M	$CaCl_2$	soluble	TcO_4^-
0.1 M	NaOH	bound	TcO_4
0.25%	NaOCl	reduced	Tc
6N	H_2SO_4	residual	

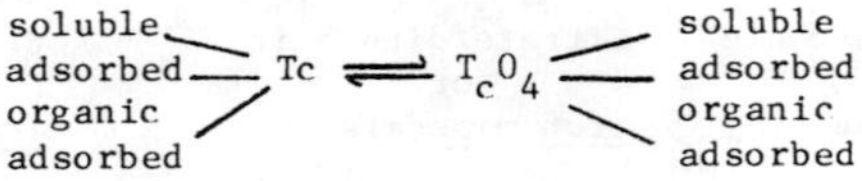

There appears to be all kinds of technetium. Perhaps it is a matter of redox potential. I know it is difficult to measure redox potentials. But if you want to compare different data from different soils then I think it is essentail that we measure the redox potential. I suggest that it will help in interpreting differences.

The second site specific presentation was by Champ. He studied the chemical form of radionuclides in ground waters and found all kinds of species - polyvalent species but mainly anions. Now this is important because anions have high mobility. As we saw from this morning's session, TcO_4^- is easily taken up by plants. It is more easily taken up than cations. Perhaps we should be paying more attention to anions.

Cremers recognized the difficulty in describing the affinity for well defined plutonium compounds. He has introduced the EFAR concept -

Equilibrium Fractional Activity Ratio

- a concept based on sound thermodynamically well defined equations. This system has been developed for Eu but is applicable to other elements.

Mr. Chairman that, I think, must conclude my report.

BONDIETTI:

A small comment which is relevant to the measurement of redox potentials. It is critical if you are going to measure redox potentials that you measure the redox couples. At least several should be measured, e.g. Fe(II)/Fe(III), sulphide/sulphate and of course oxygen. At least you have a reference point for the measurements you make. Otherwise it becomes very dubious just how good that measurement is.

FRISSEL:

Yes, you propose to do these redox measurements. Yes, of course, I agree with this.

BONDIETTI:

Another comment, technetium this time. You can see Tc being reduced in soil with no free water. That is, a 20% soil moisture level. I don't know how you made your redox potential. Under those cases if you do put water in, is that the same thing? Secondly, a redox potential made with a platinum electrode measures the reversible electron transfer to platinum metal, it is not necessarily the same thing as the chemical reaction occurring between a pertechnate ion and some organic. Yes, I agree with you that we should pay attention to redox potential but I don't necessarily think that it is going to be the answer knowing how pertechnate behaves.

FRISSEL:

I didn't say that it is the answer. I only said that it will help us if we have two areas where we have different conclusions in different areas. It might help.

SMITH:

I now present the report for the third session.

The invited paper by Cooper on the influence of chemical form on gut absorption set the scene for the session on gastro-intestinal absorption. He reminded us of the limited understanding we have of the mechanisms involved in the uptake of materials through the gut and despite extensive physiological and biochemical studies over several decades, there is still a large uncertainty about element transfer and the relative roles of passive and active transport across the absorptive cells; on factors allowing leakage past tight junctions and on the role of enzymes at the brush border. He did indicate that there are species differences in enzyme concentrations and that these could lead to results which could be misleading when we extrapolate from GI uptake studies in animals to man. I'll refer to that later when I come to

Dr. Bhattacharyya's paper. There is obviously a need in this particular area of physiology for further studies. Cooper did remind us that there were mechansims which could regulate release of elements from naturally occurring complexing agents and could result in the hydrolysis of species, such as Pu(IV), to insoluble compounds under the alkaline conditions that exist in the digestive juices of the small intestine. This process, however, could be counteracted by competing reactions which favour the formation of soluble compounds, the release, for example, of amino acids by the digestion of proteins. For those elements that exhibit more than one oxidation state, the presence of oxidising or reducing agents in the food or in the food products may influence the bioavailability of these elements.

We then have the proffered papers which did give us some encouragement for solving some of these problems. I refer first of all to the paper by Metivier who indicated the importance of reductive capacity in the digestive juices. In the baboon it was shown that neptunium (V) and plutonium (V) as their nitrates are absorbed to the same extent (f_1 *ca* 10^{-2}). These masses I must stress, (400 g kg^{-1}) are high. When he used very low masses of neptunium and plutonium (5 g kg^{-1}), again administered in the pentavalent state, he found that the plutonium and neptunium were absorbed to about the same extent and he quoted f_1 values of about 10^{-4}. He suggested that reduction in the gut of small amounts of pentavalent actinide to the poorly absorbed tetravalent state was responsible for the lowering in f_1. Pu(V) is more readily reduced than Np(V) therefore this f_1 value starts to get smaller at higher mass values than for Np(V). He put forward as the explanation of the difference between the relative amounts taken up under natural circumstances in the baboon gut.

Then his colleague, Fritsch, described a rather clever *in vivo* perfusion study profusion study of isolated loops of rat small intestine. Preliminary results indicate that neptunium (V) is transferred across the absorptive cells merely by passive diffusion, possibly in the presence of water molecules. No damage to the gut mucosa was observed at concentrations of neptunium similar to those used by Metivier. Damage to the intestinal mucosa could, therefore, be eliminated as contributing to a GI-uptake of 1% for Np(V). At lower masses, the author's experiments support the view that the reduction of Np(V) to Np(IV) could account for a decreased uptake of Np. Fritsch could not rule out the possibility that absorption onto the GIT contents and complexation of Np by components of food stuffs suppressed uptake.

Kargacin stated that the bioavailability of strontium, lead, cadmium and mercury to neonate and young adult rats was the result of complex interactions at sites of absorption. Supplementing standard rat chow with milk caused a significantly increased uptake of these elements in the neonate and, to a lesser extent, in young adults. Adult rats fed a diet supplemented with meat, bread, beans or potatoes also displayed a significantly higher body retention of the four elements studied. While adding extra amounts of fish meal, sunflower oil, alfalfa, cane molasses and vitamin mineral supplement to a high meat diet reduced body retention. Mechanisms postulated for influencing absorption included transport across the absorptive cells and competing complexation reactions in the digestive juices. This type of experiment points the way to a future approach, namely, the need to identify dietary factors that modify mechanisms of uptake and subsequent tissue distribution of elements.

The papers presented by Ralston and Bhattacharyya reported some important observations in the baboon. Ralston studied the metabolism of neptunium and protactinium after their intravenous or intragastric administration in different oxidation states. The author found that tissue distribution of both neptunium and protactinium was similar irrespective of the mode of administrative or chemical form. He reported retention half-times in bone and liver that were markedly different from those used by ICRP (a few years compared to 100 years for the skeleton and 100 days compared to 40 years for liver). The f_1 values he recommended for "soluble" forms of these two elements were 1% for neptunium and 0.1% for protactinium although the higher value for neptunium was obtained in fasted animals gavaged with neptunium (V) bicarbonate. The value in fed animals was 1 to 2 orders of magnitude less.

Bhattacharyya evaluated the effect of chemical form and feeding regimen on gastrointestinal absorption of plutonium in mice gavaged with low masses of plutonium. Her particular concern was that chlorinated drinking water may contain plutonium (VI).

She concluded that plutonium oxidation state (VI compared to IV) and the form in which the plutonium is administered (bicarbonate or nitrate or citrate) had little effect on gastrointestinal absorption in mice. Values of about 0.2% were obtained in fasted animals. The formation of plutonium (IV) polymers or administering food decreased the f_1 values by about an order of magnitude. Most importantly, she observed that the gastrointestinal absorption of plutonium (VI) bicarbonate solutions in both fed and fasted adult baboons appeared to be similar to that observed in mice.

These data provide further evidence of the need to reconsider the f_1 values recommended by ICRP for workers (who are exposed to inorganic forms of plutonium) when recommending f_1 values for members of the public who ingest plutonium in foodstuffs.

So Mr. Chairman, in summary, the papers presented in this particular session point the way in my opinion to a future research programme. There is certainly a need for a concerted effort to identify potential binding agents in food and for biochemical and physiological studies to characterize the effects of digestive processes. In this respect I would suggest there is a need for a carefully planned human volunteer experiments to work out appropriate f_1 values for elements biologically incorporated into foodstuffs. We heard from the session that I chaired that two or three laboratories in the UK are certainly interested in feeding radioactive materials to human volunteers to find out how these are absorbed into the bloodstream. I think there is a need to use stable and non-radioactive analogues of some of our fission and activation products to do some of these studies. Just to digress, it is possible for us to collect samples of human intestine under physiological conditions if you have a good doctor or clinician to work with and one of the approaches at NRPB is to actually take biopsy samples from duodenum and jejenum and relate this to the biochemistry and physiology of the gut. Finally, I feel that the CEC have an important role to play in any proposed research programme. I feel there is a need for a Task Group to be set up to make sure that laboratories do not duplicate each other's work, waste time and effort, and so I put it to Dr. Myttenaere that there is need for this type of Task Group to look into the future programme.

COOPER:

Robert Bulman opened the session. He emphasised the complex chemistry of iodine arising from its multiplicity of oxidation states. Iodine can be biologically active and Dr. Bulman posed the question, "Could micro-organisms have a role in the cycling of iodine from the sea surface microfilm". The second speaker, Dr. Behrens, also emphasised the role played by microorganisms or rather extra-cellular enzymes produced by them in determining the speciation of iodine. He showed that in aerated surface and soil water extra-cellular peroxydase of microbial origin catalyse the conversion of iodide to organically bound iodine. The reactions may be reversible and under anerobic conditions iodide is stable and appears to be generated from organically bound iodine. The binding of iodine by organics in soil is clearly important in determining bioavailability and the migration of iodine through the soil.

The third speaker in the session was John Howe from MAFF at Weybridge who described an intriguing piece of scientific detective work. They found anomously high levels of iodine-125 in the thyroids of swans inhabiting major rivers in Britain. They traced this back to I-125 possibly from radio-immunoassay tests discharged by hospitals and research laboratories via sinks and sewerage systems into the rivers. He also had evidence that humans are absorbing I-125 via drinking water abstracted from those rivers. The doses arising were however, thankfully, low.

The fourth speaker, Dr. Guenot, from France, described work on the absorption of iodine species onto leaves. They looked in particular at hypoiodous acid which is a form of iodine observed in a humid atmosphere and which has not previously been studied in this way. He found that this form of iodine was unlike elemental iodine, only taken up by the leaf through the stomatal pores. This may be important because it would then be an inner leaf tissue and would not be washed off by rain and may, therefore, have a longer residence time.

The final speaker in the session was Dr. Cattaldo who also discussed foliar uptake of iodine but also mentioned the other side of the coin, namely volatilisation of iodine from leaf surfaces and also from soil. Their work shows that I-129 from nuclear fuel reprocessing may behave differently to stable iodine in the vicinity of reprocessing plants. The plant-to-soil concentration factors for I-129 were higher than for stable iodine. The suggested mechanism was volatilisation from the soils and absorption by the leaf surfaces, the recently deposited iodine being preferentially volatilised.

In conclusion, clearly microorganisms have an important role in determining iodine speciation. Nearly all speakers mentioned micro-organisms at some point, but if I could just introduce a contentious point perphaps. Interest in iodine speciation might be of limited importance because the major iodine isotope produced by the nuclear fuel cycle has such a long half-life that it will eventually give rise to doses to man no matter what rate it migrates through the environment.

CREMERS:

The session on speciation of radionuclides in fresh water bodies covered four papers dealing with the possible effects of speciation on transfer to biota and geochemical behaviour, the two areas for which speciation is generally considered to be relevant.

The first paper (Vanderborght), covering a state of the art on the subject, deals in part with a discussion of some of the difficulties, inherent to speciation studies in fresh water environments. However, the main emphasis in this paper is being put on the scarcity of solid evidence for the effect of speciation - most often taken for granted - on the bioavailability. In particular, the need was stressed to divert our interests and scale up our research efforts in the study of possible links between speciation and bioavailability.

In the paper by Bierkens et al., an attempt is made to demonstrate possible effects of speciation on availability by injecting intragastrically americium in various forms and pH conditions in crayfish. Both pH and chemical forms appear to induce effects on uptake but it appears not readily feasible to sort out the effects of the two parameters separately.

Effects of speciation on geochemical behaviour of radionuclides are more readily evidenced, as shown in the paper by Nelson and Orlandini. In particular, it was demonstrated that the level of dissolved organic carbon (DOC) in the liquid phase has a dramatic influence on the solid-liquid distribution of radionuclides, illustrating the competitive effect of DOC - most likely through a process of complex formation - with the sorptive particulate phase. The most striking features of this study are the quasi-identical effect of DOC on the decrease in K_D for ^{241}Am, $^{239,240}Pu$ and ^{232}Th and the pronounced differences obtained with DOC of different origin. This particular feature illustrates the fact that the use of DOC values may not be a fully adequate measure of expressing the effect of dissolved organic matter, a point which was also raised by Vanderborght. Quite possibly, such differences arise from differences in complexing capacity and molecular weight, these differences being also likely operative in freshwater and seawater bodies.

In the paper by Picat et al., it was shown that cobalt, when introduced in the Loire river, evolves rather slowly - over a period of days, depending on pH and hydrological regime - into particulate forms. The phases, which are the most likely candidates for the trapping of cobalt were inferred to be calcium carbonate and phosphate, the process responsible being co-precipitation. A particularly striking feature in this paper is the large extent of pH fluctuations, 1 to 2 units, sometimes over very short timescales.

In regard to the concept of bioavailability, questions were raised as to how one should define such property. Formally, it is generally expressed in terms of concentration factors (CF) values, very often on a whole body basis. Such practice is evidently far from ideal and is in many respects similar to what is done in other areas of environmental research, such as:

1. K_D values of radionuclides in sediments.
2. K_D values of hydrophobic pollutants such as polychlorobiphenyls in soils and sediment-aquifer systems.
3. CF values of hydrophobic pollutants in biota.

The common idea, underlying this practice, is that such processes can be viewed as equilibrium partition phenomena, the distribution coefficient being the ratio of sorption (accumulation) and desorption rates. However, in the three areas just referred to, the partition phenomenon can be rationalised by a normalisation step, which takes care of the dilution effects of the inactive phases, and which is made possible by the clear-cut prevalence of just one, or perhaps just two sub-phrases which are generally present at very small concentrations. In particular, we refer to (1) the normalisation of K_D with respect to some given reference element (taking care of the dilution effect by the inactive phase), the expression of K_D with respect to the sediment or soil organic matter, (2) and the normalisation of the CF with respect to the lipid content of the organism, (3) the lipid fraction being the sink for the hydrophobics. Such procedure is evidently not readily possible in the case which concerns us here, as evidenced by the widely varying degrees of radionuclide partitioning in the various organs of a body, such as a crustacea. Moreover, the process may further be complicated by possible effects related to the mode of entry (ingestion, membranous uptake).

Evidently, when we are dealing with cases of membrane passage, (algae, micro-organisms, plant roots), the situation becomes much more tractable and the concept of bioavailability can then more readily be quantified. In such cases, provided it can be demonstrated that we are dealing with absorption processes, perhaps the most rational way of expressing the effect of speciation on bioavailability would be in purely kinetic terms, ie. as a first order rate constant. Such an approach may well be worth pursuing in future work.

SIBLEY:

I wish to make only a few comments regarding speciation in the marine environment. There were five papers presented in the session. The invited paper by Dr. Van den Berg provided a state of the art discussion of some of the methods that are presently available to study speciation in solution. The other papers in the session on marine environment considered the behaviour of stable element analogues or transuranic elements and some of the environmental factors that affect the behaviour of these elements. Other papers that relate to the marine environment were presented by Drs. Fisher and Bonotto this morning and by others earlier in the week. I'm not going to take time to discuss what was in all of those papers but I would like to point out that the CEC Radiation Protection programme has sponsored two other seminars in the past two years, one at La Spezia, Italy, on the behaviour of long-life radionuclides in the marine evironment and one at Renesse, Netherlands, on the behaviour of long-life radionuclides in estuaries. The proceedings of each of those meetings have a fair number of papers which relate to the question of speciation in the marine environment. In addition there is a great deal of information published on trace metal speciation in seawater including the effects of speciation on bioaccumulation and toxicity. Much of this has been published in the chemical oceanography literature but it is pertinent to radioecology and we would be well advised to review that literature.

It seems to me that the radioecological questions that are of concern in the marine environment are the same as for other environments, and there are 2 general questions.

1. How will radionuclides be dispersed or contained? We might consider this to be a hydrogeological question.

and 2. What are the critical pathways by which radionuclides may be returned to man? This is clearly a more biological question.

I would like to consider the second question briefly. We heard in the sessions on the freshwater environment and on gastro-intestinal uptake that the principal route of radionuclide uptake for man is from drinking water not from aquatic food chain accumulation. Therefore, in freshwater environments speciation in solution is an important subject. It is, however, significantly less important in marine environments because we do not drink seawater. It is important to identify the potential pathways of radionuclide uptake by man from marine foodstuffs. However, I do not believe speciation is very relevant to that problem at the present time.

On the contrary, I think speciation is quite relevant to the question of dispersion and long-term availability because speciation can affect partitioning between the soluble and particulate phases. As we learned from several papers, we need to be concerned not only with speciation in solution but also in the solid phases and Dr. Cremers indicated that reliable information on solid phase speciation is much more difficult to obtain. Various selective extraction procedures have been developed to evaluate distribution in the solid phases and it appears that these methods are becoming more popular. In the methods sections of papers and in discussions at conferences it is generally recognised that these methods are not absolutely selective. Unfortunately the lack of selectivity is often ignored in the results sections of papers and conclusions are stated much more emphatically than the methods allow. I believe selective extractions can provide valuable information on radionuclide distribution in solid phases but we must remember the methodological limitations when we are interpreting the results. I would request authors to provide more detailed information on their extraction procedures especially the time of extraction and solid/-solution ratios. I have recently tried to compare results of selective extractions by different authors and have been unable to do that because often I could not determine from the methods presented in the papers how the authors did their extractions. I know editors and publishers want us to keep papers short. However, it does not take long to include that information. It would make the paper that much more valuable.

The use of stable analogues has been proposed to evaluate the long term behaviour of radionuclides, but we learned yesterday that stable Eu is partitioned in sediments differently than freshly introduced Eu-152. It would be interesting and valuable to determine how solid phase partitioning changes with time and selective extractions will be valuable to address that problem. One can argue on geochemical grounds that stable analogues will always be more tightly bound to sediments than recently introduced radionuclides. Thus, stable analogues will give the least conservative estimate of mobility. What we must do is determine how quickly introduced radionuclides approach the solid phase speciation of their stable analogues.

CREMERS:

I would just like to make this statement from the original paper of Tessiers in 1979. In his conclusions he writes "for the moment it is obvious that the distribution of a given metal between the various fractions does not necessarily reflect the relative scavenging action of discrete sediment phases but rather should be considered as operationally defined by the method of extraction".

BHATTACHARYYA:

Dr. Desmet started our session this morning with the very stimulating presentation of the many physiological pathways in plants that can be involved in the uptake and handling of radioactive fission products. These included CO_2 fixation during photosynthesis, the incorporation and decorporation of water molecules in the TCA cycle, possible incorporation of tritium in the form of HT into plants that contained hydrogenases. This set the stage for all of us in terms of thinking of the very complex ways in which we have to think about the handling by plants of the various elements that are of interest. He then gave an introduction to the technetium studies that are being carried out in his laboratory and since there were a number of papers on technetium I think I'd like to end up with a summary of those papers. I will move on now to the presentation by Jack Schubert. He showed us that when Psuedomonas aeroginosa cultures are exposed to high levels of thorium and uranium, they respond by synthesising chelating agents. Some of these are similar to those induced under conditions of iron deprivation. Some, however, appear to be made in response to thorium and uranium. He is interested in identifying these molecules and maybe synthesising them with the ultimate interest in using them as potential chelating agents for the removal and decorporation of thorium and uranium from persons exposed. It is a very ambitious and interesting approach to this problem.

Dr. Fisher presented a study of the uptake of radioactive tin by marine diatoms that have a cell wall composed of silica. He showed us that the concentration factor for tin in these organisms is as high as 10^4 and that it can be shown also that for other organisms, like picoplankton, the concentration factor is as high as 10^6. He indicated then that the incorporation into these organisms might be a way of introducing tin into the food chain or else bringing about changes in the vertical transport of tin once its been incorporated into the sea environment.

Dr. Cattaldo presented us with a very nice study of the radionuclide complexation in the xylem of plants. He showed us a system whereby he was looking at the distribution of the complexation of elements by the amino acid, organic acid, sugars and polyphosphate portions of the organic carbons present in xylem. He showed that their special components of the xylem are handling the various radioelements and that plutonium is fully complexed in the xylem and seems to be complexed mainly by the organic acids. A similar association was noted for iron but in the case of nickel and cadmium you get an association with the amino acid fraction. It was interesting to me, to see that he wasn't able to duplicate the *in vivo* patterns for these elements when he incubated the elements directly with the xylem exudates *in vitro* so that there is something very particular happening during the uptake process that can't be entirely reproduced *in vitro*.

I'll turn lastly then to the studies on technetium which were covered in a number of different papers. If we start out with the consideration of the handling of technetium by algae, it was very clear from Dr. Bonnotto's presentation that algae on the Belgian coast accumulate technetium. The concentrations of technetium that he found in the algae he analysed were such that the concentration factor that he would calculate from a technetium concentration in seawater were as high as 25,000 - 30,000 which is similar to that reported by others. He showed data on the uptake in the laboratory by two different strains of algae, one of which concentrated technetium and the other which did not. The *Borgesina* algae showed a presence of technetium in the vacuol sap after exposure to pertechnetate. The concentration factor was very low, however, less than 1, and most of the technetium could be accounted for by pertechnetate present in the vacuol sap. In the case of *Acetabularia* however, he showed that there was probably apical uptake of the pertechnetate with a concentration factor that increased with time up to a value of about 300 when the concentration of technetium was very low. When the concentration of technetium was elevated by using technetium-99 the concentration factor that he measured was much lower, about 10. His gel permeation studies showed that technetium taken up by the algae was predominantly protein-bound. Also in some peaks it looked as if it was bound to small molecules with a small fraction migrating as the pertechnetate itself.

If you move on from this algae system to the spinach system, both Dr. Vandecasteele and Dr. Van Loon presented a very nice body of data that gives us an idea of how the spinach plant handles pertechnetate when it is exposed to this form of the element. In pertechnetate-amended soils the bioavailability of the technetium initially was quite high with 40% being taken up by the plants and 85% being present either as water soluble or as exchangeable technetium, as measured by extraction methods. With time, however, this water soluble plus exchangeable fraction decreased. Initially the transfer factors of the technetium in the soil to plants was high, about 400, and with time there was a five to tenfold decrease in the transfer factor. This corresponded to the decrease in the water-soluble plus exchangeable fraction. This speaks then of the characterisation of soil. If you look at the plant uptake, Dr. Van Loon showed us an equation by which he could predict the pertechnetate concentration in leaves from the pertechnetate concentration in the growth medium and vice versa. Technetium (IV) complexed by cysteine, underwent oxidation to pertechnetate in the plant leaves. You could calculate back and figure out how much of the pertechnetate should have been in solution assuming that it was the absorbable species.

Dr. Van Loon also showed that technetium complexed by 'tetren' was taken up to a reduced extent by the plants. Both investigators showed that if you extracted the spinach leaves and looked at its migration on column chromatograph systems, the technetium as in the algae system appears to be protein-bound - perhaps as much as 50%. The other 50% seemed to be either as technetium bound to a small peptide or partly as the free unchanged pertechnetate. So I would like to conclude then and say that there are a number of questions that arise out of these very nice studies on technetium. One might be why is technetium taken up by some algae and not by others. If we knew that it would tell us something about the handling of technetium by the algae. What is happening to the chemical form of technetium with time in soil? We could see that the bioavailability is decreasing with time and what are the bound forms of technetium in algae and plants. Can we identify those species to help us understand things? The last thing that I think that needs to be done with all the data on technetium that has been presented, is to take it and decide how it can be used in terms of evaluation of risk from exposure to technetium that has been released into the environment. We see a decrease in bioavailability in time and a change in chemical species. How can all this data be used to help us to estimate the risk to man of technetium.

DESMET:

At the end of the meeting I would like to add a few ideas. This meeting was very stimulating since it poised the finger on the big uncertainty that changes of the chemical form of any radionuclide represent for a source-man food-chain. It was, however, also very obvious that only the tip of the iceberg was revealed and that no real clear image originated from the different researches on speciation. To my personal feeling, the physico-chemical problems were satisfactorily assessed, especially in the sessions on soils and aquatic ecosystems. Less salient however was the assessment of the chemical characterization of the complex chemical species, although it should be a logical step in the speciation research. This is the more true in the plant-animal transfer studies. Therefore, I am particularly interested by the proposition of Dr. Smith of NRPB, speaking about the creation of a special Task Group which would be charged to giving direction and to stimulating research in this particular area. It would be good to try to integrate somehow the results and make them accessible and available to those who have to make predictions by means of food-chain or ecological models.

Now I have one last agreeable duty to fulfil. First of all, I wish to thank on behalf of Dr. Myttenaere and myself the responsible authorities of this University who gave us the opportunity to live a few days in these historical premises, which really were impressive to me. I thank Dr. Hylton Smith for having coordinated and stimulated the organization of this seminar. I thank as well Dr. R. Bulman and Dr. J. Cooper for their contribution to the more practical points of the organization.

SMITH:

On behalf of the National Radiological Protection Board I thank the Commission for allowing this meeting to take place and for supporting it financially and I thank you all for participating. It has been a successful meeting, I am quite sure but we'll need time to think about it because we've mixed so many different disciplines, chemists and biologists etc. It has been a very difficult subject to cover. I feel it has been a useful pointer for the future and I hope you all enjoyed it. I hope you've had a chance to look round Oxford and appreciate its beauty. I wish you all a safe journey home.

PARTICIPANTS

UK

Dr Colin Bowlt	St. Bartholomews Hospital Medical College, Department of Radiation Biology, 2 Charter House Square, London EC1 M6BQ
Dr David Prime Mr Barry Frith	University of Manchester, Radiological Protection Service, Coupland III Building, Oxford Road, Manchester M13 9PL
Dr Ralph Atherton	British Nuclear Fuel PLC, Risley, Warrington, Cheshire
Dr Ronald Crawford Dr J. Gomme	UKAEA Dounreay, Lab 107, D 1200, Dounreay NPDE, Thurso, Caithness KW14 7TZ
Dr Steven Ogden Dr R A Gomme	Department of Chemistry, University of Southampton, Southampton, SO9 5NH
Dr David Horill Dr Brenda Howard Mr Franics Livens	Institute of Terrestrial Ecology, Merlewood Research Station, Grange-over-Sands, Cumbria
Mr R S B Jones Dr Brian R Bowsker	UKAEA - Atomic Energy Establishment, Winfrith, AEE Dorchester, Dorset
Miss Margaret Minski	Imperial College Reactor Centre, Silwood Park, Ascot, Berkshire, SL5 7PY
Dr Murdoch Baxter Mr Joseph Toole	University of Glasgow, Chemistry Department, Glasgow, G12 8QQ
Mrs Jennifer Higgo	Imperial College, Department of Chemistry, South Kensington, London SW7 2AZ
Dr Stephen Nicholson	Safety and Reliability Directorate - UKAEA, Wigshaw Lane, Culcheth, Warrington, WA3 4NE
Dr Neil Lynn	Royal Naval College, Greenwich, London, SE10 9NN
Dr Jacqueline Cross	University of Wales, Institute of Science and Technology, Department of Applied Chemistry, PO Box 13, Cardiff, CF1 3XF
Dr Helen Crews	MAFF, Food Sciences Laboratory, Haldin House, Queen Street, Norwich, NR2 4SX
Dr Philip Day	Department of Chemistry, University of Manchester, Manchester, M13 9PL
Mr Bernard Harvey	MAFF, Fisheries Laboratory, Lowestoft, Suffolk, NR33 6HT
Mrs J A Kirton	Associated Nuclear Services, 123 High Street, Epsom, Surrey, KT19 8EB

Dr Robert A Bulman Dr John R Cooper Mr George J Jam Dr John D Harrison Dr Donald S Popplewell Dr Jane Smith-Briggs Dr Hylton Smith Dr Bernard Wilkins	NRPB, Chilton, Didcot, Oxfordshire. OX11 ORQ
Mr John R Howe	Central Veterinary Laboratory, MAFF, New Haw, Weybridge, Surrey, KT15 3NB
Dr J D F Ramsay Dr P M Pollard Mr R J Russell Mr J W McMillan Mr D C Pryke Dr S J Williams Mrs H Thomason Mr Roger M.Brown Dr Chris Pickford Dr John Hislop	AERE, Harwell, Didcot, Oxfordshire, OX11 ORA
Dr C M G Van den Berg	Department of Oceanography, University of Liverpool, Liverpool, L69 3BX
Mr John Wrench	Department of Environmental Sciences, University of Lancaster, Lancaster, LA1 4YQ
Dr I R Hall	H.M. Industrial Pollution Inspectorate, Pentland House, Robb's Loan, Edinburgh

AUSTRALIA

Dr Terry F Hamilton	University of Melbourne, School of Chemistry, Parkville, Victoria, 3052
Dr Richard T Lawson	Australian Atomic Energy Commission, Lucas Heights Research Laboratories, Private Mailbag, Sutherland, N.S.W. 2232

BELGIUM

Dr Silvano Bonotto Prof Oscar Vanderborght Dr Johan Bierkens Dr Jozef Vangenechten	C E N /S C K, Dept. of Radiobiology, B-2400 MOL
Prof Adrien Cremers	Laboratorium voor Colloîdale Scheikun de, K.U. Leuven, Kard. Mercierlaan 92, B-3030 Heverlee, Leuven
Dr Gilbert Desmet Dr Constant Myttenaere	Commission of the European Communities, Directorate General for Science Research and Development, Rue de La Loi 200, B.1049 Brussels

Dr Thomas Sibley	Unité de physiologic Végetale, K.U. Leuven, Place Croix de Sud 4, B.1348, Leuven

CANADA

Dr Doug Champ	Atomic Energy of Canada Ltd., Chalk River Nuclear Laboratories, Chalk River, Ontario K08 1J0
Dr Bryan E Imber,	Dobrocky Seatech, 9865 West Saanich Road, P.O.Box 6500, Sidney, B C, U8L 4M7

DENMARK

Dr Henning Dahlgaard	Health Physics Department, Risø National Laboratory, DK-4000 Roskilde

FRANCE

Dr Guenot	Centre d' Etudes Nucliaires, DERS/SERE, B.P. No. 6, 92260 Fontenay-Aux-Roses
Dr Louis Farges	Commisariat à l'Energie Atomique, CEN/FAR BP6-92260, Fontenay aux Roses.
Dr Henri Metivier Dr Paul Fritsch	C E A - IPSN - DPS - Section de Toxicologie et Dr. Cancerologie Expérimentale - BP No. 12-91680, Bruyeres-le-Châtel
Dr Philippe Picat	CEA/IPSN CEN Cadarche, BP. No.1., 13115 Saint Paul Lez Durance
Dr A Thomas Dr P Nirel	Laboratorie de geologic, Ecole Normale Supérieure, 46 Rue D'Ulm, 75230 Paris, Cedex 05
Dr Pierre Gueguenìat Dr Dominique Boust	C.E.A IPSN-DERS-SERE-SRTCM, Laboratorie de Radioécologic, Marine Centre de la Hag - B P 270, F-50107 Cherbourg

WEST GERMANY

Dr J R Duffield	IGT, Kernforschungszentrum, Postfach 3640, D7500 Karlsruhe
Prof Hans Bonka	Lehrgebiet Strahlenschutz in der Kerntechnik, RWTH Aachen, Templergraben 55, D5100, Aachen
Dr Klaus Fischer	D W K - Büro Wackersdorf, Postfach 62 D-8464 Wackersdorf
Dr Heinz Schüttelkopf	Kernforschungszentrum Karlshrue GmbH, Postach 3640, D-7500 Karlsruhe
Dr Angela Erzberger Dr Viggenhauser	Bundesgesundheitsamt, Ingolstädter Str.1, D-8042 Neuherberg

Dr Mario F Bernkopf,	GSF-IFT, Thesdor-Heuss-Str.4, 3300 Braunschweig
Dr Heinz Müller J E Johnson	GSF-Institut für Strahlenschutz, Ingolstadter Dr. Landstr 7, D-8042 Neuherberg
Dr Gerhard Pröle	Gesellschaft für Strahlen - und Umweltforschung, Ingolstadter Land Str. 1, D-8042 Neuherberg
Dr Horst Behrens	Gessellschaft für Strahlen - und Umweltforschung, mbH München, Institut für Radiohydrometric, D-8042 Neuherberg
Prof G Marx Dr A R Flambard	Freie Universitat Berlin, Institute für Anorganische und Analytische Chemie, Fabeck Str. 34-36, 1000 Berlin 33

ITALY

Dr W Martinotti,	ENEL, Rubartino 54, Milan 20134
Dr Giovanni Ciceri	CISE, P.O. Box 12081, I-20134 Milan

MONACO

Dr Nicholas Fisher Dr Elis Holm	International Laboratory of Marine Radioactivity, IAEA Muséé Oceanographique, MC98000 Monaco

NETHERLANDS

Dr.M M Rutgers Van Der Loeff	Netherlands Institute for Sea Research, P.O. Box 59, 1790 AB Den Burg
Dr Martin J Frissel	Laboratory for Radiation Research, RIVM, P O Box 1, 3720 BA, Bilthoven
Dr R Luc Van Loon	Stichting Ital, Keyenbergseweg 6, P.O. 48, 67000 AA Wageningen

NORWAY

Professor Alexis Pappas Dr Brit Salbu	Department of Chemistry, University of Oslo, P.O. Box 1033, Blindern, 0315 OLSO-3

SWEDEN

Dr Åke Eriksson	Department of Radioecology, Swedish University of Agricultural Sciences, S-750 07 Uppsala
Dr Per Ostlund	Department of Geology, University of Stockholm, S-106 91 Stockholm

SWITZERLAND

Dr Hans-Jurgen Pfeiffer Dr Hans Wanner	HSK, CH-5303, Wuereulingen

USA

Dr Dominic Cataldo Dr Ray Wildung Dr David E.Robertson	Battelle, Pacific Northwest Laboratories, P O Box 999, Richland, Washington, 99352
Dr Maryka Bhattacharyya Dr Donald Nelson	Argonne National Laboratory, 97005 S Cass Avenue, Argonne, IL 60439.
Dr David La Touche	Oregon State University, Department of General Science, Corvallis, Oregon, 97330
Mr Lowell Ralston	NYU Medical Centre, Long Meadow Rd, Tuxedo, NY 10987
Prof Jack Schubert	Department of Chemistry, University of Maryland, Baltimore County, Catonsville, MD 21228
Dr Ernest A Bondietti	Oak Ridge National Laboratory, P O Box X, Oak Ridge, Tennessee 37831

INDEX